Introduction to Vibrations and Waves

T0256837

Introduction to Vibrations and Waves

H.J. PAIN, D.Sc.
Emeritus Reader, Department of Physics,
Imperial College, London, UK

and

PROFESSOR P. RANKIN
Department of Physics, University of Colorado, Boulder, Colorado, USA

This edition first published 2015 © 2015 John Wiley & Sons, Ltd

Registered office
John Wiley & Sons Ltd, The Atrium, Southern Gate, Chichester, West Sussex, PO19 8SQ, United Kingdom

For details of our global editorial offices, for customer services and for information about how to apply for permission to reuse the copyright material in this book please see our website at www.wiley.com.

Library of Congress Cataloging-in-Publication Data is available

Hardback ISBN: 9781118441107
Paperback ISBN: 9781118441084

A catalogue record for this book is available from the British Library.

Set in 10.5/12.5pt Times LT Std by SPi Publisher Services, Pondicherry, India

1   2015

Contents

Acknowledgement

The authors wish to acknowledge the work of Dr. Youfang Hu in preparing the manuscript of the work and to thank him for it.

About the companion website

This book is accompanied by a companion website:

Go to http://booksupport.wiley.com where you will be prompted to enter the book ISBN to access the following resources:

- Solutions to chapter problems
- PowerPoint slides of all the figures from the book for you to download

Preface

The co-authors first met nearly 40 years ago when, as an undergraduate, Patricia Rankin joined the Physics Department at Imperial College, London, where John Pain was teaching "Vibrations & Waves". She still has her copy of an earlier edition of this book. As an undergraduate she was the top student in her year in a large class, overwhelmingly male, and won the most prestigious of awards, the Governors' Prize. She stayed on to do her Ph.D. before moving to the United States where as an experimental particle physicist she was awarded a Sloan Fellowship and recognized as an Outstanding Junior Investigator by the U.S. Department of Energy. She is a tenured professor in the Physics Department at Boulder, Colorado, with Nobel Prize winners as colleagues. Her extensive experience in teaching physics is reflected in this revised edition, where well-tried principles of the past are matched with the needs of the present generation of students.

John Pain is an Emeritus Reader in Physics from Imperial College.

Introduction

Students, unless highly motivated, tend to bypass introductions and it is left to the lecturer to find out what a book hopes to achieve. This book is based on *The Physics of Vibrations and Waves* (Wiley) which first appeared in 1968. The principles of that book were stated in its introduction and remain valid today. The theme is that a medium through which energy is transmitted via wave propagation behaves essentially as a continuum of coupled oscillations. A simple oscillator is characterized by three parameters, two of which are capable of storing and exchanging energy, while the third is energy dissipating. This is equally true of any medium. The product of the energy storing parameters determines the velocity of wave propagation through the medium and, in the absence of the third parameter, their ratio governs the impedance which the medium presents to the waves. The energy dissipating parameter introduces a loss term into the impedance, energy is absorbed from the wave system and it attenuates.

This viewpoint allows a discussion of simple harmonic, damped, forced and coupled oscillators which leads seamlessly to the behaviour of transverse waves on a string, longitudinal waves in a gas and a solid, to current and voltage waves on a transmission line and electromagnetic waves. All are amenable to this common treatment and it is the wide validity of relatively few physical principles which this book seeks to demonstrate.

What has changed since 1968 has not been in the students' favour. At that time the required mathematics had been done at school or was covered in the first weeks of a university course. This is no longer true and a major effort has been made in this book to address this change. Great emphasis is placed on the fact that a single mathematical principle covers a wide range of physical situations. There are three major principles used continuously throughout the book which may not be familiar to every student. They are:

(1) The square root of minus 1 (i).
(2) The exponential series (and its connection with the binomial theorem).
(3) Taylor's theorem which appears at the end of Chapter 4.

As principles (1) and (2) are already needed in Chapter 2, it has been decided to provide numbered working plans on these topics and sufficient examples at the beginning of Chapter 2 to help students over any difficulties, particularly if guided by a tutor. Where helpful, references to these are made in the text. A detailed derivation of the binomial theorem and its connection with the exponential series can be found at the end of the book together with an explanation of the Taylor series.

Each chapter contains a number of worked examples on which problems may be based.

Table of Constants

Charge on electron	1.602×10^{-19}	coulombs
Rest mass of electron	9.1×10^{-31}	kilograms
Atomic mass unit	1.66×10^{-27}	kilograms
1 newton	10^5	dynes
1 electron volt	1.6×10^{-19}	joules
Planck's constant h	6.62×10^{-34}	joule sec
Boltzmann's constant k	1.38×10^{-23}	joules/degree
	8.61×10^{-5}	electron volt/degree
Avogadro's number	6.022×10^{23}	per mole
Velocity of light c	3×10^8	metres/sec
Permeability of free space μ_0	$4\pi \times 10^{-7}$	henries/metre
Permittivity of free space ϵ_0	$(36\pi \times 10^9)^{-1}$	farads/metre

Table of Energy Storing Processes

Table showing how energy storing processes in a medium govern the wave velocity and the impedance. Potential energy is stored in medium via parameter C and kinetic or inductive energy is stored by ρ or L.

Type of Wave	(Velocity)^2	Impedance	Symbols	
transverse on string	T/ρ	ρc	T	tension
			ρ	linear density
			c	wave velocity
longitudinal in gas	$\gamma P/\rho = B/\rho = (\rho C)^{-1}$	$\rho c = \sqrt{\rho/C}$	γ	specific heat ratio
			P	gas pressure
			B	bulk modulus
			C	compressibility
			c	wave velocity
voltage and current on transmission line	$(L_0 C_0)^{-1}$	$\sqrt{L_0/C_0}$	L_0 inductance $\}$ per unit C_0 capacitance $\}$ length	
electromagnetic waves in a dielectric	$(\mu\epsilon)^{-1}$	$\sqrt{\mu/\epsilon}$	μ	permeability (henries/metre)
			ϵ	permittivity (farads/metre)

1

Simple Harmonic Motion

Notes to the students

After reading this chapter and completing the problems, you will understand:

- What a simple harmonic oscillator is.
- How a simple harmonic oscillator is described mathematically.
- How to use the equations describing simple harmonic motion to extract quantities of physical interest.
- The wide variety of physical systems that behave as simple harmonic oscillators. This will include all the oscillators and waves in this book except those in the last chapter.
- One of the most important things to learn as a physicist is that many seemingly different systems can be described in the same mathematical terms.

At first sight the eight physical systems in Figure 1.1 appear to have little in common.

1.1(a) is a mass fixed to a wall via a spring of stiffness s sliding to and fro in the x direction on a frictionless plane.

1.1(b) is a simple pendulum, a mass m swinging at the end of a light rigid rod of length l.

1.1(c) is a flat disc supported by a rigid wire through its centre and oscillating through small angles in the plane of its circumference.

1.1(d) is a mass m at the centre of a light string of length $2l$ fixed at both ends under a constant tension T. The mass vibrates in the plane of the paper.

1.1(e) is a frictionless U-tube of constant cross-sectional area containing a length l of liquid, density ρ, oscillating about its equilibrium position of equal levels in each limb.

1.1(f) is an open flask of volume V and a neck of length l and constant cross-sectional area A in which the air of density ρ vibrates as sound passes across the neck.

Introduction to Vibrations and Waves, First Edition. H. J. Pain and Patricia Rankin.
© 2015 John Wiley & Sons, Ltd. Published 2015 by John Wiley & Sons, Ltd.
Companion website: http://booksupport.wiley.com

1.1(g)   is a hydrometer, a body of mass m floating in a liquid of density ρ with a neck of constant cross-sectional area cutting the liquid surface. When depressed slightly from its equilibrium position it performs small vertical oscillations.

1.1(h)   is an electrical circuit, an inductance L connected across a capacitance C carrying a charge q.

All of these systems are simple harmonic oscillators which, when slightly disturbed from their equilibrium or rest postion, will oscillate with simple harmonic motion. This is the most fundamental vibration of a single particle or one-dimensional system. A small displacement x from its equilibrium position sets up a restoring force which is proportional to x acting in a direction towards the equilibrium position.

Thus, this restoring force F in Figure 1.1(a) may be written

$$F = -sx$$

where s, the constant of proportionality, is called the stiffness and the negative sign shows that the force is acting against the direction of increasing displacement and back towards the equilibrium position. A constant value of the stiffness restricts the displacement x to small values (this is Hooke's Law of Elasticity). The stiffness s is obviously the restoring force per unit distance (or displacement) and has the dimensions

$$\frac{\text{force}}{\text{distance}} \equiv \frac{MLT^{-2}}{L} \quad \text{where } T \text{ is time}$$

The equation of motion of such a disturbed system is given by the dynamic balance between the forces acting on the system, which by Newton's Law is

$$\text{mass times acceleration} = \text{restoring force}$$

or

$$m\ddot{x} = -sx$$

where the acceleration

$$\ddot{x} = \frac{\mathrm{d}^2 x}{\mathrm{d}t^2}$$

This gives

$$m\ddot{x} + sx = 0$$

or

$$\ddot{x} + \frac{s}{m}x = 0$$

where the dimensions of

$$\frac{s}{m} \quad \text{are} \quad \frac{MLT^{-2}}{ML} = T^{-2} = \nu^2$$

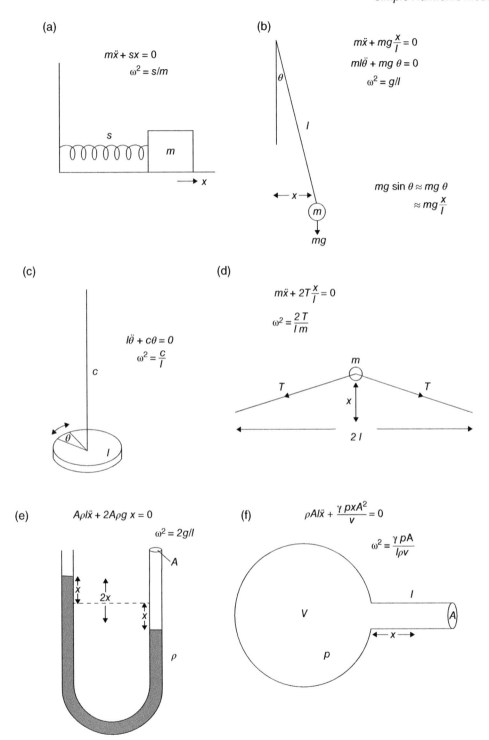

Figure 1.1 *Simple harmonic oscillators with their equations of motion and angular frequencies ω of oscillation. (a) A mass on a frictionless plane connected by a spring to a wall. (b) A simple pendulum. (c) A torsional pendulum. (d) A mass at the centre of a string under constant tension T. (e) A fixed length of non-viscous liquid in a U-tube of constant cross-section. (f) An acoustic Helmholtz resonator.*

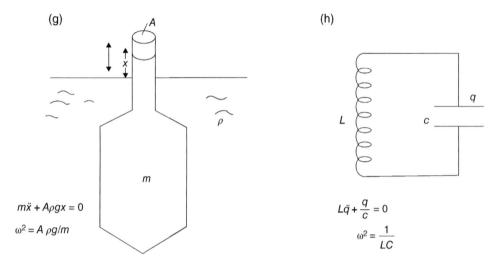

Figure 1.1 *(Continued)*   (g) A hydrometer mass m in a liquid of density ρ. (h) An electrical L C resonant circuit.

Here T is a time, or period of oscillation, the reciprocal of ν which is the frequency with which the system oscillates.

However, when we solve the equation of motion we shall find that the behaviour of x with time has a sinusoidal or cosinusoidal dependence, and it will prove more appropriate to consider, not ν, but the angular frequency $\omega = 2\pi\nu$ so that the period

$$T = \frac{1}{\nu} = 2\pi\sqrt{\frac{m}{s}}$$

where s/m is now written as ω^2. Thus the equation of simple harmonic motion

$$\ddot{x} + \frac{s}{m}x = 0$$

becomes

$$\boxed{\ddot{x} + \omega^2 x = 0} \tag{1.1}$$

1.1   Displacement in Simple Harmonic Motion

The behaviour of a simple harmonic oscillator is expressed in terms of its displacement x from equilibrium, its velocity $\dot{x}$, and its acceleration $\ddot{x}$ at any given time. If we try the solution

$$x = A\cos\omega t$$

where A is a constant with the same dimensions as x, we shall find that it satisfies the equation of motion

$$\ddot{x} + \omega^2 x = 0$$

for

$$\dot{x} = -A\omega \sin \omega t$$

and

$$\ddot{x} = -A\omega^2 \cos \omega t = -\omega^2 x$$

Another solution

$$x = B \sin \omega t$$

is equally valid, where B has the same dimensions as A, for then

$$\dot{x} = B\omega \cos \omega t$$

and

$$\ddot{x} = -B\omega^2 \sin \omega t = -\omega^2 x$$

The complete or general solution of equation (1.1) is given by the addition or superposition of both values for x so we have

$$x = A \cos \omega t + B \sin \omega t \qquad (1.2)$$

with

$$\ddot{x} = -\omega^2 \left(A \cos \omega t + B \sin \omega t\right) = -\omega^2 x$$

where A and B are determined by the values of x and $\dot{x}$ at a specified time. If we rewrite the constants as

$$A = a \sin \phi \quad \text{and} \quad B = a \cos \phi$$

where ϕ is a constant angle, then

$$A^2 + B^2 = a^2 (\sin^2 \phi + \cos^2 \phi) = a^2$$

so that

$$a = \sqrt{A^2 + B^2}$$

and

$$x = a \sin\phi \cos\omega t + a\cos\phi \sin\omega t$$
$$= a\sin(\omega t + \phi)$$

The maximum value of $\sin(\omega t + \phi)$ is unity so the constant a is the maximum value of x, known as the amplitude of displacement. The limiting values of $\sin(\omega t + \phi)$ are ± 1 so the system will oscillate between the values of $x = \pm a$ and we shall see that the magnitude of a is determined by the total energy of the oscillator.

The angle ϕ is called the 'phase constant' for the following reason. Simple harmonic motion is often introduced by reference to 'circular motion' because each possible value of the displacement x can be represented by the projection of a radius vector of constant length a on the diameter of the circle traced by the tip of the vector as it rotates in a positive anticlockwise direction with a constant angular velocity ω. Each rotation, as the radius vector sweeps through a phase angle of 2π rad, therefore corresponds to a complete vibration of the oscillator. In the solution

$$x = a\sin(\omega t + \phi)$$

the phase constant ϕ, measured in radians, defines the position in the cycle of oscillation at the time $t = 0$, so that the position in the cycle from which the oscillator started to move is

$$x = a\sin\phi$$

The solution

$$x = a\sin\omega t$$

defines the displacement only of that system which starts from the origin $x = 0$ at time $t = 0$ but the inclusion of ϕ in the solution

$$x = a\sin(\omega t + \phi)$$

where ϕ may take all values between zero and 2π allows the motion to be defined from any starting point in the cycle. This is illustrated in Figure 1.2 for various values of ϕ.

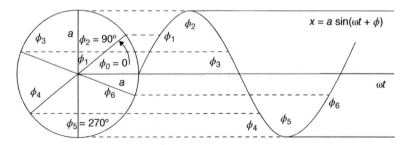

Figure 1.2   *Sinusoidal displacement of simple harmonic oscillator with time, showing variation of starting point in cycle in terms of phase angle ϕ.*

Worked Examples

Show that $x = A \cos \omega t + B \sin \omega t$ may be written as $x = a \left(\cos \omega t + \phi \right)$.

$x = a \left(\cos \omega t + \phi \right) = a \cos \omega t \cos \phi - a \sin \omega t \sin \phi$

$\therefore A = a \cos \phi$ and $B = -a \sin \phi$

The pendulum in Figure 1.1(b) swings with a displacement amplitude a. If its starting point from rest is (a) $x = a$, (b) $x = -a$, what are the values of ϕ in the solution $x = a \sin(\omega t + \phi)$?

$x = a \sin(\omega t + \phi) = a$ at $t = 0$ requires $\phi = \frac{\pi}{2}$

$x = a \sin(\omega t + \phi) = -a$ at $t = 0$ requires $\phi = -\frac{\pi}{2}$

If $x = a$ at $t = 0$ with $\phi = \frac{\pi}{2}$, at what values of ωt will $x = \frac{a}{\sqrt{2}}, \frac{a}{2}$ and $x = 0$?

Answers: $\omega t = \frac{\pi}{4}, \omega t = \frac{\pi}{3}, \omega t = \frac{\pi}{2}$.

1.2   Velocity and Acceleration in Simple Harmonic Motion

The values of the velocity and acceleration in simple harmonic motion for

$$x = a \sin(\omega t + \phi)$$

are given by

$$\frac{dx}{dt} = \dot{x} = a\omega \cos(\omega t + \phi)$$

and

$$\frac{d^2 x}{dt^2} = \ddot{x} = -a\omega^2 \sin(\omega t + \phi)$$

The maximum value of the velocity $a\omega$ is called the velocity amplitude and the acceleration amplitude is given by $a\omega^2$.

From Figure 1.2 we see that a positive phase angle of $\pi/2$ rad converts a sine into a cosine curve. Thus the velocity

$$\dot{x} = a\omega \cos(\omega t + \phi)$$

leads the displacement

$$x = a \sin(\omega t + \phi)$$

by a phase angle of $\pi/2$ rad and its maxima and minima are always a quarter of a cycle ahead of those of the displacement; the velocity is a maximum when the displacement is zero and is zero at maximum displacement. The acceleration is 'anti-phase' (π rad) with respect to the displacement, being maximum positive when the displacement is maximum negative and vice versa. These features are shown in Figure 1.3.

Often, the relative displacement or motion between two oscillators having the same frequency and amplitude may be considered in terms of their phase difference $\phi_1 - \phi_2$ which can have any value

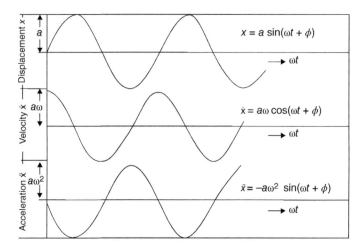

Figure 1.3 *Variation with time of displacement, velocity and acceleration in simple harmonic motion. Displacement lags velocity by $\pi/2$ rad and is π rad out of phase with the acceleration. The initial phase constant ϕ is taken as zero.*

because one system may have started several cycles before the other and each complete cycle of vibration represents a change in the phase angle of $\phi = 2\pi$. When the motions of the two systems are diametrically opposed; that is, one has $x = +a$ whilst the other is at $x = -a$, the systems are 'anti-phase' and the total phase difference

$$\phi_1 - \phi_2 = n\pi \text{ rad}$$

where n is an odd integer. Identical systems 'in phase' have

$$\phi_1 - \phi_2 = 2n\pi \text{ rad}$$

where n is any integer. They have exactly equal values of displacement, velocity and acceleration at any instant.

1.2.1 Non-linearity

If the stiffness s is constant, then the restoring force $F = -sx$, when plotted versus x, will produce a straight line and the system is said to be linear. The displacement of a linear simple harmonic motion system follows a sine or cosine behaviour. Non-linearity results when the stiffness s is not constant but varies with displacement x (see the beginning of Chapter 14).

1.3 Energy of a Simple Harmonic Oscillator

The fact that the velocity is zero at maximum displacement in simple harmonic motion and is a maximum at zero displacement illustrates the important concept of an exchange between kinetic and potential energy. In an ideal case the total energy remains constant but this is never realized in practice. If no energy is dissipated then all the potential energy becomes kinetic energy and vice versa, so that the values of

(a) the total energy at any time, (b) the maximum potential energy and (c) the maximum kinetic energy will all be equal; that is

$$E_{\text{total}} = \text{KE} + \text{PE} = \text{KE}_{\text{max}} = \text{PE}_{\text{max}}$$

The solution $x = a\sin(\omega t + \phi)$ implies that the total energy remains constant because the amplitude of displacement $x = \pm a$ is regained every half cycle at the position of maximum potential energy; when energy is lost the amplitude gradually decays as we shall see later in Chapter 2. The potential energy is found by summing all the small elements of work $sx \cdot dx$ (force sx times distance dx) done by the system against the restoring force over the range zero to x where $x = 0$ gives zero potential energy.

Thus the potential energy $=$

$$\int_0^x sx \cdot dx = \frac{1}{2}sx^2$$

The kinetic energy is given by $\frac{1}{2}m\dot{x}^2$ so that the total energy

$$\boxed{E = \frac{1}{2}m\dot{x}^2 + \frac{1}{2}sx^2}$$

Since E is constant we have

$$\frac{\mathrm{d}E}{\mathrm{d}t} = (m\ddot{x} + sx)\dot{x} = 0$$

giving again the equation of motion

$$m\ddot{x} + sx = 0$$

Worked Example

In Figure 1.1(g) the energy equation is

$$E = \frac{1}{2}m\dot{x}^2 + A\rho g x^2$$

$$\frac{\mathrm{d}E}{\mathrm{d}t} = (m\ddot{x} + 2A\rho g x)\dot{x} = 0$$

$$m\ddot{x} + sx = 0$$

with

$$s = 2A\rho g$$

$$\omega^2 = \frac{2A\rho g}{m} = \frac{2g}{l}$$

The maximum potential energy occurs at $x = \pm a$ and is therefore

$$\text{PE}_{\text{max}} = \frac{1}{2}sa^2$$

The maximum kinetic energy is

$$\mathrm{KE}_{\max} = \left(\frac{1}{2}m\dot{x}^2\right)_{\max} = \frac{1}{2}ma^2\omega^2 \left[\cos^2(\omega t + \phi)\right]_{\max}$$
$$= \frac{1}{2}ma^2\omega^2$$

when the cosine factor is unity.

But $m\omega^2 = s$ so the maximum values of the potential and kinetic energies are equal, showing that the energy exchange is complete.

The total energy at any instant of time or value of x is

$$E = \frac{1}{2}m\dot{x}^2 + \frac{1}{2}sx^2$$
$$= \frac{1}{2}ma^2\omega^2 \left[\cos^2(\omega t + \phi) + \sin^2(\omega t + \phi)\right]$$
$$= \frac{1}{2}ma^2\omega^2$$
$$= \frac{1}{2}sa^2$$

as we should expect.

Figure 1.4 shows the distribution of energy versus displacement for simple harmonic motion. Note that the potential energy curve

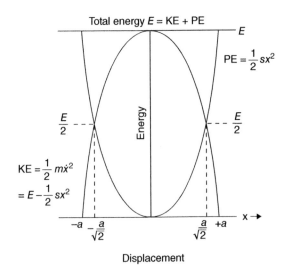

Figure 1.4 *Parabolic representation of potential energy and kinetic energy of simple harmonic motion versus displacement. Inversion of one curve with respect to the other shows a 90° phase difference. At any displacement value the sum of the ordinates of the curves equals the total constant energy E.*

$$PE = \frac{1}{2}sx^2 = \frac{1}{2}ma^2\omega^2 \sin^2(\omega t + \phi)$$

is parabolic with respect to x and is symmetric about $x = 0$, so that energy is stored in the oscillator both when x is positive and when it is negative, e.g. a spring stores energy whether compressed or extended, as does a gas in compression or rarefaction. The kinetic energy curve

$$KE = \frac{1}{2}m\dot{x}^2 = \frac{1}{2}ma^2\omega^2 \cos^2(\omega t + \phi)$$

is parabolic with respect to both x and $\dot{x}$. The inversion of one curve with respect to the other displays the $\pi/2$ phase difference between the displacement (related to the potential energy) and the velocity (related to the kinetic energy).

For any value of the displacement x the sum of the ordinates of both curves equals the total constant energy E.

Worked Example

A particle oscillates with simple harmonic motion along the x axis with a displacement amplitude a and spends a time dt in moving from x to $x + dx$. Show that the probability of finding it between x and $x + dx$ is given by

$$\frac{dx}{\pi(a^2 - x^2)^{\frac{1}{2}}}$$

Let

$$x = a\sin(\omega t + \phi)$$

then

$$dt = \frac{dx}{v}$$

where

$$v = \dot{x} = a\omega(\cos\omega t + \phi)$$

Particle is at same dx twice per oscillation.
∴

$$\text{probability} = \eta = \frac{2dt}{T}$$

where

$$T = \text{period} = \frac{2\pi}{\omega}$$

$$\therefore$$

$$\eta = \frac{2dt}{T}$$

$$= \frac{2\omega dx}{2\pi a\omega(\cos\omega t + \phi)}$$

$$= \frac{dx}{\pi a \cos(\omega t + \phi)}$$

$$= \frac{dx}{\pi a(1 - \sin^2(\omega t + \phi))^{\frac{1}{2}}}$$

$$= \frac{dx}{\pi(a^2 - x^2)^{\frac{1}{2}}}$$

1.4   Simple Harmonic Oscillations in an Electrical System

So far we have discussed the simple harmonic motion of the mechanical and fluid systems of Figure 1.1, chiefly in terms of the inertial mass stretching the weightless spring of stiffness s. The stiffness s of a spring defines the difficulty of stretching; the reciprocal of the stiffness, the compliance C (where $s = 1/C$) defines the ease with which the spring is stretched and potential energy stored. This notation of compliance C is useful when discussing the simple harmonic oscillations of the electrical circuit of Figure 1.1(h) and Figure 1.5, where an inductance L is connected across the plates of a capacitance C. The force equation of the mechanical and fluid examples now becomes the voltage equation (balance of voltages) of the electrical circuit, but the form and solution of the equations and the oscillatory behaviour of the systems are identical.

In the absence of resistance the energy of the electrical system remains constant and is exchanged between the magnetic field energy stored in the inductance and the electric field energy stored between the plates of the capacitance. At any instant, the voltage across the inductance is

$$V = -L\frac{\mathrm{d}I}{\mathrm{d}t} = -L\frac{\mathrm{d}^2q}{\mathrm{d}t^2}$$

where I is the current flowing and q is the charge on the capacitor, the negative sign showing that the voltage opposes the increase of current. This equals the voltage q/C across the capacitance so that

$$L\ddot{q} + q/C = 0 \qquad \text{(Kirchhoff's Law)}$$

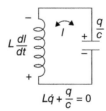

$$L\ddot{q} + \frac{q}{C} = 0$$

Figure 1.5   *Electrical system which oscillates simple harmonically. The sum of the voltages around the circuit is given by Kirchhoff's law as $L\,\mathrm{d}I/\mathrm{d}t + q/C = 0$.*

or

$$\ddot{q} + \omega^2 q = 0$$

where

$$\omega^2 = \frac{1}{LC}$$

The circuit in Figure 1.5 is very useful in producing oscillators with fixed frequencies in the range 30 Hz (low frequency acoustic) to 50 MHz (high frequency stereo). e.g. $L = 1\mu$H and $C = 25$pF oscillates at a frequency of 31.26 MHz.

The energy stored in the magnetic field or inductive part of the circuit throughout the cycle, as the current increases from 0 to I, is formed by integrating the power at any instant with respect to time; that is

$$E_L = \int VI \cdot dt$$

(where V is the magnitude of the voltage across the inductance).

So

$$E_L = \int VIdt = \int L\frac{dI}{dt} I \, dt = \int_0^I LIdI$$

$$= \frac{1}{2}LI^2 = \frac{1}{2}L\dot{q}^2$$

The potential energy stored mechanically by the spring is now stored electrostatically by the capacitance and equals

$$\frac{1}{2}CV^2 = \frac{q^2}{2C}$$

Comparison between the equations for the mechanical and electrical oscillators

$$\text{mechanical (force)} \rightarrow m\ddot{x} + sx = 0$$

$$\text{electrical (voltage)} \rightarrow L\ddot{q} + \frac{q}{C} = 0$$

$$\text{mechanical (energy)} \rightarrow \frac{1}{2}m\dot{x}^2 + \frac{1}{2}sx^2 = E$$

$$\text{electrical (energy)} \rightarrow \frac{1}{2}L\dot{q}^2 + \frac{1}{2}\frac{q^2}{C} = E$$

shows that magnetic field inertia (defined by the inductance L) controls the rate of change of current for a given voltage in a circuit in exactly the same way as the inertial mass controls the change of velocity for a given force. Magnetic inertial or inductive behaviour arises from the tendency of the magnetic flux threading a circuit to remain constant and reaction to any change in its value generates a voltage and hence a current which flows to oppose the change of flux. This is the physical basis of Fleming's right-hand rule.

1.5   Superposition of Two Simple Harmonic Vibrations in One Dimension

(1) Vibrations Having Equal Frequencies

In the following chapters we shall meet physical situations which involve the superposition of two or more simple harmonic vibrations on the same system.

We have already seen how the displacement in simple harmonic motion may be represented in magnitude and phase by a constant length vector rotating in the positive (anticlockwise) sense with a constant angular velocity ω. To find the resulting motion of a system which moves in the x direction under the simultaneous effect of two simple harmonic oscillations of equal angular frequencies but of different amplitudes and phases, we can represent each simple harmonic motion by its appropriate vector and carry out a vector addition.

If the displacement of the first motion is given by

$$x_1 = a_1 \cos (\omega t + \phi_1)$$

and that of the second by

$$x_2 = a_2 \cos (\omega t + \phi_2)$$

then Figure 1.6 shows that the resulting displacement amplitude R is given by

$$R^2 = (a_1 + a_2 \cos \delta)^2 + (a_2 \sin \delta)^2$$
$$= a_1^2 + a_2^2 + 2a_1a_2 \cos \delta$$

where $\delta = \phi_2 - \phi_1$ is constant.

The phase constant θ of R is given by

$$\tan \theta = \frac{a_1 \sin \phi_1 + a_2 \sin \phi_2}{a_1 \cos \phi_1 + a_2 \cos \phi_2}$$

so the resulting simple harmonic motion has a displacement

$$x = R \cos(\omega t + \theta)$$

an oscillation of the same frequency ω but having an amplitude R and a phase constant θ.

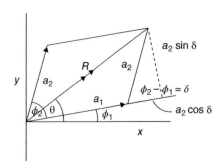

Figure 1.6   *Addition of vectors, each representing simple harmonic motion along the x axis at angular frequency ω to give a resulting simple harmonic motion displacement $x = R\cos(\omega t + \theta)$— here shown for $t = 0$.*

Worked Example

If $a_1 = a_2 = a$ in Figure 1.6, show that $R^2 = 4a^2 \cos^2 \frac{\delta}{2}$.

$$R^2 = 2a^2 + 2a^2 \cos \delta = 2a^2 + 2a^2 \left(2\cos^2 \frac{\delta}{2} - 1 \right) = 4a^2 \cos^2 \frac{\delta}{2}$$

(2) Vibrations Having Different Frequencies

Suppose we now consider what happens when two vibrations of equal amplitudes but different frequencies are superposed. If we express them as

$$x_1 = a \sin \omega_1 t$$

and

$$x_2 = a \sin \omega_2 t$$

where

$$\omega_2 > \omega_1$$

then the resulting displacement is given by

$$x = x_1 + x_2 = a(\sin \omega_1 t + \sin \omega_2 t)$$
$$= 2a \sin \frac{(\omega_1 + \omega_2)t}{2} \cos \frac{(\omega_2 - \omega_1)t}{2}$$

This expression is illustrated in Figure 1.7. It represents a sinusoidal oscillation at the average frequency $(\omega_1 + \omega_2)/2$ having a displacement amplitude of $2a$ which modulates; that is, varies between $2a$ and

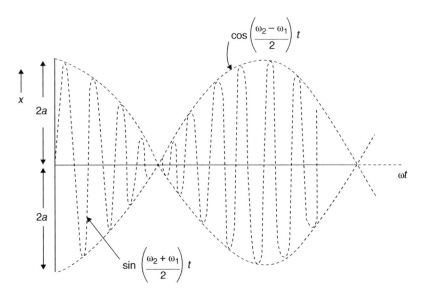

Figure 1.7   *Superposition of two simple harmonic displacements $x_1 = a \sin\omega_1 t$ and $x_2 = a \sin\omega_2 t$ when $\omega_2 > \omega_1$. The slow $\cos[(\omega_2 - \omega_1)/2]t$ envelope modulates the $\sin [(\omega_2 + \omega_1)/2]t$ curve between the values $x = \pm 2a$.*

zero under the influence of the cosine term of a much slower frequency equal to half the difference $(\omega_2 - \omega_1)/2$ between the original frequencies.

When ω_1 and ω_2 are almost equal the sine term has a frequency very close to both ω_1 and ω_2 whilst the cosine envelope modulates the amplitude $2a$ at a frequency $(\omega_2 - \omega_1)/2$ which is very slow.

Acoustically this growth and decay of the amplitude is registered as 'beats' of strong reinforcement when two sounds of almost equal frequency are heard. The frequency of the 'beats' is $(\omega_2 - \omega_1)$, the difference between the separate frequencies (not half the difference) because the maximum amplitude of $2a$ occurs twice in every period associated with the frequency $(\omega_2 - \omega_1)/2$. We shall meet this situation again when we consider the coupling of two oscillators in Chapter 4 and the wave group of two components in Chapter 6.

Problem 1.1.   The equation of motion

$$m\ddot{x} = -sx \quad \text{with} \quad \omega^2 = \frac{s}{m}$$

applies directly to the system in Figure 1.1(a).

If the pendulum bob of Figure 1.1(b) is displaced a small distance x show that the stiffness (restoring force per unit distance) is mg/l and that $\omega^2 = g/l$ where g is the acceleration due to gravity. Now use the small angular displacement θ instead of x and show that ω is the same.

In Figure 1.1(c) the angular oscillations are rotational so the mass is replaced by the moment of inertia I of the disc and the stiffness by the restoring couple of the wire which is C rad^{-1} of angular displacement. Show that $\omega^2 = C/I$.

In Figure 1.1(d) show that the stiffness is $2T/l$ and that $\omega^2 = 2T/lm$.

In Figure 1.1(e) show that the stiffness of the system is $2\rho Ag$, where A is the area of cross section and that $\omega^2 = 2g/l$ where g is the acceleration due to gravity.

In Figure 1.1(f) only the gas in the flask neck oscillates, behaving as a piston of mass ρAl. If the pressure changes are calculated from the equation of state use the adiabatic relation $pV^\gamma = $ constant and take logarithms to show that the pressure change in the flask is

$$dp = -\gamma p \frac{dV}{V} = -\gamma p \frac{Ax}{V},$$

where x is the gas displacement in the neck. Hence show that $\omega^2 = \gamma pA/l\rho V$. Note that γp is the stiffness of a gas (see Chapter 7).

In Figure 1.1(g), if the cross-sectional area of the neck is A and the hydrometer is a distance x above its normal floating level, the restoring force depends on the volume of liquid displaced (Archimedes' principle). Show that $\omega^2 = g\rho A/m$.

Check the dimensions of ω^2 for each case.

Problem 1.2.   Show by the choice of appropriate values for A and B in equation (1.2) that equally valid solutions for x are

$$x = a\cos(\omega t + \phi)$$
$$x = a\sin(\omega t - \phi)$$
$$x = a\cos(\omega t - \phi)$$

and check that these solutions satisfy the equation

$$\ddot{x} + \omega^2 x = 0$$

Problem 1.3.   The pendulum in Figure 1.1(a) swings with a displacement amplitude a. If its starting point from rest is

$$(a)\, x = a$$
$$(b)\, x = -a$$

find the different values of the phase constant ϕ for the solutions

$$x = a\sin(\omega t + \phi)$$
$$x = a\cos(\omega t + \phi)$$
$$x = a\sin(\omega t - \phi)$$
$$x = a\cos(\omega t - \phi)$$

For each of the different values of ϕ, find the values of ωt at which the pendulum swings through the positions

$$x = +a/\sqrt{2}$$
$$x = a/2$$

and

$$x = 0$$

for the first time after release from

$$x = \pm a$$

Problem 1.4.   When the electron in a hydrogen atom bound to the nucleus moves a small distance from its equilibrium position, a restoring force per unit distance is given by

$$s = e^2/4\pi\epsilon_0 r^3$$

where $r = 0.05$ nm may be taken as the radius of the atom. Show that the electron can oscillate with a simple harmonic motion with

$$\omega_0 \approx 4.5 \times 10^{16} \mathrm{rad\,s}^{-1}$$

If the electron is forced to vibrate at this frequency, in which region of the electromagnetic spectrum would its radiation be found?

$$e = 1.6 \times 10^{-19} \mathrm{C}, \text{ electron mass } m_e = 9.1 \times 10^{-31} \mathrm{kg}$$
$$\epsilon_0 = 8.85 \times 10^{-12} \mathrm{N}^{-1}\mathrm{m}^{-2}\mathrm{C}^2$$

Problem 1.5.   Show that the values of ω^2 for the three simple harmonic oscillations (a), (b), (c) in the diagram are in the ratio 1: 2: 4.

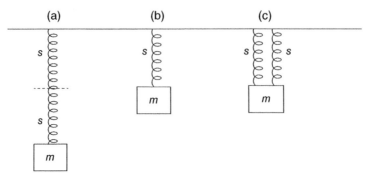

Problem 1.6.   The displacement of a simple harmonic oscillator is given by

$$x = a \sin(\omega t + \phi)$$

If the oscillation started at time $t = 0$ from a position x_0 with a velocity $\dot{x} = v_0$ show that

$$\tan \phi = \omega x_0 / v_0$$

and

$$a = \left(x_0^2 + v_0^2 / \omega^2 \right)^{1/2}$$

Problem 1.7.   Following the 2nd worked example in section 1.3 Energy of a Simple Harmonic Oscillator, show that if the charge on the capacitor of Figure 1.5 is given by $q = q_0 \cos \omega t$ the probability of q having a value between q and $q + dq$ is

$$\eta = \frac{1}{\pi} \frac{dq}{\left(q_0^2 - q^2 \right)^{\frac{1}{2}}}$$

Use this to write down the equation for the probability of the current in the inductance L to lie between I and $I + dI$ where $I = \dot{q}$.

Problem 1.8.   Many identical simple harmonic oscillators are equally spaced along the x axis of a medium and a photograph shows that the locus of their displacements in the y direction is a sine curve. If the distance λ separates oscillators which differ in phase by 2π radians, what is the phase difference between two oscillators a distance x apart?

Problem 1.9.   A mass stands on a platform which vibrates simple harmonically in a vertical direction at a frequency of 5 Hz. Show that the mass loses contact with the platform when the displacement exceeds 10^{-2} m.

Problem 1.10.   A mass M is suspended at the end of a spring of length l and stiffness s. If the mass of the spring is m and the velocity of an element dy of its length is proportional to its distance y from the fixed end of the spring, show that the kinetic energy of this element is

$$\frac{1}{2}\left(\frac{m}{l}dy\right)\left(\frac{y}{l}v\right)^2$$

where v is the velocity of the suspended mass M. Hence, by integrating $y^2 dy$ over the length of the spring, show that its total kinetic energy is $\frac{1}{6}mv^2$ and, from the total energy of the oscillating system, show that the frequency of oscillation is given by

$$\omega^2 = \frac{s}{M + m/3}$$

Problem 1.11.   The general form for the energy of a simple harmonic oscillator is

$$E = \frac{1}{2}\text{mass (velocity)}^2 + \frac{1}{2}\text{stiffness (displacement)}^2$$

Set up the energy equations for the oscillators in Figure 1.1(a), (b), (c), (d), (e), (f) and (g), and use the expression

$$\frac{dE}{dt} = 0$$

to derive the equation of motion in each case.

Problem 1.12.   The displacement of a simple harmonic oscillator is given by $x = a\sin\omega t$. If the values of the displacement x and the velocity $\dot{x}$ are plotted on perpendicular axes, eliminate t to show that the locus of the points $(x, \dot{x})$ is a circle. Show that this circle represents a path of constant energy.

Problem 1.13.   In Chapter 12 the intensity of the pattern when light from two slits interfere (Young's experiment) will be seen to depend on the superposition of two simple harmonic oscillations of equal amplitude a and phase difference δ. Show that the intensity

$$I = R^2 \propto 4a^2 \cos^2 \delta/2$$

Between what values does the intensity vary?

Problem 1.14.   The electrical circuit of Figure 1.5 has $L = 250$ mH and $C = 100$ μF. Show that it oscillates with a frequency $\nu \approx 31.2$ Hz (the low end of human acoustic range). If $L = 25$ μH and $C = 4$ μF, show that $\nu = 16$ kHz (the top end of the human acoustic range).

Problem 1.15.   The equation for an ellipse is

$$\frac{x^2}{a^2} + \frac{y^2}{b^2} = 1$$

where a is the horizontal semi axis and b is the vertical semi axis. The coordinates of the displacement of a particle of mass m are given by

$$x = a\sin\omega t$$
$$y = b\cos\omega t$$

Eliminate t to show that the particle follows an elliptical path and show by adding its kinetic and potential energy at any position x, y that the ellipse is a path of constant energy equal to the sum of the separate energies of the simple harmonic vibrations.

Prove that the quantity $m(x\dot{y} - y\dot{x})$ is also constant. This quantity represents the angular momentum which changes sign when the motion reverses its direction.

Problem 1.16.   If, in Figure 1.7, $\omega_1 \approx \omega_2 = \omega$ and $\omega_1 - \omega_2 = \Delta\omega$, show that the number of rapid oscillations between two consecutive zeros of the slow frequency envelope is $\frac{\omega}{\Delta\omega}$.

2

Damped Simple Harmonic Motion

Introduction

The introduction to this book states that the behaviour of any oscillator is governed by three parameters, two of which store and exchange energy while the third parameter causes energy loss. The first chapter dealt with the two energy storing parameters. The energy of simple harmonic motion is constant. This chapter introduces the third parameter which dissipates the energy and changes the behaviour of the simple harmonic oscillator. One of the changes is the decay of the oscillations - a decay which is known as exponential.

Where the force s is a *force per unit distance* the resistive dissipating force r is a *force per unit velocity* which acts to oppose the direction of motion. The rate at which the energy decays is determined by the relative strength of two forces r and s. There are three regions of relative strength:

(1) $r/2$ is greater than s – which is non oscillatory.
(2) $r/2$ is equal to s – which is non oscillatory.
(3) $r/2$ is less than s – which is oscillatory.

The region $r/2$ is greater than s is of least interest to us. It is called 'dead beat'.

The region $r/2$ is equal to s is important because it explains the behaviour of all shock absorbers. It is known as critical damping and describes the response of a system initially at rest which is subject to a sudden jolt and is required to return to equilibrium in the minimum possible time.

Region 3, $r/2$ is less than s, is where the amplitude of the oscillator is gradually reduced as energy is lost due to the action of r. Two particular methods which describe this damping are the logarithmic decrement and the Q value of the system.

This chapter requires two mathematical techniques with which you may not be familiar. These are:

(a) the use of i equals the square root of minus one, and
(b) the exponential series which describes the laws of natural growth (compound interest) and decay (damped oscillations).

Introduction to Vibrations and Waves, First Edition. H. J. Pain and Patricia Rankin.
© 2015 John Wiley & Sons, Ltd. Published 2015 by John Wiley & Sons, Ltd.
Companion website: http://booksupport.wiley.com

The formal derivations of the Binomial Theorem, the exponential series and Taylor's series appear at the end of the book, in Appendices 1 and 2, but a working plan of six points is presented here which covers all the aspects of the exponential series which you will meet in this book. They are numbered and are referred to in the text where appropriate. They are offered as immediate help in this chapter and for reference in later use. Ask your tutor for help (or discuss in a study group) if you have any difficulty in following the examples and look out for applications in the text.

2.1   Complex Numbers

The algebra of complex numbers is straightforward. A complex number has two parts, one real and one imaginary. It is written $z = a + ib$ where a is real and ib is imaginary because $i = \sqrt{-1}$.

(i)   If $z_1 = a + ib$ and $z_2 = c + id$ and $z_1 = z_2$ then $a = c$ and $b = d$ (real parts are equal and imaginary parts are equal.)

(ii) $z_1 + z_2 = (a + c) + i(b + d)$

(iii) $z_1 \times z_2 = (a + ib)(c + id) = (ac - bd) + i(ad + bc)$ for $i^2 = -1$

(iv) z^* is the complex conjugate of z which changes i to $-i$ so $z_1^* = a - ib$ is the complex conjugate of $z_1 = a + ib$

(v) $|z_1|$ is the magnitude of $\sqrt{z_1 z_1^*} = [(a + ib)(a - ib)]^{\frac{1}{2}} = (a^2 + b^2)^{\frac{1}{2}}$

(vi)

$$\frac{z_1}{z_2} = \frac{z_1 z_2^*}{z_2 z_2^*} = \frac{(a + ib)(c - id)}{(c + id)(c - id)} = \frac{(ac + bd) + i(bc - ad)}{c^2 + d^2}$$

(vii)   If $b > a$ then $\sqrt{a - b}$ is imaginary, written $\sqrt{i^2 a - i^2 b} = \pm i\sqrt{b - a}$ (this is used in the section on damped oscillations).

2.2   The Exponential Series

(1) The **exponential series** is written

$$e^x = 1 + x + \frac{x^2}{2!} + \frac{x^3}{3!} - \cdots \frac{x^n}{n!} \quad \text{where} \quad n \to \infty \quad \text{and}$$

$$n! \text{ (called } n \text{ factorial)} = 1 \cdot 2 \cdot 3 \cdot 4 \cdots n.$$

Both e^x and e^{-x} are shown in Figure 2.1, respectively as growth and decay processes. Note that e^x and e^{-x} equal 1 for $x = 0$.

$$\frac{de^x}{dx} = 1 + x + \frac{x^2}{2!} + \frac{x^3}{3!} \cdots = e^x$$

so the function equals its gradient at x.

$$\frac{d^2 e^x}{dx} = 1 + x + \frac{x^2}{2!} + \frac{x^3}{3!} \cdots = e^x$$

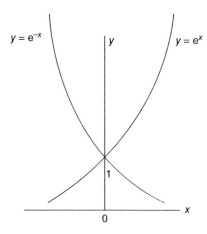

Figure 2.1   *The behaviour of the exponential series $y = e^x$ and $y = e^{-x}$.*

(2) When $x = 1$

$$e^x = e = 1 + 1 + \frac{1}{2!} + \frac{1}{3!} + \frac{1}{4!} + \cdots$$

which is the base of natural logarithms.

Logarithms are first met with a base 10. The logarithm of a number is the power to which the base must be raised in order to equal the given number. Thus

$$\log_{10} 100 = 2 \quad \log_{10} 10 = 1 \quad \log_{10} 1 = 0$$

($\log 1$ to any base is zero.)

2.2.1   The Exponential Series and the Law of Compound Interest

The law of compound interest may be written as $(1 + 1/n)^n$ where the bracket $(1 + 1/n)$ is multiplied by itself n times. Within the bracket the digit one represents the unit of currency being saved, a dollar, pound sterling or euro, and the term $1/n$ represents the fraction of that unit pound as interest. As a percentage of interest we write $100/n\%$. There are n payments of interest (the index outside the bracket) and each time interest is paid the savings increase by a factor $(1 + 1/n)$.

We consider three rates of interest to observe the growth of capital. If $n = 20$ the rate of interest is $100/20 = 5\%$ and after 20 payments of interest the capital has grown by a factor of 2.63. If $n = 40$ the rate of interest is $100/40 = 2\frac{1}{2}\%$ and after 40 payments of interest the capital has grown by a factor of 2.685. If $n = 100$ the rate of interest is $100/100 = 1\%$ and after 100 payments the capital has grown by 2.705. In the limit as $n \to \infty$ as $1/n \to 0$ we have

$$\lim_{n \to \infty} \left(1 + \frac{1}{n}\right)^n = 2.718 = e$$

the base of natural logarithms, which describes the infinitesimal growth (or decay with a negative sign) of natural systems. Finally we note that $[(1+1/n)^n]^2 = e \cdot e = e^2 : (\text{for } n \to \infty) [(1+1/n)^n]^3 = e \cdot e \cdot e = e^3$ and $[(1+1/n)^n]^x = e^x = 1 + x + x^2/2! + x^3/3! \cdots$

(3) $e^{\pm \alpha x}$

Often we shall meet expressions such as $e^{\pm \alpha x}$ where α is real. The index αx is a power of the base and is a pure number, so α has the dimensions of $1/x$ which often gives us a great deal of physical information. (See for example note 6 on Relaxation Time.)

$$e^{\alpha x} = 1 + \alpha x + \frac{\alpha^2 x^2}{2!} + \frac{\alpha^3 x^3}{3!} + \cdots$$

$$\frac{d e^{\alpha x}}{dx} = \alpha + \frac{2\alpha^2}{2!}x + \frac{3\alpha^3}{3!}x^2 = \alpha \left(1 + \alpha x + \frac{\alpha^2}{2!}x^2 + \frac{\alpha^3}{3!}x^3\right) = \alpha e^{\alpha x}$$

$$\text{Similarly} \quad \frac{d^2 e^{\alpha x}}{dx} = \alpha^2 e^{\alpha x}$$

(4) e^{ix}

In this case the index is imaginary and yields a great deal of information.

$$e^{ix} = 1 + ix + \frac{i^2 x^2}{2!} + \frac{i^3 x^3}{3!} + \frac{i^4 x^4}{4!} + \cdots = 1 + ix - \frac{x^2}{2!} - \frac{ix^3}{3!} + \frac{x^4}{4!} + \cdots$$

$$= 1 - \frac{x^2}{2!} + \frac{x^4}{4!} + \cdots + i \left(x - \frac{x^3}{3!} - \frac{x^5}{5!} + \cdots\right)$$

$$= \cos x + i \sin x$$

$$\cos x = (e^{ix} + e^{-ix})/2 \quad \text{and} \quad \sin x = (e^{ix} - e^{-ix})/2i$$

We see also that

$$\frac{d}{dx} e^{ix} = i e^{ix} = i \cos x - \sin x$$

Often we shall represent a sine or cosine oscillation in the form e^{ix} and recover the original form by taking that part of the solution preceded by i in the case of the sine and the real part of the solution when the oscillation is that of a cosine.

Let us consider the expression $e^{i\omega t}$ and try $x = a e^{i\omega t} = a(\cos\omega t + i\sin\omega t)$ as a solution to the simple harmonic motion where a is a constant length (the amplitude) and ω is the constant angular frequency.

$$\frac{dx}{dt} = \dot{x} = i\omega a e^{i\omega t} = i\omega x$$

$$\frac{d^2 x}{dt^2} = \ddot{x} = i^2 \omega^2 a e^{i\omega t} = -\omega^2 x$$

$$\therefore \ddot{x} + \omega^2 x = 0$$

and $x = a(\cos\omega t + i\sin\omega t)$ is a complete solution of the simple harmonic motion equation.

(5) $\int \frac{dy}{y} = \log_e y + \text{constant}$

If $y = e^x$ then $x = \log_e y$ and

$$\frac{dy}{dx} = \log_e y = \frac{1}{\frac{dx}{dy}} = e^x = y$$

$$\therefore dx = \frac{dy}{y} \quad \text{and} \quad \int dx = \int \frac{dy}{y} = \log_e y + \text{constant}$$

This results in the Differential Form of the Exponential Series

$$\text{Putting } \frac{dy}{y} = \frac{dN}{N} = \pm \alpha dx$$

where α is a real constant and $\frac{dN}{N}$ is a constant fraction of growth or decay with x (or t).

$$\text{we have} \int_{N_0}^{N} \frac{dN}{N} = \pm \alpha \int dx \quad \therefore \log_e N - \log_e N_0 = \pm \alpha x$$

$$\therefore N = N_0 e^{\pm \alpha x}$$

where $N = N_0$ is the original value of N at $x = 0$.

(6) e^{-1} Relaxation Time

Theoretically the time for an exponential to decay to zero is infinite. The convention for comparison is to choose the time, known as the **relaxation time**, for the system to decay to e^{-1}, that is $1/e$, of its original value.

Example

A capacitor C discharges q through a resistance R. The voltage equation around the circuit is given

$$IR + q/C = 0 \quad \text{where } I = dq/dt$$

$$\therefore R\frac{dq}{dt} = \frac{-q}{C} \quad \therefore \frac{dq}{q} = -\frac{dt}{RC} \quad \text{and} \quad \int_{q_0}^{q} \frac{dq}{q} = -\int_{t_0}^{t} \frac{dt}{RC}$$

giving $q = q_0 e^{-t/RC}$ where q_0 is the original charge on C.
At $t = RC$, $q = q_0 e^{-1}$ so the relaxation time is $t = RC$. Check that RC has the dimensions of time t.

2.2.2 Note on the Binomial Theorem

The expansion of $(1 + 1/n)^n$ used in deriving the exponential e is close to that of the binomial series which is explained in detail in Appendix 1, showing the connection.

For our purpose here the binomial series is written $(1 + x)^n$ where $-1 < x < 1$ and x can be integral or a fraction. Expanding this expression, we have

$$(1 + x)^n = 1 + nx + \frac{n(n-1)x^2}{2!} + \frac{n(n-1)(n-2)x^3}{3!}$$

$$\text{e.g. } (1 + x)^2 = 1 + 2x + x^2 \approx 1 + 2x \text{ neglecting } x^2$$

which is a very small error when $|x| < 1$.

It is a very common practice to take the first two terms of the expansion as an approximation – the error being of the order of the first term which is dropped.

The square root form $(1 + x)^{1/2}$ is often written $1 + \frac{1}{2}x$. This will be used and pointed out in the text. Initially we discussed the case of ideal simple harmonic motion where the total energy remained constant and the displacement followed a sine curve, apparently for an infinite time. In practice some energy is always dissipated by a resistive or viscous process; for example, the amplitude of a freely swinging pendulum will always decay with time as energy is lost. The presence of resistance to motion means that another force is active, which is taken as being proportional to the velocity. The frictional force acts in the direction opposite to that of the velocity (see Figure 2.2) and so applying Newton's Second Law, the equation of motion, becomes

$$m\ddot{x} = -sx - r\dot{x}$$

where r is the constant of proportionality and has the dimensions of force per unit of velocity. The presence of such a term will always result in energy loss.

The problem now is to find the behaviour of the displacement x from the equation

$$m\ddot{x} + r\dot{x} + sx = 0 \tag{2.1}$$

where the coefficients m, r and s are constant.

When these coefficients are constant a solution of the form $x = Ce^{\alpha t}$ can always be found. Since an exponential term is always nondimensional, C has the dimensions of x (a length, say) and α has the dimensions of inverse time, T^{-1}. We shall see that there are three possible forms of this solution, each

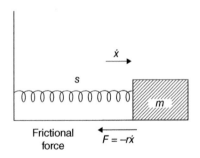

Figure 2.2   *Simple harmonic motion system with a damping or frictional force $r\dot{x}$ acting against the direction of motion. The equation of motion is $m\ddot{x} + r\dot{x} + sx = 0$.*

describing a different behaviour of the displacement x with time. In the first of these solutions C appears explicitly as a constant length, but in the second and third cases it takes the form

$$C = A + Bt \qquad *$$

where A is a length, B is a velocity and t is time, giving C the overall dimensions of a length, as we expect. From our point of view the first case is not the most important.

Taking C as a constant length gives $\dot{x} = \alpha C e^{\alpha t}$ and $\ddot{x} = \alpha^2 C e^{\alpha t}$, so that equation (2.1) may be rewritten

$$C e^{\alpha t} (m\alpha^2 + r\alpha + s) = 0$$

so that either

$$x = C e^{\alpha t} = 0 \quad \text{(which is trivial)}$$

or

$$m\alpha^2 + r\alpha + s = 0$$

Solving the quadratic equation in α gives

$$\alpha = \frac{-r}{2m} \pm \sqrt{\frac{r^2}{4m^2} - \frac{s}{m}} \quad \text{where} \quad \frac{s}{m} = \omega_0^2$$

Note that $r/2m$ and $(\omega_0^2)^{1/2}$, and therefore, α, all have the dimensions of inverse time, that is a frequency T^{-1}, which we expect from the form of $e^{\alpha t}$.

The displacement can now be expressed as

$$x_1 = C_1 e^{-rt/2m + (r^2/4m^2 - \omega_0^2)^{1/2}t}, \quad x_2 = C_2 e^{-rt/2m - (r^2/4m^2 - \omega_0^2)^{1/2}t}$$

or the sum of both these terms

$$x = x_1 + x_2 = C_1 e^{-rt/2m + (r^2/4m^2 - \omega_0^2)^{1/2}t} + C_2 e^{-rt/2m - (r^2/4m^2 - \omega_0^2)^{1/2}t} \tag{2.2}$$

The bracket $(r^2/4m^2 - \omega_0^2)$ can be positive, zero or negative depending on the relative magnitude of the two terms inside it. Each of these conditions gives one of the three possible solutions referred to earlier and each solution describes a particular kind of behaviour. We shall discuss these solutions in order of increasing significance from our point of view; the third solution is the one we shall concentrate upon throughout the rest of this book.

The conditions are:

(1) Bracket positive $(r^2/4m^2 > \omega_0^2)$. Here the damping resistance term $r^2/4m^2$ dominates the stiffness term s/m, and heavy damping results in a dead beat system.

---

*The number of constants allowed in the general solution of a differential equation is always equal to the order (that is, the highest differential coefficient) of the equation. The two values A and B are allowed because equation (2.1) is second order. The values of the constants are adjusted to satisfy the initial conditions.

(2) Bracket zero ($r^2/4m^2 = \omega_0^2$). The balance between the two terms results in a critically damped system.

Neither (1) nor (2) gives oscillatory behaviour.

(3) Bracket negative ($r^2/4m^2 < \omega_0^2$). The system is lightly damped and gives oscillatory damped simple harmonic motion.

2.2.3   *Region 1. Heavy Damping* ($r^2/4m^2 > \omega_0^2$)

We can write equation 2.2 as

$$C_1 e^{-\alpha_1 t} + C_2 e^{-\alpha_2 t} = 0$$

where

$$\alpha_1 \equiv \left[\frac{r}{2m} + \left(\frac{r^2}{4m^2} - \omega_0^2 \right)^{\frac{1}{2}} \right]$$

and

$$\alpha_2 \equiv \left[\frac{r}{2m} - \left(\frac{r^2}{4m^2} - \omega_0^2 \right)^{\frac{1}{2}} \right]$$

α_1 and α_2 are positive and we see that $-\alpha_2$ with three negative coefficients to t is dominant in terms of the decay after the system is displaced from equilibrium.

Note that

$$\alpha_1 > \frac{r}{2m} > \omega_0 \quad \text{and} \quad \alpha_1 \alpha_2 = \omega_0^2$$

so

$$\alpha_2 < \omega_0$$

This is an important result when we examine region 2.

When $r^2/4m^2 \gg \omega_0^2$

$$\alpha_1 \approx \frac{r}{m}$$

and

$$\alpha_2 = \frac{r}{2m} - \frac{r}{2m}\left(1 - \frac{4\omega_0^2 m^2}{r^2}\right)^{\frac{1}{2}}$$

$$\approx \frac{r}{2m} - \frac{r}{2m}\left(1 - \frac{2\omega_0^2 m^2}{r^2}\right)$$

$$\approx \frac{\omega_0^2 m}{r}$$

(using the Binomial Theorem with $n = 1/2$).

In the above C_1 and C_2 are arbitrary in value but have the same dimensions as C. Two separate values of C are allowed because the differential equation 2.1 is of second order.

Figure 2.3 illustrates heavily damped behaviour when a system is disturbed from equilibrium by a sudden impulse (that is, given a velocity at $t = 0$). It will return to zero displacement quite slowly without oscillating about its equilibrium position. More advanced mathematics shows that the value of the velocity dx/dt vanishes only once so that there is only one value of maximum displacement.

Worked Example

A dead beat (heavily damped) system is displaced a distance A from equilibrium and released from rest. Its subsequent motion is given by

$$x = A\,e^{-\omega_0^2 mt/r}$$

Show by considering the relaxation time that its decay is slowed down by the resistive force r acting against the stiffness force s.
From

$$e^{-\omega_0^2 mt/r}$$

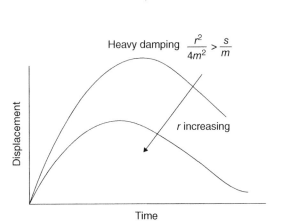

Heavy damping $\dfrac{r^2}{4m^2} > \dfrac{s}{m}$

r increasing

Displacement

Time

Figure 2.3   *Non-oscillatory behaviour of damped simple harmonic system with heavy damping (where $r^2/4m^2 > s/m$) after the system has been given an impulse from a rest position $x = 0$.*

we have

$$t = \frac{r}{\omega_0^2 m} = \frac{r}{m}\frac{m}{s} = \frac{r}{s}$$

as the relaxation time, so s has to work against r.

(Permission to use eq. 3.15 page 39, eq. 3.17 and 3.18 page 41, I. C. Main, *Vibrations and Waves in Physics* (1978) (C. U. P.) is gratefully acknowledged.)

2.2.4   *Region 2. Critical Damping* $(r^2/4m^2 = \omega_0^2)$

Using the notation of Region 1, we see that $r^2/4m^2 = \omega_0^2$ and that $x = e^{-\omega_0 t}(C_1 + C_2)$. This is, in fact, the limiting case of the behaviour of Region 1 as the bracket $(r^2/4m^2 - \omega_0^2)$ changes from positive to negative. In this case the quadratic equation in α has equal roots, which, in a differential equation solution, demands that C must be written $C = A + Bt$, where A is a constant length and B a given velocity which depends on the boundary conditions. We now prove that the value

$$x = (A + Bt)e^{-rt/2m} = (A + Bt)e^{-\omega_0 t} \tag{2.3}$$

satisfies $m\ddot{x} + r\dot{x} + sx = 0$ when $r^2/4m^2 = \omega_0^2$.

Worked Example

We write

$$m\ddot{x} + r\dot{x} + sx = 0 = \ddot{x} + 2\omega_0\dot{x} + \omega_0^2 x \tag{2.4}$$

using $(r/2m = \omega_0)$

The first term $A\,e^{-\omega_0 t}$ in equation 2.3 gives $A(\omega_0^2 - 2\omega_0^2 + \omega_0^2)e^{-\omega_0 t} = 0$ in equation 2.4. The second term $Bt\,e^{-\omega_0 t}$ in equation 2.3 gives, in equation 2.4, three terms

$$\ddot{x} = B(\omega_0^2 t - \omega_0 - \omega_0)\,e^{-\omega_0 t}$$
$$+\frac{r}{m}\dot{x} = 2\omega_0\dot{x} = 2\omega_0 B(1 - \omega_0 t)\,e^{-\omega_0 t} \quad \text{and the third term}$$
$$+\frac{s}{m}x = \omega_0^2 x = \omega_0^2 Bt\,e^{-\omega_0 t}$$

The sum of these three terms is zero in equation 2.4 proving that $(A + Bt)\,e^{-\omega_0 t}$ is a solution of equation 2.4.

Application to a Damped Mechanical Oscillator

Critical damping is of practical importance in mechanical oscillators which experience sudden impulses and are required to return to zero displacement in the minimum time. Suppose such a system has zero displacement at $t = 0$ and receives an impulse which gives it an initial velocity V.

Then $x = 0$ (so that $A = 0$) and $\dot{x} = V$ at $t = 0$. However,

$$\dot{x} = B[(-\omega_0 t)e^{-\omega_0 t} + e^{-\omega_0 t}] = B \text{ at } t = 0$$

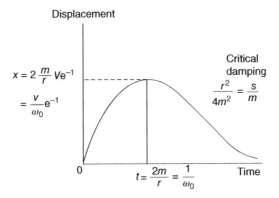

Figure 2.4 *Limiting case of non-oscillatory behaviour of damped simple harmonic system where $r^2/4m^2 = \omega_0^2$ (critical damping).*

so that $B = V$ and the complete solution is

$$x = Vt\,e^{-\omega_0 t}$$

The maximum displacement x occurs when the system comes to rest before returning to zero displacement. At maximum displacement

$$\dot{x} = Ve^{-\omega_0 t}(1 - \omega_0 t) = 0$$

thus giving $(1 - \omega_0 t) = 0$, i.e. $t = 1/\omega_0$.
At this time the displacement is therefore

$$x = Vt\,e^{-\omega_0 t} = \frac{V}{\omega_0}e^{-1}$$

The curve of displacement versus time is shown in Figure 2.5; the return to zero in a critically damped system is reached in minimum time.

Note that the relaxation time in critical damping is $1/\omega_0$ which is faster than that of $1/\alpha_2$ in the heavy damping case. The value of $V = 0$ at $t = 1/\omega_0$ so the resistive force $r = 0$ and the restoring stiffness force s is unopposed at a maximum, whereas s is opposed by r in the heavy damping case.

2.2.5   Region 3. Damped Simple Harmonic Motion ($r^2/4m^2 < \omega_0^2$)

When $r^2/4m^2 < \omega_0^2$ the damping is light, and this gives from the present point of view the most important kind of behaviour, oscillatory damped simple harmonic motion.

The expression $(r^2/4m^2 - \omega_0^2)^{1/2}$ is an imaginary quantity, the square root of a negative number, which can be rewritten

$$\pm\left(\frac{r^2}{4m^2} - \omega_0^2\right)^{1/2} = \pm\sqrt{-1}\left(\omega_0^2 - \frac{r^2}{4m^2}\right)^{1/2}$$

$$= \pm i\left(\omega_0^2 - \frac{r^2}{4m^2}\right)^{1/2} \quad (\text{where } i = \sqrt{-1})$$

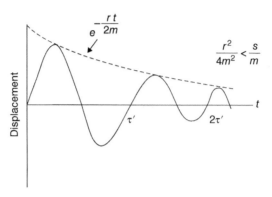

Figure 2.5 *Damped oscillatory motion where $s/m > r^2/4m^2$. The amplitude decays with $\mathrm{e}^{-rt/2m}$, and the reduced angular frequency is given by $\omega'^2 = \omega_0^2 - r^2/4m^2$.*

so the displacement

$$x = C_1\, \mathrm{e}^{-rt/2m}\mathrm{e}^{+\mathrm{i}(\omega_0^2 - r^2/4m^2)^{1/2}t} + C_2\, \mathrm{e}^{-rt/2m}\mathrm{e}^{-\mathrm{i}(\omega_0^2 - r^2/4m^2)^{1/2}t}$$

The graph of this expression is shown in Figure 2.5, a $\sin \omega t$ curve with an amplitude which decays exponentially as $\mathrm{e}^{-rt/2m}$. To reconcile the graph and its expression note that

(a) The factor $\mathrm{e}^{-rt/2m}$ in both the C_1 and C_2 terms of the expression is the envelope of the decay of the amplitude in Figure 2.5.

(b) The bracket has the dimensions of inverse time; that is, of frequency, and can be written $(\omega_0^2 - r^2/4m^2)^{1/2} = \omega'$, so that the second exponential becomes $\mathrm{e}^{\mathrm{i}\omega't} = \cos \omega't + \mathrm{i}\sin \omega't$. This shows that the behaviour of the displacement x is oscillatory with a new frequency $\omega' < \omega = (\omega_0^2)^{1/2}$, the frequency of ideal simple harmonic motion. To compare the behaviour of the damped oscillator with the ideal case we should like to express the solution in a form similar to $x = A\sin(\omega't + \phi)$ as in the simple harmonic case, where ω has been replaced by ω'.

We can do this by writing

$$x = \mathrm{e}^{-rt/2m}(C_1\, \mathrm{e}^{\mathrm{i}\omega't} + C_2\, \mathrm{e}^{-\mathrm{i}\omega't})$$

If we now choose

$$C_1 = \frac{A}{2\mathrm{i}}\mathrm{e}^{\mathrm{i}\phi}$$

and

$$C_2 = -\frac{A}{2\mathrm{i}}\mathrm{e}^{-\mathrm{i}\phi}$$

where A and ϕ (and thus $\mathrm{e}^{\mathrm{i}\phi}$) are constants which depend on the motion at $t = 0$, we find after substitution

$$x = A\mathrm{e}^{-rt/2m}\frac{[\mathrm{e}^{\mathrm{i}(\omega't + \phi)} - \mathrm{e}^{-\mathrm{i}(\omega't + \phi)}]}{2\mathrm{i}}$$
$$= A\mathrm{e}^{-rt/2m}\sin(\omega't + \phi)$$

This procedure is equivalent to imposing the boundary condition $x = A \sin \phi$ at $t = 0$ upon the solution for x. The displacement therefore varies sinusoidally with time as in the case of simple harmonic motion, but now has a new frequency

$$\omega' = \left(\omega_0^2 - \frac{r^2}{4m^2} \right)^{1/2}$$

and its amplitude A is modified by the exponential term $e^{-rt/2m}$, a term which decays with time.

If $x = 0$ at $t = 0$ then $\phi = 0$; Figure 2.5 shows the behaviour of x with time, its oscillations gradually decaying with the envelope of maximum amplitudes following the dotted curve $e^{-rt/2m}$. The constant A is obviously the value to which the amplitude would have risen at the first maximum if no damping were present.

The presence of the force term $r\dot{x}$ in the equation of motion therefore introduces a loss of energy which causes the amplitude of oscillation to decay with time as $e^{-rt/2m}$.

Worked Example

The amplitude of a vibrating mass of 200 grams decays because of the pressure of a resistive force r. What value of r will reduce the amplitude A to $A\,e^{-1}$ in 5 seconds?

$$A\,e^{-1} = A\,e^{\frac{-t}{2m/r}} \quad \text{when} \quad 5 = \frac{.4}{r} \quad \therefore r = \frac{.4}{5} = .08\ \mathrm{kg\,s}^{-1}$$

2.3   Methods of Describing the Damping of an Oscillator

Earlier in Chapter 1 we saw that the energy of an oscillator is given by

$$E = \frac{1}{2} m a^2 \omega^2 = \frac{1}{2} s a^2$$

that is, proportional to the square of its amplitude.

We have just seen that in the presence of a damping force $r\dot{x}$ the amplitude decays with time as

$$e^{-rt/2m}$$

so that the energy decay will be proportional to

$$\left(e^{-rt/2m} \right)^2$$

that is, $e^{-rt/m}$. The larger the value of the damping force r the more rapid the decay of the amplitude and energy. Thus we can use the exponential factor to express the rates at which the amplitude and energy are reduced.

2.3.1   Logarithmic Decrement

This measures the rate at which the amplitude dies away. Suppose in the expression

$$x = A e^{-rt/2m} \sin(\omega' t + \phi)$$

we choose

$$\phi = \pi/2$$

and we write

$$x = A_0 \, e^{-rt/2m} \cos \omega' t$$

with $x = A_0$ at $t = 0$. Its behaviour will follow the curve in Figure 2.6.

If the period of oscillation is τ' where $\omega' = 2\pi/\tau'$, then one period later the amplitude is given by

$$A_1 = A_0 \, e^{(-r/2m)\tau'}$$

so that

$$\frac{A_0}{A_1} = e^{r\tau'/2m} = e^\delta$$

where

$$\delta = \frac{r}{2m}\tau' = \log_e \frac{A_0}{A_1}$$

is called the logarithmic decrement. (Note that this use of δ differs from that in Figure 1.6.) The logarithmic decrement δ is the logarithm of the ratio of two amplitudes of oscillation which are separated by one period, the larger amplitude being the numerator since $e^\delta > 1$.

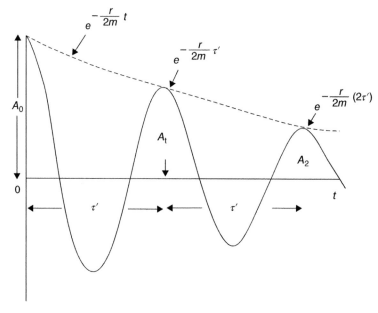

Figure 2.6   *The logarithmic ratio of any two amplitudes one period apart is the logarithmic decrement, defined as $\delta = \log_e (A_n/A_{n+1}) = r\tau'/2m$.*

Similarly

$$\frac{A_0}{A_2} = e^{r(2\tau')/2m} = e^{2\delta}$$

and

$$\frac{A_0}{A_n} = e^{n\delta}$$

Experimentally, the value of δ is best found by comparing amplitudes of oscillations which are separated by n periods. The graph of

$$\log_e \frac{A_0}{A_n}$$

versus n for different values of n has a slope δ.

Worked Example

Show that the reduction of a damped amplitude to half its value takes $1.39m/r$ seconds

$$\log_e 2 = \frac{t}{2}\frac{r}{m} = 0.693 \quad \therefore t = 1.386\frac{m}{r} \sec$$

2.3.2 Relaxation Time or Modulus of Decay

Another way of expressing the damping effect is by means of the time taken for the amplitude to decay to

$$e^{-1} = 0.368$$

of its original value A_0. This time is called the relaxation time or modulus of decay and the amplitude

$$A_t = A_0\, e^{-rt/2m} = A_0\, e^{-1}$$

at a time $t = 2m/r$.

Measuring the natural decay in terms of the fraction e^{-1} of the original value is a very common procedure in physics. The time for a natural decay process to reach zero is, of course, theoretically infinite.

2.3.3 The Quality Factor or Q-value of a Damped Simple Harmonic Oscillator

This measures the rate at which the energy decays. Since the decay of the amplitude is represented by

$$A = A_0\, e^{-rt/2m}$$

the decay of energy is proportional to

$$A^2 = A_0^2\, e^{(-rt/2m)^2}$$

and may be written

$$E = E_0 \, e^{(-r/m)t}$$

where E_0 is the energy value at $t = 0$.

The time for the energy E to decay to $E_0 \, e^{-1}$ is given by $t = m/r$ seconds during which time the oscillator will have vibrated through $\omega' m/r$ rad.

We define the quality factor

$$Q = \frac{\omega' m}{r}$$

as the number of radians through which the damped system oscillates as its energy decays to

$$E = E_0 \, e^{-1}$$

If r is small, then Q is very large and

$$\frac{s}{m} \gg \frac{r^2}{4m^2}$$

so that

$$\omega' \approx \omega_0 = \left(\frac{s}{m} \right)^{1/2}$$

Thus, we write, to a very close approximation,

$$Q = \frac{\omega_0 m}{r}$$

which is a constant of the damped system.

Since r/m now equals ω_0/Q we can write

$$E = E_0 \, e^{(-r/m)t} = E_0 \, e^{-\omega_0 t/Q}$$

The fact that Q is a constant $(= \omega_0 m/r)$ implies that the ratio

$$\frac{\text{energy stored in system}}{\text{energy lost per cycle}}$$

is also a constant, for

$$\frac{Q}{2\pi} = \frac{\omega_0 m}{2\pi r} = \frac{\nu_0 m}{r}$$

is the number of cycles (or complete oscillations) through which the system moves in decaying to

$$E = E_0 \, e^{-1}$$

and if

$$E = E_0 \, e^{(-r/m)t}$$

the energy lost per cycle is

$$-\Delta E = \frac{\mathrm{d}E}{\mathrm{d}t}\Delta t = \frac{-r}{m}E\frac{1}{\nu'}$$

where $\Delta t = 1/\nu' = \tau'$, the period of oscillation.
  Thus, the ratio

$$\frac{\text{energy stored in system}}{\text{energy lost per cycle}} = \frac{E}{-\Delta E} = \frac{\nu'm}{r} \approx \frac{\nu_0 m}{r}$$

$$= \frac{Q}{2\pi}$$

  In the next chapter we shall meet the same quality factor Q in two other roles, the first as a measure of the power absorption bandwidth of a damped oscillator driven near its resonant frequency and again as the factor by which the displacement of the oscillator is amplified at resonance.

Worked Example

When an electron in an excited atom radiates light it behaves as a damped simple harmonic oscillator. The wavelength of the radiation is $500 \cdot 10^{-9}$ m and its intensity decays to e^{-1} in 10^{-8} sec. What is the Q of the system? The length of the radiated wavetrain is called the coherence length $l = ct$. How long is the coherence length and how many waves does it contain?

Solution

$$\frac{E}{-\mathrm{d}E} = Q = \omega_0 t = 2\pi\nu t = \frac{2\pi ct}{\lambda} = \frac{2\pi \times 3 \times 10^8 \times 10^{-8}}{5 \times 10^{-7}} \approx 4 \times 10^7$$

$$l = ct = 3 \times 10^8 \times 10^{-8} = 3 \text{ metres}$$

$$\text{No. of waves} = 3 \times \frac{10^9}{500} = 6 \times 10^6$$

2.3.4   Energy Dissipation

We have seen that the presence of the resistive force reduces the amplitude of oscillation with time as energy is dissipated.
  The total energy remains the sum of the kinetic and potential energies

$$E = \frac{1}{2}m\dot{x}^2 + \frac{1}{2}sx^2$$

Now, however, $\mathrm{d}E/\mathrm{d}t$ is not zero but negative because energy is lost, so that

$$\frac{\mathrm{d}E}{\mathrm{d}t} = \frac{\mathrm{d}}{\mathrm{d}t}\left(\frac{1}{2}m\dot{x}^2 + \frac{1}{2}sx^2\right) = \dot{x}(m\ddot{x} + sx)$$

$$= \dot{x}(-r\dot{x}) \quad \text{for} \quad m\ddot{x} + r\dot{x} + sx = 0$$

i.e. $\mathrm{d}E/\mathrm{d}t = -r\dot{x}^2$, which is the rate of doing work against the frictional force (dimensions of force $\times$ velocity = force $\times$ distance/time).

Worked Example

Show that the average value of energy loss per cycle $= \frac{1}{2}r\omega^2 a^2$ when $x = a\sin\omega t$.

If $x = a\sin\omega t$, $\dot{x} = \omega a\cos\omega t$ with an average value per cycle of $\dot{x}^2 = \frac{1}{2}\omega^2 a^2$.

$$\therefore \text{Average loss} = \frac{1}{2}r\omega^2 a^2$$

2.3.5   Damped SHM in an Electrical Circuit

The force equation in the mechanical oscillator is replaced by the voltage equation in the electrical circuit of inductance, resistance and capacitance (Figure 2.7).

We have, therefore,

$$L\frac{dI}{dt} + RI + \frac{q}{C} = 0$$

or

$$L\ddot{q} + R\dot{q} + \frac{q}{C} = 0$$

and by comparison with the solutions for x in the mechanical case we know immediately that the charge

$$q = q_0\, e^{-Rt/2L \pm (R^2/4L^2 - 1/LC)^{1/2}t}$$

which, for $1/LC > R^2/4L^2$, gives oscillatory behaviour at a frequency

$$\omega^2 = \frac{1}{LC} - \frac{R^2}{4L^2}$$

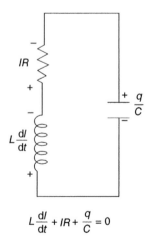

$$L\frac{dI}{dt} + IR + \frac{q}{C} = 0$$

Figure 2.7   *Electrical circuit of inductance, capacitance and resistance capable of damped simple harmonic oscillations. The sum of the voltages around the circuit is given from Kirchhoff's Law as $L\frac{dI}{dt} + RI + \frac{q}{C} = 0$.*

From the exponential decay term we see that R/L has the dimensions of inverse time T^{-1} or ω, so that ωL has the dimensions of R; that is, ωL is measured in ohms.

Similarly, since $\omega^2 = 1/LC$, $\omega L = 1/\omega C$, so that $1/\omega C$ is also measured in ohms. We shall use these results in the next chapter.

Worked Example

In Figure 2.6 show that critical damping occurs when $R = 2\omega_0 L$ where $\omega_0^2 = \frac{1}{LC}$. Using its mechanical equivalent show that $Q = \omega_0 L/R$. The system is very heavily damped when $R \gg L$ so that L may be ignored. Show that the relaxation time is RC for such a circuit.

Solution

$$\text{Critical dumping:} \quad \frac{1}{LC} = \frac{R^2}{4L^2} = \omega_0^2 \quad \therefore R = 2\omega_0 L$$

$$\text{Analogy gives } L \rightarrow m \text{ and } R \rightarrow r \quad \therefore Q = \omega_0 \frac{m}{r} \rightarrow \frac{\omega_0 L}{R}$$

Without L, 2.6 is an RC circuit with $q = q_0 e^{-t/RC}$ and a relaxation time $= RC$. (Note that $RC \equiv$ mechanical r/s. See end of section 2.2.3 Region 1 Heavy Damping.)

Problem 2.1. A critically damped vibrating mechanical system is displaced a distance A from equilibrium and released from rest. Use the boundary condition $x = A$ and $\dot{x} = 0$ at $t = 0$ in equation 2.3 to show that its subsequent displacement is given by $x = A(1 + \omega_0 t)e^{-\omega_0 t}$.

Problem 2.2. A damped simple harmonic oscillator has a mass of 5 kg, an oscillation frequency of 0.5 Hz and a logarithmic decrement of 0.02. Calculate the values of the stiffness force s and the resistive force r of the oscillator.

Problem 2.3. A critically damped mechanical system consist of a pan hanging from a spring with a damping. What is the value of the damping force r if a mass extends the spring by 10 cm without overshoot. The mass is 5 kg. ($g = 9.81$ m s^{-2})

Problem 2.4. A capacitance C with a charge q_0 at $t = 0$ discharges through a resistance R. Use the voltage equation $q/C + IR = 0$ to show that the relaxation time of this process is RC s; that is,

$$q = q_0 e^{-t/RC}$$

(Note that t/RC is non-dimensional.)

Problem 2.5. The frequency of a damped simple harmonic oscillator is given by

$$\omega'^2 = \frac{s}{m} - \frac{r^2}{4m^2} = \omega_0^2 - \frac{r^2}{4m^2}$$

(a) If $\omega_0^2 - \omega'^2 = 10^{-6}\omega_0^2$ show that $Q = 500$ and that the logarithmic decrement $\delta = \pi/500$.
(b) If $\omega_0 = 10^6$ and $m = 10^{-10}$ kg show that the stiffness of the system is 100 N m^{-1}, and that the resistive constant r is 2×10^{-7} N $\cdot$ s m^{-1}.
(c) If the maximum displacement at $t = 0$ is 10^{-2} m, show that the energy of the system is 5×10^{-3} J and the decay to e^{-1} of this value takes 0.5 ms.
(d) Show that the energy loss in the first cycle is $2\pi \times 10^{-5}$ J.

Problem 2.6.   Show that the fractional change in the resonant frequency ω_0 ($\omega_0^2 = s/m$) of a damped simple harmonic mechanical oscillator is $\approx (8Q^2)^{-1}$ where Q is the quality factor.

Problem 2.7.   The maximum displacement of a simple harmonic oscillator $x = a\sin\omega t$ occurs when $\omega t = \pi/2$. In a damped oscillator the maximum occurs at ωt slightly less than $\pi/2$. Show that the maximum is advanced an angle $\Phi \approx \frac{1}{2Q}$, where $Q = \omega_0 m/r$.

Problem 2.8.   A plasma consists of an ionized gas of ions and electrons of equal number densities ($n_i = n_e = n$) having charges of opposite sign $\pm e$, and masses m_i and m_e, respectively, where $m_i > m_e$. Relative displacement between the two species sets up a restoring

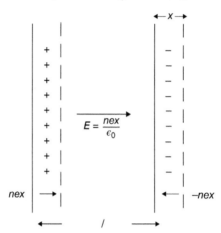

electric field which returns the electrons to equilibrium, the ions being considered stationary. In the diagram, a plasma slab has all its electrons displaced a distance x to give a restoring electric field $E = nex/\varepsilon_0$, where ε_0 is constant. Show that the restoring force on each electron is $-ne^2x/\varepsilon_0$ and that they oscillate simple harmonically with angular frequency $\omega_e^2 = ne^2/m_e\varepsilon_0$. This frequency is called the electron plasma frequency, and only those radio waves of frequency $\omega > \omega_e$ will propagate in such an ionized medium. Hence the reflection of such waves from the ionosphere.

Problem 2.9.   When the string of an instrument is plucked the sound intensity (energy) decreases by a factor of 2 after 4 seconds. The natural frequency of the string is 330 Hz. Calculate the relaxation time, the Q of the system and the fractional energy loss per cycle.

3

The Forced Oscillator

Introduction

In Chapter 1 we saw how many different physical systems could be described by the same equation of motion. Here we extend that principle to more complicated systems, those driven by outside influences which increase their energy. A tuned radio circuit is rarely associated with a forced mechanical oscillator but we shall see that the equations which govern the behaviour of both oscillators are identical so the solutions to their equations are applicable to each system. Moreover the electrical force from an electromagnetic wave operating on a charged electron in an atom combines both types of oscillator and reveals many optical properties of matter.

Essentially there is no new mathematics in this chapter but in showing the similarity between electrical and mechanical oscillators we shall highlight how much more information is gained by the use of $i = \sqrt{-1}$ as a vector operator particularly in the cases of impedance and phase.

3.1 The Operation of i upon a Vector

We have already seen that a harmonic oscillation can be conveniently represented by the form $e^{i\omega t}$. In addition to its mathematical convenience i can also be used as a vector operator of physical significance. We say that when i precedes or operates on a vector the direction of that vector is turned through a positive angle (anticlockwise) of $\pi/2$, i.e. i acting as an operator advances the phase of a vector by 90°. The operator $-i$ rotates the vector clockwise by $\pi/2$ and retards its phase by 90°. The mathematics of i as an operator differs in no way from its use as $\sqrt{-1}$ and from now on it will play both roles.

The vector $\boldsymbol{r} = \boldsymbol{a} + i\boldsymbol{b}$ is shown in Figure 3.1, where the direction of b is perpendicular to that of a because it is preceded by i. The magnitude or modulus of r is written

$$r = |\boldsymbol{r}| = (a^2 + b^2)^{1/2}$$

Introduction to Vibrations and Waves, First Edition. H. J. Pain and Patricia Rankin.
© 2015 John Wiley & Sons, Ltd. Published 2015 by John Wiley & Sons, Ltd.
Companion website: http://booksupport.wiley.com

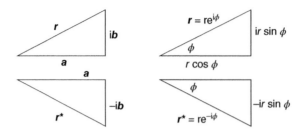

Figure 3.1   *Vector representation using i operator and exponential index. Star superscript indicates complex conjugate where −i replaces i.*

and

$$r^2 = (a^2 + b^2) = (a + ib)(a - ib) = rr^*,$$

where $(a - ib) = r^*$ is defined as the complex conjugate of $(a + ib)$; that is, the sign of i is changed.

The vector $r^* = a - ib$ is also shown in Figure 3.1.

The vector r can be written as a product of its magnitude r (scalar quantity) and its phase or direction in the form (Figure 3.1)

$$r = r\,e^{i\phi} = r(\cos\phi + i\sin\phi)$$
$$= a + ib$$

showing that $a = r\cos\phi$ and $b = r\sin\phi$.

It follows that

$$\cos\phi = \frac{a}{r} = \frac{a}{(a^2 + b^2)^{1/2}}$$

and

$$\sin\phi = \frac{b}{r} = \frac{b}{(a^2 + b^2)^{1/2}}$$

giving $\tan\phi = b/a$.

Similarly

$$r^* = r\,e^{-i\phi} = r(\cos\phi - i\sin\phi)$$
$$\cos\phi = \frac{a}{r}, \quad \sin\phi = \frac{-b}{r} \quad \text{and} \quad \tan\phi = \frac{-b}{a} \quad \text{(Figure 3.1)}$$

The reader should confirm that the operator i rotates a vector by $\pi/2$ in the positive direction (as stated in the first paragraph of section 3.1 The Operation of i upon a Vector) by taking $\phi = \pi/2$ in the expression

$$r = r\,e^{i\phi} = r(\cos\pi/2 + i\sin\pi/2)$$

Note that $\phi = -\pi/2$ in $r = r\,e^{-i\pi/2}$ rotates the vector in the negative direction.

3.2 Vector Form of Ohm's Law

We begin our discussion on forced oscillators with a very common electrical equation.

Ohm's Law is first met as the scalar relation $V = IR$, where V is the voltage across the resistance R and I is the current through it. Its scalar form states that the voltage and current are always in phase. Both will follow a $\sin \omega t$ or a $\cos \omega t$ curve.

However, the presence of either or both of the other two electrical components, inductance L and capacitance C, will introduce a phase difference between voltage and current, and Ohm's Law takes the vector form

$$V = IZ_e,$$

Where Z_e, called the impedance, replaces the resistance, and is the vector sum of the effective resistances of R, L and C in the circuit.

When an alternating voltage $V_0 e^{i\omega t}$ of frequency ω is applied across a resistance, inductance and capacitor in series as in Figure 3.2a, the balance of voltages is given by

$$IR + L\frac{dI}{dt} + q/C = V_0 e^{i\omega t} = V_0(\cos \omega t + i \sin \omega t) \tag{3.1}$$

and the current through the circuit is given by $I = I_0 e^{i\omega t}$. The voltage across the inductance

$$V_L = L\frac{dI}{dt} = L\frac{d}{dt}I_0 e^{i\omega t} = i\omega L I_0 e^{i\omega t} = i\omega L I$$

But ωL, as we saw at the end of the last chapter, has the dimensions of ohms, being the value of the effective resistance presented by an inductance L to a current of frequency ω. The product $\omega L I$ with dimensions of ohms times current, i.e. volts, is preceded by i; this tells us that the phase of the voltage across the inductance is 90° ahead of that of the current through the circuit.

Similarly, the voltage across the capacitor is

$$\frac{q}{C} = \frac{1}{C}\int I dt = \frac{1}{C}I_0\int e^{i\omega t}dt = \frac{1}{i\omega C}I_0 e^{i\omega t} = -\frac{iI}{\omega C}$$

(since $1/i = -i$).

Again $1/\omega C$, measured in ohms, is the value of the effective resistance presented by the capacitor to the current of frequency ω. Now, however, the voltage $I/\omega C$ across the capacitor is preceded by $-i$ and

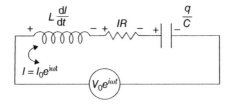

Figure 3.2a *An electrical forced oscillator. The voltage $V_0 e^{i\omega t}$ is applied to the series LCR circuit giving $V_0 e^{i\omega t} =$ $LdI/dt + IR + q/C$.*

therefore lags the current by 90°. Equation 3.1 becomes

$$i\omega L I + IR - iI/\omega C = V_0\, e^{i\omega t}$$

or

$$\left[R + i\left(\omega L - \frac{1}{\omega C}\right)\right] I_0\, e^{i\omega t} = V_0\, e^{i\omega t}$$

The voltage and current across the resistance are in phase and Figure 3.2b shows that the vector form of Ohm's Law may be written $V = IZ_e = I[R + i(\omega L - 1/\omega C)]$, where the impedance $Z_e = R + i(\omega L - 1/\omega C)$. The quantities ωL and $1/\omega C$ are called reactances because they introduce a phase relationship as well as an effective resistance, and the bracket $(\omega L - 1/\omega C)$ is often written X_e, the reactive component of Z_e.

The magnitude, in ohms, i.e. the value of the impedance, is

$$Z_e = \left[R^2 + \left(\omega L - \frac{1}{\omega C}\right)^2\right]^{1/2}$$

and the vector Z_e may be represented by its magnitude and phase as

$$\mathbf{Z_e} = Z_e e^{i\phi} = Z_e(\cos\phi + i\,\sin\phi)$$

so that

$$\cos\phi = \frac{R}{Z_e},\ \sin\phi = \frac{X_e}{Z_e}$$

and

$$\tan\phi = X_e/R,$$

where ϕ is the phase difference between the total voltage across the circuit and the current through it.

The value of ϕ can be positive or negative depending on the relative value of ωL and $1/\omega C$: when $\omega L > 1/\omega C$, ϕ is positive, but the frequency dependence of the components shows that ϕ can change both sign and size.

The magnitude of $\mathbf{Z_e}$ is also frequency dependent and has its minimum value $Z_e = R$ when $\omega L = 1/\omega C$ with I at a maximum when $\omega^2 = 1/LC$.

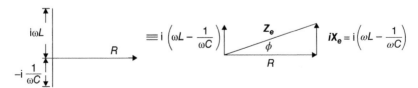

Figure 3.2b   *Vector addition of resistance and reactances to give the electrical impedance $\mathbf{Z_e} = R + i(\omega L - 1/\omega C)$.*

In the vector form of Ohm's Law, $\mathbf{V} = \mathbf{I}\mathbf{Z}_e$. If $\mathbf{V} = V_0\,\mathrm{e}^{\mathrm{i}\omega t}$ and $\mathbf{Z}_e = Z_e\,\mathrm{e}^{\mathrm{i}\phi}$, then we have

$$I = \frac{V_0\mathrm{e}^{\mathrm{i}\omega t}}{Z_e\mathrm{e}^{\mathrm{i}\phi}} = \frac{V_0}{Z_e}\,\mathrm{e}^{\mathrm{i}(\omega t - \phi)}$$

giving a current of amplitude V_0/Z_e which lags the voltage by a phase angle ϕ.

In such an equation the real part of the left-hand side matches the real part of the right-hand side and the imaginary part of the left-hand side matches the imaginary part of the right-hand side.

So for $V_0\mathrm{e}^{\mathrm{i}(\omega t - \phi)}$, the real part $V_0\cos(\omega t - \phi)$ gives $I_0\cos(\omega t - \phi)$ and the imaginary part (i) $V_0\sin(\omega t - \phi)$ gives (i) $I_0\sin(\omega t - \phi)$.

3.3 The Tuned LCR Circuit

Our first example of a lightly damped forced oscillator forms one of the early stages of an electronic system such as a radio receiver. It is a series circuit of an inductance L, a resistance R and a variable capacitor C. Figure 3.3.

Changing the value of C changes the resonant frequency $\omega_0^2 = 1/LC$ of the circuit and allows us to select and amplify the different frequencies ω_0 transmitted by radio stations in the region. The value of V_0 the transmitted signal is constant and is independent of the current drawn by the circuit. At a given resonant frequency ω_0 the impedance of the receiver is R and for an input voltage $V_0\cos\omega t$, for all practical purposes, the voltages across the circuit elements are at a maximum with $V_R = V_0$ and $V_L - V_C = 0$ because they are anti-phase. The maximum values of V_L and V_C may be written as QV_0 where Q is the amplification factor at resonance. To find the amplification of V_C at resonance which is passed on to the next stage of the receiver we consider the voltage equation taken around Figure 3.3 as

$$L\frac{\mathrm{d}I}{\mathrm{d}t} + IR + \frac{q}{c} = V_0\,\mathrm{e}^{\mathrm{i}\omega t} \quad \text{or} \quad \left(\mathrm{i}\omega L + R - \frac{1}{\omega C}\right)I_0\,\mathrm{e}^{\mathrm{i}\omega t} = V_0\,\mathrm{e}^{\mathrm{i}\omega t}$$

giving

$$I_0 = \frac{V_0}{\left[R^2 + (\omega L - \frac{1}{\omega C})^2\right]^{1/2}} = \frac{V_0}{\frac{L}{\omega}\left[\frac{\omega^2 R^2}{L^2} + (\omega^2 - \omega_0^2)^2\right]^{1/2}} \quad \text{where} \quad \omega_0^2 = \frac{1}{LC}$$

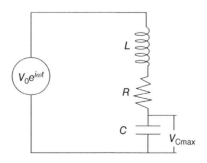

Figure 3.3 *Tuned LCR circuit with* V_{Cmax} *connected to next stage at resonance.*

The resonance value of the capacitance voltage $V_{C\max}$ results from the maximum current $I_{0\max}$ flowing through the capacitance impedance $Z_C = 1/\omega C$ so

$$V_{C\max} = I_{0\max}/\omega C = \frac{V_0}{\frac{\omega LC}{\omega}\left[\frac{\omega^2 R^2}{L^2} + (\omega^2 - \omega_0^2)^2\right]^{1/2}}$$

$$\therefore \frac{V_{C\max}}{V_0} = \frac{1}{LC[\omega^2 R^2/L^2 + (\omega^2 - \omega_0^2)^2]^{1/2}} = \frac{\omega_0^2}{[\omega^2 R^2/L^2 + (\omega^2 - \omega_0^2)^2]^{1/2}}$$

Plotting $V_{C\max}/V_0$ against ω gives the curve of Figure 3.4 with a peak at the input frequency $\omega = \omega_0$. The peak value is written Q which measures the amplification at that frequency.

Worked Example

At ω_0, $Z_e = R$ and $I = V_0/R$

$$V_L = L\frac{dI}{dt} = L\frac{\omega_0 V_0}{R}$$

$$\therefore \frac{V_L}{V_0} = \frac{\omega_0 L}{R} = Q$$

At ω_0, $V_{C\max} = V_L$

$$\therefore \frac{V_{C\max}}{V_0} = Q$$

Tuning the circuit selects one station at a time and the curve of Figure 3.4 is translated along the horizontal ω axis peaking at successive values of ω_0 in the transmitting region. The narrower the upper part of the curve the sharper the tuning and the increase in selectivity, free from interference from signals from nearby frequencies.

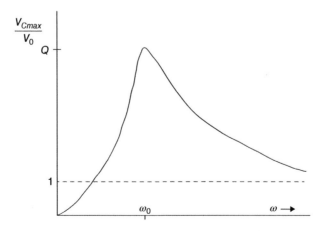

Figure 3.4   *Maximum $V_{C\max}$ at ω_0 resonance of tuned LCR circuit. $V_{C\max}/V_0 = Q$, the amplification factor.*

The height of the peak measures Q at which stage the voltage across the capacitor equals QV_0 (equal and opposite to the voltage across the inductance) while V_R across the resistance equals V_0 the input signal voltage.

The upper part of the curve in Figure 3.4 is asymmetric but the curve of the power $VI = I^2 R$ absorbed by the circuit from the transmitted signal generator is symmetric. This is discussed in section 3.5 (The Q Value in Terms of the Resonance Absorption Bandwidth) which confirms that the amplification factor $Q = \omega_0 L/R$. Since L/R is constant then $Q = \omega_0 L/R \rightarrow 0$ as ω_0 decreases. This is shown in Figure 3.4, which also shows that for a fixed value of ω_0 $V_{C\,\mathrm{max}}/V_0 \rightarrow 1$ as ω the transmitted frequency increases.

In conventional radio circuits at frequencies of a megacycle Q values are of the order of a few hundred; at higher radio frequencies resonant copper cavities have Q values of 500,000. Optical absorption in crystals and nuclear magnetic resonances are often described in terms of Q values. The Mössbauer effect in nuclear physics involves Q values of 10^{10}.

3.4 Power Supplied to Oscillator by the Input Voltage

In order to maintain the steady state oscillations of a system the input voltage must replace the energy lost in each cycle because of the presence of the resistance. *We shall now derive the most important result that*:

'in the steady state the amplitude and phase of a driven oscillator adjust themselves so that the average power supplied by the input voltage just equals that being dissipated by the resistance'.

The instantaneous power P supplied is equal to the product of the instantaneous applied voltage and the instantaneous current; that is,

$$P = VI = V_0 \cos \omega t \frac{V_0}{Z_e} \cos(\omega t - \phi)$$

$$= \frac{V_0^2}{Z_e} \cos \omega t \cos(\omega t - \phi)$$

The average power

$$P_{\mathrm{av}} = \frac{\text{total work per oscillation}}{\text{oscillation period}}$$

$$\therefore P_{\mathrm{av}} = \int_0^T \frac{P \, \mathrm{d}t}{T} \text{ where } T = \text{oscillation period}$$

$$= \frac{V_0^2}{Z_e T} \int_0^T \cos \omega t \cos(\omega t - \phi) \mathrm{d}t$$

$$= \frac{V_0^2}{Z_e T} \int_0^T [\cos^2 \omega t \cos \phi + \cos \omega t \sin \omega t \sin \phi) \mathrm{d}t$$

$$= \frac{V_0^2}{2Z_e} \cos \phi$$

because

$$\int_0^T \cos \omega t \times \sin \omega t \, \mathrm{d}t = 0$$

and

$$\frac{1}{T} \int_0^T \cos^2 \omega t \, dt = \frac{1}{2}$$

The power supplied by V_0 is not stored in the system, but dissipated as the power $VI = I^2 R$ lost across the resistance R.

Now

$$I^2 R = R \frac{V_0^2}{Z_e^2} \cos^2(\omega t - \phi)$$

and the average value of this over one period of oscillation is

$$\frac{1}{2} \frac{R V_0^2}{Z_e^2} = \frac{1}{2} \frac{V_0^2}{Z_e} \cos \phi \quad \text{because} \quad \frac{R}{Z_e} = \cos \phi$$

This proves the initial statement that the power supplied equals the power dissipated.

The power supplied is given by $VI \cos \phi$, where V and I are the instantaneous r.m.s. values of voltage and current and $\cos \phi$ is known as the power factor.

$$VI \cos \phi = \frac{V^2}{Z_e} \cos \phi = \frac{V_0^2}{2Z_e} \cos \phi$$

since

$$V = \frac{V_0}{\sqrt{2}}$$

the r.m.s. value for an alternating voltage.

Note that in $\frac{V^2}{Z_e} \cos \phi$ when Z_e is a reactance the factor i in Z_e means that ϕ is $90°$ between V and I and no power is consumed.

The power supplied (and absorbed) is at a maximum at the resonant frequency ω_0 (where $\omega_0^2 = 1/LC$) of the circuit, that is when $Z_e = R$ its minmum value.

These features are displayed in Figure 3.5 where $P_{av}(\text{maximum}) = V_0^2/2R$, and $\cos \phi = 1$ because $\phi = 0$ when $Z_e = R$ and V and I are in phase. Note that the Absorption Resonance Curve is symmetric about ω_0.

3.5   The Q-Value in Terms of the Resonance Absorption Bandwidth

In the last chapter we discussed the quality factor of an oscillator system in terms of energy decay. We may derive the same parameter in terms of the curve of Figure 3.5, where the sharpness of the resonance is precisely defined by the ratio

$$Q = \frac{\omega_0}{\omega_2 - \omega_1} = \frac{\omega_0}{\Delta \omega},$$

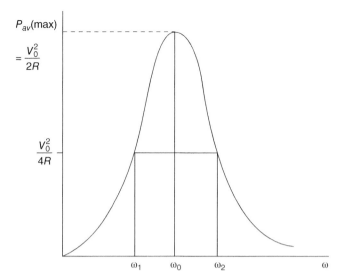

Figure 3.5   *Graph of average power versus ω supplied to an oscillator by the input voltage. Bandwidth $\omega_2 - \omega_1$ of resonance curve defines response in terms of the quality factor, $Q = \omega_0/(\omega_2 - \omega_1)$, where $\omega_0^2 = 1/LC$. Note the curve symmetry about ω_0.*

where ω_2 and ω_1 are those frequencies at which the power supplied

$$P_{\text{av}} = \frac{1}{2}P_{\text{av}}(\text{maximum})$$

The frequency difference $\omega_2 - \omega_1$ is often called the bandwidth.
  Now

$$P_{\text{av}} = RV_0^2/2Z_e^2 = \frac{1}{2}P_{\text{av}}(\text{maximum}) = \frac{1}{2}V_0^2/2R = \frac{1}{4}\frac{V_0^2}{R}$$

when

$$Z_e^2 = 2R^2$$

that is, when

$$R^2 + \left(\omega L - \frac{1}{\omega C}\right)^2 = 2R^2 \quad \text{or} \quad \left(\omega L - \frac{1}{\omega C}\right) = \pm R.$$

If $\omega_2 > \omega_1$, then

$$\omega_2 L - \frac{1}{\omega_2 C} = +R$$

and

$$\omega_1 L - \frac{1}{\omega_1 C} = -R$$

Eliminating C between these equations gives the bandwidth frequency at half resonance absorption as

$$\omega_2 - \omega_1 = R/L$$

so that

$$Q = \omega_0 L / R$$

Worked Example

The purpose of this example is to show how Q of the resonance absorption curve, associated with a bandwidth $\Delta\omega$, is the same Q as the energy decay Q of Chapter 2. The process of energy absorption and decay are the mirror images of each other. The resistance force in damped simple harmonic motion is exactly that force responsible for the energy absorption of a driven oscillator. An electron in an atom absorbs energy when excited by an electromagnetic wave and will radiate that energy when free to do so. The bandwidth $\Delta\omega$ enters the scene because, as we shall see in Chapter 6, the decaying radiated wavetrain from the atom has a finite length. Only a wavetrain of infinite length may be represented by a single frequency. Shorter wavetrains need a frequency range $\Delta\omega$ to describe them; the shorter the wavetrain the wider the range $\Delta\omega$.

Problem

A peak in the absorption spectrum of an atom occurs at $\lambda = 550\,\mathrm{nm}$ and has a width of $\Delta\lambda = 1.2 \times 10^{-5}\,\mathrm{nm}$. Calculate the lifetime of the excited atom and the length of the radiated wavetrain.

Solution

$$\omega = 2\pi\nu = 2\pi c/\lambda \quad \therefore \Delta\omega = -2\pi c\Delta\lambda/\lambda^2 \quad \text{where} \quad \Delta\lambda = 1.2 \times 10^{-5}\,\mathrm{nm}.$$

$$Q = \omega_0\tau \ (\text{where } \tau \text{ is the lifetime}) = \omega_0 \frac{m}{r} = \omega_0/\Delta\omega$$

$$\frac{(550 \times 10^{-9})^2}{2\pi \times 3 \times 10^8 \times 1.2 \times 10^{-14}} = \frac{1}{\Delta\nu} = 1.3 \times 10^{-8}\,\mathrm{sec} = \tau$$

$$\text{Length of radiated wavetrain} = c\tau = 3 \times 1.3 = 3.9\,\mathrm{metres}$$

3.6 The Forced Mechanical Oscillator

We are now in a position to discuss the physical behaviour of a mechanical oscillator of mass m, stiffness s and resistance r being driven by an alternating force $F_0 \cos\omega t$, where F_0 is the amplitude of the force (Figure 3.6). This is the mechanical equivalent of the series electrical LCR circuit (Figure 3.3).

The mechanical equation of motion, i.e. the dynamic balance of forces, is given by

$$m\ddot{x} + r\dot{x} + sx = F_0 \cos\omega t$$

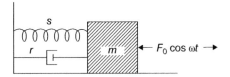

Figure 3.6 *Mechanical forced oscillator with force $F_0 \cos \omega t$ applied to damped mechanical circuit of Figure 2.2.*

In the voltage equation $q \equiv x$, $\dot{q} \equiv \dot{x} = I$ and $\ddot{q} \equiv \frac{\mathrm{d}I}{\mathrm{d}t} = \ddot{x}$, so we have

$$L\ddot{q} + R\dot{q} + q/C = V_0 \cos \omega t$$

In the mechanical oscillator our chief focus will be on the variable x, the displacement of the oscillator from equiliblium and not on $\dot{x}$, the velocity which is the equivalent to the current I in the electrical case. The identical form of the force and voltage equations means that a number of the results we obtained from the electrical equation may be carried over directly to the mechanical equation.

(a) Impedance

The impedance of the mechanical oscillator is defined as the force required to produce unit velocity, that is, $Z_m = F/v$ or $F = vZ_m$. (c.f. electrical $V = IZ_e$) Comparing the constant coefficients of the variables in the force and voltage equations, we can write $Z_m = r + i(\omega m - s/m) = r + iX_m$ where $X_m = (\omega m - s/\omega)$ (c.f. electrical $Z_e = R + i(\omega L - 1/\omega C)$). As with Z_e we can write $Z_m = Z_m \mathrm{e}^{\mathrm{i}\phi}$ where $\cos \phi = r/Z_m$, $\sin \phi = X_m/Z_m$ and $\tan \phi = X_m/r$.

(b) The power supplied by the force to maintain steady state mechanical oscillations is given by

$$\frac{F_0^2}{2Z_m} \cos \phi \quad \left(\text{c.f. electrical } \frac{V_0^2}{2Z_e} \cos \phi \right)$$

This power replaces the loss due to the work rate done against the frictional force, that is, force times velocity

$$(r\dot{x})\dot{x} = r\dot{x}^2 = r\frac{F_0^2}{Z_m^2} \cos^2(\omega t - \phi)$$

which averages over one period of oscillation

$$\frac{1}{2}\frac{F_0^2 r}{Z_m^2} = \frac{1}{2}\frac{F_0^2}{Z_m} \cos \phi \quad \text{where} \quad \cos \phi = \frac{r}{Z_m} \left(\text{c.f. electrical } \frac{1}{2}\frac{V_0^2}{Z_e} \cos \phi \right)$$

(c) Variation of P_{av} with ω versus absorption resonance curve (Figure 3.5)

In the mechanical case the average power supplied $P_{\mathrm{av}} = F_0^2 \cos \phi / 2Z_m$ is a maximum when $\cos \phi = 1$ and $(\omega m - \frac{s}{m}) = 0$ at $\omega_0^2 = s/m$. The force and velocity ($\equiv$ voltage and current) are then in phase and Z_m has its minimum value equal to r. Thus $P_{\mathrm{av}}(\max) = F_0^2/2r$.

(d) The Q value in terms of the resonance absorption bandwidth

Replace R by r: V_0 by F_0 and Z_e by Z_m.

$$Z_m^2 = r^2 + \left(\omega m - \frac{s}{m}\right)^2 = 2r^2 \text{ at } \frac{1}{2}P_{\text{av}}(\text{max})$$

when

$$\left(\omega m - \frac{s}{\omega}\right)^2 = r^2 \quad \text{i.e.} \quad \left(\omega m - \frac{s}{\omega}\right) = \pm r$$

for $\omega_2 > \omega_1$,

$$\omega_2 m - \frac{s}{\omega_2} = +r \quad \text{and} \quad \omega_1 m - \frac{s}{\omega_1} = -r$$

Eliminate s between these equations to give

$$\omega_2 - \omega_1 = \frac{r}{m} \quad \text{so} \quad Q = \frac{\omega_0}{\omega_2 - \omega_1} = \frac{\omega_0 m}{r} \quad \left(\text{c.f. } Q = \frac{\omega_0 L}{R}\right)$$

Returning to the equation of motion for the forced mechanical oscillator

$$m\ddot{x} + r\dot{x} + sx = F_0 \cos \omega t$$

the complete solution for x in the equation of motion consists of two terms:

(1)   a 'transient' term which dies away with time and is, in fact, the solution to the equation $m\ddot{x} + r\dot{x} + sx = 0$ discussed in Chapter 2. This contributes the term

$$x = Ce^{-rt/2m} e^{i(s/m - r^2/4m^2)^{1/2}t}$$

which decays with $e^{-rt/2m}$. The second term
(2)   is called the 'steady state' term, and describes the behaviour of the oscillator after the transient term has died away.

Both terms contribute to the solution initially, but for the moment we shall concentrate on the 'steady state' term which describes the ultimate behaviour of the oscillator.

To do this we shall rewrite the force equation in vector form and represent $\cos \omega t$ by $e^{i\omega t}$ as follows:

$$m\ddot{x} + r\dot{x} + sx = F_0 e^{i\omega t} \tag{3.2}$$

Solving for the vector x will give both its magnitude and phase with respect to the driving force $F_0 e^{i\omega t}$. Initially, let us try the solution $x = A e^{i\omega t}$, where A may be complex, so that it may have components in and out of phase with the driving force.

The velocity

$$\dot{x} = i\omega A e^{i\omega t} = i\omega x$$

so that the acceleration

$$\ddot{x} = i^2 \omega^2 x = -\omega^2 x$$

and equation (3.2) becomes

$$(-A\omega^2 m + i\omega A r + As)\, e^{i\omega t} = F_0\, e^{i\omega t}$$

which is true for all t when

$$A = \frac{F_0}{i\omega r + (s - \omega^2 m)}$$

or, after multiplying numerator and denominator by $-i$

$$A = \frac{-iF_0}{\omega[r + i(\omega m - s/\omega)]} = \frac{-iF_0}{\omega Z_m}$$

Hence

$$x = A\, e^{i\omega t} = \frac{-iF_0\, e^{i\omega t}}{\omega Z_m} = \frac{-iF_0\, e^{i\omega t}}{\omega Z_m\, e^{i\phi}}$$
$$= \frac{-iF_0\, e^{i(\omega t - \phi)}}{\omega Z_m}$$

where

$$Z_m = \left[r^2 + (\omega m - s/\omega)^2\right]^{1/2}$$

This vector form of the steady state behaviour of x gives three pieces of information and completely defines the magnitude of the displacement x and its phase with respect to the driving force after the transient term dies away. It tells us

(1) **That the phase difference ϕ exists between x and the force because of the reactive part $(\omega m - s/\omega)$ of the mechanical impedance.**
(2) **That an extra difference is introduced by the factor $-i$ and even if ϕ were zero the displacement x would lag the force $F_0 \cos \omega t$ by $90°$.**
(3) **That the maximum amplitude of the displacement x is $F_0/\omega Z_m$. We see that this is dimensionally correct because the velocity x/t has dimensions F_0/Z_m.**

Having used $F_0\, e^{i\omega t}$ to represent its real part $F_0 \cos \omega t$, we now take the real part of the solution

$$x = \frac{-iF_0\, e^{i(\omega t - \phi)}}{\omega Z_m}$$

to obtain the actual value of x. (If the force had been $F_0 \sin \omega t$, we would now take that part of x preceded by i.)

Now

$$x = -\frac{iF_0}{\omega Z_m} e^{i(\omega t - \phi)}$$

$$= -\frac{iF_0}{\omega Z_m} [\cos (\omega t - \phi) + i \sin(\omega t - \phi)]$$

$$= -\frac{iF_0}{\omega Z_m} \cos (\omega t - \phi) + \frac{F_0}{\omega Z_m} \sin(\omega t - \phi)$$

The value of x resulting from $F_0 \cos \omega t$ is therefore

$$x = \frac{F_0}{\omega Z_m} \sin(\omega t - \phi)$$

[the value of x resulting from $F_0 \sin \omega t$ would be $-F_0 \cos (\omega t - \phi)/\omega Z_m$].

Note that both of these solutions satisfy the requirement that the total phase difference between displacement and force is ϕ plus the $-\pi/2$ term introduced by the $-i$ factor. When $\phi = 0$ the displacement $x = F_0 \sin \omega t/\omega Z_m$ lags the force $F_0 \cos \omega t$ by exactly $90°$.

To find the velocity of the forced oscillation in the steady state we write

$$v = \dot{x} = (i\omega)\frac{(-iF_0)}{\omega Z_m} e^{i(\omega t - \phi)}$$

$$= \frac{F_0}{Z_m} e^{i(\omega t - \phi)}$$

We see immediately that

(1)   There is no preceding i factor so that the velocity v and the force differ in phase only by ϕ, and when $\phi = 0$ the velocity and force are in phase.
(2)   The amplitude of the velocity is F_0/Z_m, which we expect from the definition of mechanical impedance $Z_m = F/v$.

Again we take the real part of the vector expression for the velocity, which will correspond to the real part of the force $F_0 e^{i\omega t}$. This is

$$v = \frac{F_0}{Z_m} \cos(\omega t - \phi)$$

Thus, the velocity is always exactly $90°$ ahead of the displacement in phase and differs from the force only by a phase angle ϕ, where

$$\tan \phi = \frac{\omega m - s/\omega}{r} = \frac{X_m}{r}$$

so that a force $F_0 \cos \omega t$ gives a displacement

$$x = \frac{F_0}{\omega Z_m} \sin(\omega t - \phi)$$

and a velocity

$$\dot{x} = \omega x = v = \frac{F_0}{Z_m}\cos(\omega t - \phi)$$

and an acceleration

$$\ddot{x} = -\omega^2 x = -\frac{\omega F_0}{Z_m}\sin(\omega t - \phi) \qquad \text{(cf. Figure 1.3)}$$

Frequencies at which x_{max}, $\dot{x}_{max}$ and $\ddot{x}_{max}$ occur

The amplitude of $x = F_0/\omega Z_m$ so x_{max} occurs at a minimum of ωZ_m that is when

$$\frac{d}{d\omega}\omega Z_m = 0 = \frac{d}{d\omega}\omega[r^2 + (\omega m - s/m)^2]^{\frac{1}{2}}$$

i.e. when

$$2\omega r^2 + 4\omega m(\omega^2 m - s) = 0$$

or

$$2\omega[r^2 + 2m(\omega^2 m - s)] = 0$$

giving

$$\text{either } \omega = 0 \text{ or } \omega^2 = \frac{s}{m} - \frac{r^2}{2m^2} = \omega_0^2 - \frac{r^2}{2m^2}$$

The amplitude of $\dot{x} = \omega x = F_0/Z_m$ so $\dot{x}_{max}$ occurs at $Z_m(\text{minimum})$, i.e.

$$\left(\omega m - \frac{s}{m}\right) = 0 \quad \text{or} \quad \omega_0^2 = \frac{s}{m}$$

Worked Example

Prove that the acceleration $\ddot{x}_{max}$ occurs at

$$\frac{\omega_0^2}{\omega^2} = \left(1 - \frac{r^2}{2\omega_0^2 m^2}\right) \quad \text{where} \quad \omega_0^2 = \frac{s}{m}$$

The amplitude of

$$\ddot{x}_{max} = |\omega^2 x| = \left|\frac{F_0}{Z_m/\omega}\right|$$

so $\ddot{x}_{max}$ occurs at the minimum of

$$\frac{1}{Z_m/\omega}$$

that is when

$$\frac{d}{d\omega} \frac{1}{Z_m/\omega} = 0$$

i.e. when

$$\frac{d}{d\omega} \left(\frac{r^2}{\omega^2} + \frac{\omega^2}{\omega^2}m^2 - \frac{2m\omega_0^2}{\omega^2} + \frac{\omega_0^4 m^2}{\omega^4} \right) = 0$$

or

$$-2r^2 \frac{\omega}{\omega^3} + 4m^2\omega_0^2 \frac{\omega}{\omega^3} - 4\omega_0^4 m^2 \frac{\omega^3}{\omega^5} = 0$$

or

$$m^2 \frac{\omega}{\omega^3} \left[\frac{-r^2}{2\omega_0^2 m^2} + 1 \right] = m^2\omega_0^2 \frac{\omega^3}{\omega^5} = m^2 \frac{\omega_0^2}{\omega^2} \frac{\omega}{\omega^3}$$

i.e. when

$$\frac{\omega_0^2}{\omega^2} = \left[1 - \frac{r^2}{2\omega_0^2 m^2} \right]$$

3.7   Behaviour of Velocity v in Magnitude and Phase versus Driving Force Frequency ω

The velocity amplitude is

$$\frac{F_0}{Z_m} = \frac{F_0}{[r^2 + (\omega m - s/\omega)^2]^{1/2}}$$

so that the magnitude of the velocity will vary with the frequency ω because Z_m is frequency dependent.

At low frequencies, the term $-s/\omega$ is the largest term in Z_m and the impedance is said to be stiffness controlled. At high frequencies ωm is the dominant term and the impedance is mass controlled. At a frequency ω_0 where $\omega_0 m = s/\omega_0$, the impedance has its minimum value $Z_m = r$ and is a real quantity with zero reactance.

The velocity F_0/Z_m then has its maximum value $v = F_0/r$, and ω_0 is said to be the frequency of velocity resonance. Note that $\tan \phi = 0$ at ω_0, the velocity and force being in phase.

The variation of the magnitude of the velocity with driving frequency, ω, is shown in Figure 3.7, the height and sharpness of the peak at resonance depending on r, which is the only effective term of Z_m at ω_0.

The expression

$$v = \frac{F_0}{Z_m} \cos(\omega t - \phi)$$

where

$$\tan \phi = \frac{\omega m - s/\omega}{r}$$

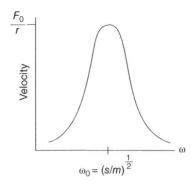

Figure 3.7   *Velocity of forced oscillator versus driving frequency ω. Maximum velocity $v_{\max} = F_0/r$ at $\omega_0^2 = s/m$.*

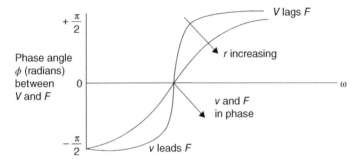

Figure 3.8   *Variation of phase angle ϕ versus driving frequency, where ϕ is the phase angle between the velocity of the forced oscillator and the driving force. $\phi = 0$ at velocity resonance. Each curve represents a fixed resistance value.*

shows that for positive ϕ; that is, $\omega m > s/\omega$, the velocity v will lag the force because $-\phi$ appears in the argument of the cosine. When the driving force frequency ω is very high and $\omega \to \infty$, then $\phi \to 90°$ and the velocity lags the force by that amount.

When $\omega m < s/\omega$, ϕ is negative, the velocity is ahead of the force in phase, and at low driving frequencies as $\omega \to 0$ the term $s/\omega \to \infty$ and $\phi \to -90°$.

Thus, at low frequencies the velocity leads the force (ϕ negative) and at high frequencies the velocity lags the force (ϕ positive).

At the frequency ω_0, however, $\omega_0 m = s/\omega_0$ and $\phi = 0$, so that velocity and force are in phase. Figure 3.8 shows the variation of ϕ with ω for the velocity, the actual shape of the curves depending upon the value of r.

3.8   Behaviour of Displacement x versus Driving Force Frequency ω

The phase of the displacement

$$x = \frac{F_0}{\omega Z_m} \sin(\omega t - \phi)$$

is at all times exactly 90° behind that of the velocity. Whilst the graph of ϕ versus ω remains the same, the total phase difference between the displacement and the force involves the extra 90° retardation introduced by the $-i$ operator. Thus, at very low frequencies, where $\phi = -\pi/2$ rad and the velocity leads the force, the displacement and the force are in phase as we should expect. At high frequencies the displacement lags the force by π rad and is exactly out of phase, so that the curve showing the phase angle between the displacement and the force is equivalent to the ϕ versus ω curve, displaced by an amount equal to $\pi/2$ rad. This is shown in Figure 3.9.

The amplitude of the displacement $x = F_0/\omega Z_m$, and at low frequencies $Z_m = [r^2 + (\omega m - s/\omega)^2]^{1/2} \rightarrow s/\omega$, so that $x \approx F_0/(\omega s/\omega) = F_0/s$.

At high frequencies $Z_m \rightarrow \omega m$, so that $x \approx F_0/(\omega^2 m)$, which tends to zero as ω becomes very large. At very high frequencies, therefore, the displacement amplitude is almost zero because of the mass-controlled or inertial effect.

The velocity resonance occurs at $\omega_0^2 = s/m$, where the denominator Z_m of the velocity amplitude is a minimum, but the displacement resonance will occur, since $x = (F_0/\omega Z_m) \sin(\omega t - \phi)$, when the denominator ωZ_m is a minimum. We saw just before the worked example at the end of section 3.6 that this takes place when

$$\omega^2 = \frac{s}{m} - \frac{r^2}{2m^2} = \omega_0^2 - \frac{r^2}{2m^2}$$

Thus the displacement resonance occurs at a frequency slightly less than ω_0, the frequency of velocity resonance. For a small damping constant r or a large mass m these two resonances, for all practical purposes, occur at the frequency ω_0.

Denoting the displacement resonance frequency by

$$\omega_r = \left(\frac{s}{m} - \frac{r^2}{2m^2} \right)^{1/2}$$

we can write the maximum displacement as

$$x_{\max} = \frac{F_0}{\omega_r Z_m}$$

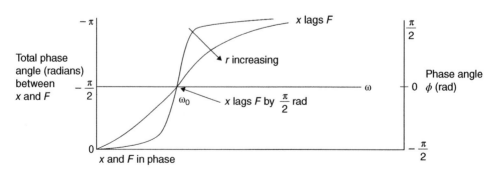

Figure 3.9 *Variation of total phase angle between displacement and driving force versus driving frequency ω. The total phase angle is $-\phi - \pi/2$ rad.*

Worked Example

We now prove that the value of $\omega_r Z_m$ at ω_r is equal to $\omega' r$ where

$$\omega'^2 = \frac{s}{m} - \frac{r^2}{4m^2} = \omega_0^2 - \frac{r^2}{4m^2}$$

$$x_{\max} = \frac{F_0}{\omega Z_m} \quad \text{at} \quad \omega^2 = \omega_0^2 - \frac{r^2}{2m^2}$$

i.e.

$$\omega^2 Z_m^2 = \omega^2 \left[r^2 + \left(\omega m - \frac{s}{\omega} \right)^2 \right] \quad \text{with} \quad \omega^2 = \omega_0^2 - \frac{r^2}{2m^2}$$

$$= r^2 \left(\omega_0^2 - \frac{r^2}{2m^2} \right) + \left[\left(\omega_0^2 m - \frac{r^2}{2m} \right) - s \right]^2 \quad \text{where} \quad \frac{s}{m} = \omega_0^2$$

$$= r^2 \left(\omega_0^2 - \frac{r^2}{2m^2} \right) + \left(\frac{r^2}{2m} \right)^2$$

$$= r^2 \left(\omega_0^2 - \frac{r^2}{2m^2} + \frac{r^2}{4m^2} \right)$$

$$= r^2 \left(\omega_0^2 - \frac{r^2}{4m^2} \right)$$

$$= \omega'^2 r^2$$

Since $x_{\max} = F_0/\omega' r$ the amplitude at resonance is left low by increasing r with x and the variation of x with ω for different values of r is shown in Figure 3.10. Keeping the resonance amplitude low is the principle of vibration insulation.

3.9   The Q-Value as an Amplification Factor

At low frequencies ($\omega \to 0$) the displacement has a value $x_0 = F_0/s$, so that

$$\left(\frac{x_{\max}}{x_0} \right)^2 = \frac{F_0^2}{\omega'^2 r^2} \frac{s^2}{F_0^2} = \frac{m^2 \omega_0^4}{r^2 [\omega_0^2 - r^2/4m^2]}$$

$$= \frac{\omega_0^2 m^2}{r^2 [1 - 1/4Q^2]^{1/2}} = \frac{Q^2}{[1 - 1/4Q^2]^{1/2}}$$

Hence:

$$\frac{x_{\max}}{x_0} = \frac{Q}{[1 - 1/4Q^2]^{1/2}} \approx Q \left[1 + \frac{1}{8Q^2} \right] \approx Q$$

for large Q.

Thus the displacement at low frequencies is amplified by a factor of Q at displacement resonance.

3.10   Significance of the Two Components of the Displacement Curve

Any single curve of Figure 3.10 is the superposition of the two component curves (a) and (b) in Figure 3.11, for the displacement x may be rewritten

$$x = \frac{F_0}{\omega Z_m} \sin(\omega t - \phi) = \frac{F_0}{\omega Z_m}(\sin \omega t \cos \phi - \cos \omega t \sin \phi)$$

or, since

$$\cos \phi = \frac{r}{Z_m} \quad \text{and} \quad \sin \phi = \frac{X_m}{Z_m}$$

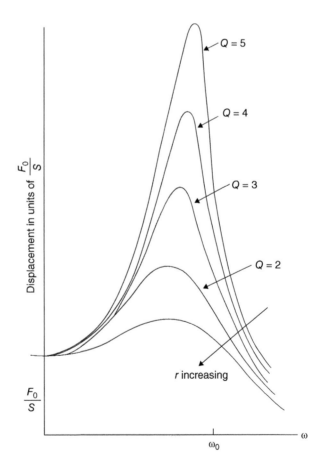

Figure 3.10   *Curves of displacement vesus frequency given in terms of the quality factor Q of the system, where Q is amplification at resonance of low frequency response $x_0 = F_0/s$.*

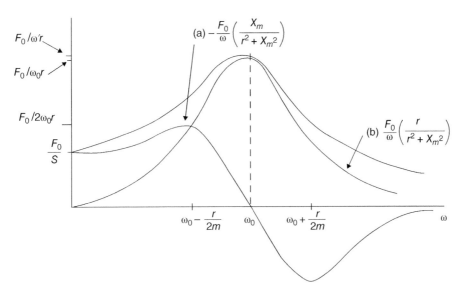

Figure 3.11 *A typical curve of Figure 3.10 resolved into its 'anti-phase' component (curve (a)) and its '90° out of phase' component (curve (b)). Curve (b) represents the resistive fraction of the impedance and curve (a) the reactive fraction. Curve (b) corresponds to absorption and curve (a) to anomalous dispersion of an electromagnetic wave in a medium having an atomic or molecular resonant frequency equal to the frequency of the wave.*

as

$$x = \frac{F_0}{\omega Z_m}\frac{r}{Z_m}\sin\omega t - \frac{F_0}{\omega Z_m}\frac{X_m}{Z_m}\cos\omega t$$

The $\cos\omega t$ component (with a negative sign) is exactly anti-phase with respect to the driving force $F_0\cos\omega t$. Its amplitude, plotted as curve (a) may be expressed as

$$-\frac{F_0}{\omega}\frac{X_m}{Z_m^2} = \frac{F_0 m(\omega_0^2 - \omega^2)}{m^2(\omega_0^2 - \omega^2)^2 + \omega^2 r^2} \tag{3.3}$$

where $\omega_0^2 = s/m$ and ω_0 is the frequency of velocity resonance.

The $\sin\omega t$ component lags the driving force $F_0\cos\omega t$ by 90°. Its amplitude plotted as curve (b) becomes

$$\frac{F_0}{\omega}\frac{r}{r^2 + X_m^2} = \frac{F_0\omega r}{m^2(\omega_0^2 - \omega^2)^2 + \omega^2 r^2}$$

We see immediately that at ω_0 curve (a) is zero and curve (b) is near its maximum but they combine to give a maximum at ω where

$$\omega^2 = \omega_0^2 - \frac{r^2}{2m^2}$$

the resonant frequency for amplitude displacement.

These curves are particularly familiar in the study of optical dispersion where the forced oscillator is an electron in an atom and the driving force is the oscillating field vector of an electromagnetic wave of frequency ω. When ω is the resonant frequency of the electron in the atom, the atom absorbs a large amount of energy from the electromagnetic wave and curve (b) is the shape of the characteristic absorption curve. Note that curve (b) represents the dissipating or absorbing fraction of the impedance

$$\frac{r}{(r^2 + X_m^2)^{1/2}}$$

and that part of the displacement which lags the driving force by $90°$. The velocity associated with this component will therefore be in phase with the driving force and it is this part of the velocity which appears in the energy loss term $r\dot{x}^2$ due to the resistance of the oscillator and which gives rise to absorption.

On the other hand, curve (a) represents the reactive or energy storing fraction of the impedance

$$\frac{X_m}{(r^2 + X_m^2)^{1/2}}$$

and the reactive components in a medium determine the velocity of the waves in the medium which in turn governs the refractive index n. In fact, curve (a) is a graph of the value of n^2 in a region of anomalous dispersion where the ω axis represents the value $n = 1$. These regions occur at every resonant frequency of the constituent atoms of the medium.

Worked Example

In Figure 3.11 show that for small r the maximum value of curve (a) equation 3.3 is $\approx F_0/2\omega_0 r$ at $\omega_1 = (\omega_0 - r/2m)$ and its minimum value is $-F_0/2\omega_0 r$ at $\omega_2 = (\omega_0 + r/2m)$ where $\omega_0^2 = s/m$. We write curve (a) as

$$\frac{F_0 m(\omega_0^2 - \omega^2)}{m^2(\omega_0^2 - \omega^2)} + \omega^2 r^2 = \frac{u}{v}$$

and we take

$$\frac{d}{d\omega}\left(\frac{u}{v}\right) = \frac{u'v - v'u}{v^2} \quad \left(\text{where } u' = \frac{d}{d\omega}u\right)$$

we equate $u'v - v'u$ to 0.

$$u'v - v'u = -2F_0\omega v - [2m^2(\omega_0^2 - \omega^2)(-2\omega) + 2r^2\omega]u = 0$$
$$= m(\omega_0^2 - \omega^2)^2 - \omega_0^2 r^2 = 0$$

with roots

$$\omega_1 = \left(\omega_0^2 - \omega_0\frac{r}{m}\right)^{\frac{1}{2}} \quad \text{and} \quad \omega_2 = \left(\omega_0^2 + \omega_0\frac{r}{m}\right)^{\frac{1}{2}}$$

For small r we write

$$\omega_1 = \left(\omega_0^2 - \omega_0\frac{r}{m}\right)^{\frac{1}{2}} = \left[\left(\omega_0^2 - \frac{r}{2m}\right)^2 - \frac{r^2}{4m^2}\right]^{\frac{1}{2}} \approx \omega_0 - \frac{r}{2m}$$

and

$$\omega_2 = \left(\omega_0^2 + \omega_0\frac{r}{m}\right)^{\frac{1}{2}} = \left[\left(\omega_0^2 + \frac{r}{2m}\right)^2 - \frac{r^2}{4m^2}\right]^{\frac{1}{2}} \approx \omega_0 + \frac{r}{2m}$$

The maximum and minimum values of curve (a) may be formed by inserting

$$\omega_1 = \left(\omega_0^2 - \omega_0\frac{r}{m}\right)^{\frac{1}{2}}$$

and

$$\omega_2 = \left(\omega_0^2 + \omega_0\frac{r}{m}\right)^{\frac{1}{2}}$$

in the expression for curve (a).

Inserting ω_1 in the curve (a) expression gives

$$x = \frac{F_0\omega_0 r}{\omega_0^2 r^2 + \omega_0^2 r^2 - \omega_0 r^3/m} = \frac{F_0\omega_0 r}{2\omega_0^2 r^2 - \omega_0 r^3/m} = \frac{F_0}{2\omega_0 r - \omega_0 r^2/m} \approx \frac{F_0}{2\omega_0 r}$$

which is the maximum value of curve (a).

Inserting ω_2 in the expression for curve (a) gives

$$x = \frac{-F_0\omega_0 r}{\omega_0^2 r^2 + \omega_0^2 r^2 + \omega_0 r^3/m} = \frac{-F_0\omega_0 r}{2\omega_0^2 r^2 + \omega_0 r^3/m} = \frac{-F_0}{2\omega_0 r + \omega_0 r^2/m} \approx \frac{-F_0}{2\omega_0 r}$$

which is the minimum value of curve (a).

Note that the frequency range $\omega_2 - \omega_1 = r/m$ is the bandwidth at 1/2 maximum value $F_0/\omega_0 r$ of the absorption curve (b).

3.11 Problem on Vibration Insulation

Keeping the resonance amplitude low is the principle of vibration insulation. A typical vibration insulator is shown in Figure 3.12. A heavy base is supported on a vibrating floor by a spring system of stiffness s and viscous damper r. The insulator will generally operate at the mass controlled end of the frequency spectrum and the resonant frequency is designed to be lower than the range of frequencies likely to be met. Suppose the vertical vibration of the floor is given by $x = x_0 e^{i\omega t}$ about its equilibrium position and y is the corresponding vertical displacement of the base about its rest position. The function of the insulator is to keep the ratio y/x_0 to a minimum in the steady state.

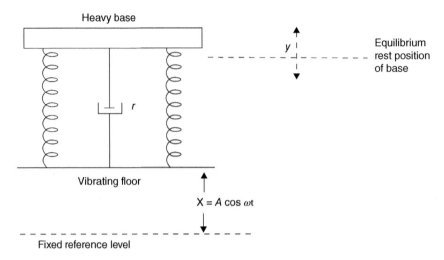

Figure 3.12   *Vibration insulator. A heavy base supported by a spring and viscous damper system on a vibrating floor.*

The equation of motion is given by

$$m\ddot{y} = -r(\dot{y} - \dot{x}) - s(y - x)$$

that is

$$m\ddot{y} + r(\dot{y} - \dot{x}) + s(y - x) = 0$$

Dividing by m we have

$$\left(\ddot{y} + \frac{r}{m}\dot{y} + \frac{s}{m}y\right) = \left(\frac{s}{m}x + \frac{r}{m}\dot{x}\right)$$

which with $s/m = \omega_0^2$, $x = x_0\,\mathrm{e}^{i\omega t}$ and $y = y_0\,\mathrm{e}^{i\omega t}$ becomes

$$(-\omega^2 + i\omega r/m + \omega_0^2)y_0\,\mathrm{e}^{i\omega t} = (\omega_0^2 + i\omega r/m)x_0\,\mathrm{e}^{i\omega t}$$

or

$$\frac{y_0}{x_0} = \frac{\omega_0^2 + i\omega r/m}{(\omega_0^2 - \omega^2) + i\omega r/m}$$

Recalling section 2.1 Complex Numbers (vi), we have

$$\left|\frac{y_0}{x_0}\right| = \left|\frac{y_0 y_0^*}{x_0 x_0^*}\right|^{\frac{1}{2}} = \left[\frac{\omega_0^4 + \omega^2 r^2/m^2}{(\omega_0^2 - \omega_0^2)^2 + \omega^2 r^2/m^2}\right]^{\frac{1}{2}}$$

$$\therefore \left|\frac{y_0}{x_0}\right| = \left[\frac{\omega_0^4 + \omega^2 r^2/m^2}{(\omega_0^4 + \omega^2 r^2/m^2) + \omega^2(\omega^2 - 2\omega_0^2)}\right]^{\frac{1}{2}}$$

Examing the denominator of $|y_0/x_0|$ we see that
(a) $y_0/x_0 = 1$ for $\omega^2 = 2s/m$
(b) $y_0/x_0 < 1$ for $\omega^2 > 2s/m$ and $y_0/x_0 > 1$ for $\omega^2 < 2s/m$
When $\omega^2 = s/m$, $y_0/x_0 > 1$ but r helps to keep y_0 low.
When $\omega^2 > 2s/m$, $y_0/x_0 < 1$ but r is unhelpful.
The value of $s/m = \omega_0^2$ should be as low as possible.
The formal derivation of y_0/x_0 is given by the sum of two simple harmonic motions

$$y = \frac{F_0}{\omega Z_m} \sin(\omega t - \phi) + A \cos \omega t \quad \text{where} \quad x = A \cos \omega t$$

and

$$F_0 = mA\omega^2$$

but, mathematically this is much more complicated.

3.12   The Effect of the Transient Term

Throughout this chapter we have considered only the steady state behaviour without accounting for the transient term mentioned in section 3.6 The Forced Mechanical Oscillator, the equation of motion, term (1). This term makes an initial contribution to the total displacement but decays with time as $e^{-rt/2m}$. Its effect is best displayed by considering the vector sum of the transient and steady state components.

The steady state term may be represented by a vector of constant length rotating anticlockwise at the angular velocity ω of the driving force. The vector tip traces a circle. Upon this is superposed the transient term vector of diminishing length which rotates anticlockwise with angular velocity $\omega' = (s/m - r^2/4m^2)^{1/2}$. Its tip traces a contracting spiral.

The locus of the magnitude of the vector sum of these terms is the envelope of the varying amplitudes of the oscillator. This envelope modulates the steady state oscillations of frequency ω at a frequency which depends upon ω' and the relative phase between ωt and $\omega' t$.

Thus, in Figure 3.13(a) where the total oscillator displacement is zero at time $t = 0$ we have the steady state and transient vectors equal and opposite in Figure 3.13(b) but because $\omega \neq \omega'$ the relative phase between the vectors will change as the transient term decays. The vector tip of the transient term is shown as the dotted spiral and the total amplitude assumes the varying lengths OA_1, OA_2, OA_3, OA_4, etc.

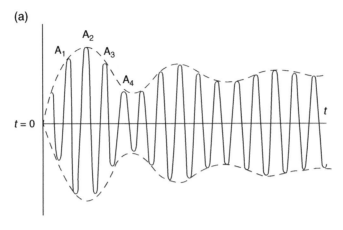

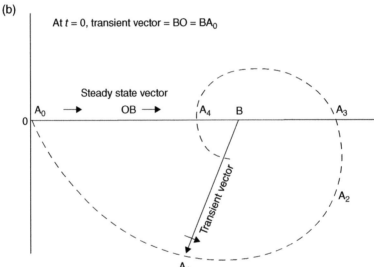

Figure 3.13   *(a) The steady state oscillation (heavy curve) is modulated by the transient term which decays exponentially with time. (b) In the vector diagram of (b) **OB** is the constant length steady state vector and **BA**$_1$ is the transient vector. Each vector rotates anticlockwise with its own angular velocity. At t = 0 the vectors **OB** and **BA**$_0$ are equal and opposite on the horizontal axis and their vector sum is zero. At subsequent times the total amplitude is the length of **OA**$_1$ which changes as A traces a contracting spiral around B. The points A$_1$, A$_2$, A$_3$ and A$_4$ indicate how the amplitude is modified in (a).*

Problem 3.1.   Show that in a resonant *LCR* series circuit the maximum potential across the capacitor occurs at a frequency

$$\omega = \omega_0 \left(1 - \frac{1}{2Q_0^2}\right)^{\frac{1}{2}}$$

where

$$\omega_0^2 = (LC)^{-1}$$

and

$$Q_0 = \omega_0 L / R.$$

See the end of section 3.6 for the frequency of x_{max}.

Problem 3.2.   In Problem 3.1 show that the maximum potential across the inductance occurs at a frequency

$$\omega = \omega_0 \left(1 - \frac{1}{2Q_0^2} \right)^{-\frac{1}{2}}.$$

See the end of section 3.6 for the frequency of $\ddot{x}_{max}$.

Problem 3.3.   A series LCR circuit has $C = 8 \times 10^{-6}$ F, $L = 2 \times 10^{-2}$ H and $R = 75\,\Omega$. It is driven by a voltage $V(t) = 15 \cos \omega t$. Find (a) the resonant frequency in Hz of the circuit and (b) the amplitude of the circuit at this frequency.

Problem 3.4.   Show that the bandwidth of the resonance absorption curve defines the phase angle range $\tan \phi = \pm 1$.

Problem 3.5.   The average power $\bar{P}$ is absorbed by a driven oscillator and its resonance curve is symmetric. $\bar{P}$ is a maximum at $\nu = 100$ Hz and its $1/2$ maximum occurs at 95 Hz. What is the value of (a) ω_0 (b) the bandwidth $\Delta \omega$ (c) Q and (d) if the force is suddenly removed after how many cycles will the energy of the system be $1/e$ of its initial value?

Problem 3.6.   Show that if $r = (sm)^{\frac{1}{2}}$ in a forced damped mechanical oscillator then the acceleration amplitude at the frequency of velocity resonance equals the limit of the acceleration amplitude at high frequencies.

Problem 3.7.   In a forced mechanical oscillator show that the following are frequency independent: (a) the displacement amplitude at low frequencies, (b) the velocity amplitude of velocity resonance and (c) the acceleration amplitude at high frequencies as $\omega \to \infty$.

Problem 3.8.   The equation $m\ddot{x} + sx = F_0 \sin \omega t$ describes the motion of an undamped simple harmonic oscillator driven by a force of frequency ω. Show, by solving the equation in vector form, that the steady state solution is given by

$$x = \frac{F_0 \sin \omega t}{m(\omega_0^2 - \omega^2)} \quad \text{where} \quad \omega_0^2 = \frac{s}{m}$$

Sketch the behaviour of the amplitude of x versus ω and note that the change of sign as ω passes through ω_0 defines a phase change of π rad in the displacement. Now show that the general solution for the displacement is given by

$$x = \frac{F_0 \sin \omega t}{m(\omega_0^2 - \omega^2)} + A \cos \omega_0 t + B \sin \omega_0 t$$

where A and B are constant.

Problem 3.9. The equation $\ddot{x} + \omega_0^2 x = (-eE_0/m)\cos \omega t$ describes the motion of a bound undamped electric charge $-e$ of mass m under the influence of an alternating electric field $E = E_0 \cos \omega t$. For an electron number density n show that the induced polarizability per unit volume (the dynamic susceptibility) of a medium

$$\chi_e = -\frac{n\,ex}{\varepsilon_0 E} = \frac{n\,e^2}{\varepsilon_0 m(\omega_0^2 - \omega^2)}$$

(The permittivity of a medium is defined as $\varepsilon = \varepsilon_0(1+\chi)$ where ε_0 is the permittivity of free space. The relative permittivity $\varepsilon_r = \varepsilon/\varepsilon_0$ is called the dielectric constant and is the square of the refractive index when E is the electric field of an electromagnetic wave.)

Problem 3.10. Repeat Problem 3.9 for the case of a damped oscillatory electron, by taking the displacement x as the component represented by curve (a) in Figure 3.11 to show that

$$\varepsilon_r = 1 + \chi = 1 + \frac{n\,e^2 m(\omega_0^2 - \omega^2)}{\varepsilon_0[m^2(\omega_0^2 - \omega^2)^2 + \omega^2 r^2]}$$

In fact, Figure 3.11(a) plots $\varepsilon_r = \varepsilon/\varepsilon_0$. Note that for

$$\omega \ll \omega_0, \quad \varepsilon_r \approx 1 + \frac{n\,e^2}{\varepsilon_0\,m\omega_0^2}$$

and for

$$\omega \gg \omega_0, \quad \varepsilon_r \approx 1 - \frac{n\,e^2}{\varepsilon_0\,m\omega^2} \quad \text{(see Figure 6.3)}$$

Problem 3.11. Light of wavelength $0.6\,\mu\text{m}$ (6000) is emitted by an electron in an atom behaving as a lightly damped simple harmonic oscillator with a Q value of 5×10^7. Show from the resonance bandwidth that the width of the spectral line from such an atom is 1.2×10^{-14} m.

Problem 3.12. The displacement of a forced oscillator is zero at time $t = 0$ and its rate of growth is governed by the rate of decay of the transient term. If this term decays to e^{-k} of its original value in a time t show that, for small damping, the average rate of growth of the oscillations is given by $x_0/t = F_0/2\,km\omega_0$ where x_0 is the maximum steady state displacement, F_0 is the force amplitude and $\omega_0^2 = s/m$.

4

Coupled Oscillations

Introduction

Until now we have dealt with oscillators in isolation. Now we couple them so that they can pass their energy to other oscillators. Such a series of oscillators forms a medium through which the energy of their simple harmonic vibrations is transmitted as waves. We begin by showing how the coupling between two atoms in a molecule leads, via the spectroscopy of their vibrations, to an estimate of the strength of the chemical bond which acts as a spring between them. Coupling with weak springs displays the concepts of normal modes, normal coordinates and degrees of freedom, that is, ways of taking up energy. Normal modes are best known as the fundamental and harmonics of strings on a musical instrument. Where the coupling takes place via spring stiffness and electrical inductance these energy storing parameters can transfer their energy without loss. Loss mechanisms will be discussed in later chapters. Finally, with identical masses equally spaced along an extended string we are able to derive the wave equation. This needs, at the end of the chapter, an introduction to the notation of partial differentiation and Taylor's series.

4.1 Stiffness (or Capacitance) Coupled Oscillators

Worked Example

Figure 4.1(a) shows two hydrogen atoms in a hydrogen molecule which are connected by a chemical bond acting as a spring. The atoms are free to vibrate along the x axis and these vibrations are antisymmetric so the centre of mass of the system is stationary. In vibration the chemical bond, stiffness s, is alternately stretched and compressed a distance of $2x$ as the atoms vibrate with amplitudes equal to x. The equation of motion of each atom is therefore given by $m\ddot{x} = -2sx$ or $\frac{m}{2}\ddot{x} + sx = 0$, a simple harmonic oscillation

Introduction to Vibrations and Waves, First Edition. H. J. Pain and Patricia Rankin.
© 2015 John Wiley & Sons, Ltd. Published 2015 by John Wiley & Sons, Ltd.
Companion website: http://booksupport.wiley.com

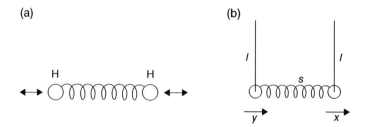

Figure 4.1   *(a) H atoms vibrating asymmetrically along the axis of the chemical bond. (b) Two identical pendulums, each a light rigid rod of length l supporting a mass m and coupled by a weightless spring of stiffness s and of natural length equal to the separation of the masses at zero displacement.*

with $\omega_0^2 = 2s/m$. This frequency is spectroscopically observed to be $\nu_0 = 1.32 \times 10^{14}$ and it lies in the infrared region of the electromagnetic spectrum. A hydrogen atom weighs 1.67×10^{-27} kg so

$$s = \frac{m}{2}\omega_0^2 = \frac{4\pi(1.32 \times 10^{14})^2(1.67 \times 10^{-27})}{2} = 574\,\text{N}\,\text{m}^{-1}.$$

The factor $m/2$ is a particular case of the 'reduced mass' where the two masses (H atoms) are equal. In general, when the masses are m_1 and m_2 the reduced mass μ is written

$$\frac{1}{\mu} = \left(\frac{1}{m_1} + \frac{1}{m_2}\right) = \frac{m_1 + m_2}{m_1 m_2}.$$

Figure 4.1(b) shows two identical pendulums, each having a mass m suspended on a light rigid rod of length l. The masses are connected by a light spring of stiffness s whose natural length equals the distance between the masses when neither is displaced from equilibrium. The small oscillations we discuss are restricted to the plane of the paper.

If x and y are the respective displacements of the masses, then the equations of motion are

$$m\ddot{x} = -mg\frac{x}{l} - s(x - y)$$

and

$$m\ddot{y} = -mg\frac{y}{l} + s(x - y)$$

These represent the normal simple harmonic motion terms of each pendulum plus a coupling term $s(x-y)$ from the spring. We see that if $x > y$ the spring is extended beyond its normal length and will act against the acceleration of x but in favour of the acceleration of y.

Writing $\omega_0^2 = g/l$, where ω_0 is the natural vibration frequency of each pendulum, gives

$$\ddot{x} + \omega_0^2 x = -\frac{s}{m}(x - y) \tag{4.1}$$

$$\ddot{y} + \omega_0^2 y = -\frac{s}{m}(y - x) \tag{4.2}$$

Instead of solving these equations directly for x and y we are going to choose two new coordinates

$$X = x + y$$
$$Y = x - y$$

The importance of this approach will emerge as this chapter proceeds. Adding equations (4.1) and (4.2) gives

$$\ddot{x} + \ddot{y} + \omega_0^2 (x + y) = 0$$

that is

$$\ddot{X} + \omega_0^2 X = 0 \qquad (4.1\text{a})$$

and subtracting (4.2) from (4.1) gives

$$\ddot{Y} + \left(\omega_0^2 + 2s/m\right) Y = 0 \qquad (4.2\text{a})$$

The motion of the coupled system is thus described in terms of the two coordinates X and Y, each of which has an equation of motion which is simple harmonic.

If $Y = 0, x = y$ at all times, so that the motion is completely described by the equation

$$\ddot{X} + \omega_0^2 X = 0$$

then the frequency of oscillation is the same as that of either pendulum in isolation and the stiffness of the coupling has no effect. This is because both pendulums are always swinging in phase (Figure 4.2a) and the light spring is always at its natural length.

If $X = 0, x = -y$ at all times, so that the motion is completely described by

$$\ddot{Y} + \left(\omega_0^2 + 2s/m\right) Y = 0$$

The frequency of oscillation is greater because the pendulums are always out of phase (Figure 4.2b) so that the spring is either extended or compressed and the coupling is effective.

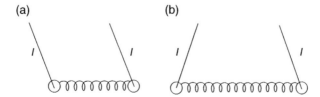

(a)   (b)

Figure 4.2 (a) The 'in phase' mode of vibration given by $\ddot{X} + \omega_0^2 X = 0$, where X is the normal coordinate $X = x + y$ and $\omega_0^2 = g/l$. (b) 'Out of phase' mode of vibration given by $\ddot{Y} + \left(\omega_0^2 + 2s/m\right)$ where Y is the normal coordinate $Y = x - y$.

4.2   Normal Modes of Vibration, Normal Coordinates and Degrees of Freedom

The significance of choosing X and Y to describe the motion is that these parameters give a very simple illustration of Normal Modes.

Equations 4.1a and 4.2a are examples of Normal Modes. A normal mode is characterized by the fact that all of its components oscillate with the same frequency. Equation 4.1a has a frequency ω_0, 4.2a has a frequency $(\omega_0^2 + 2s/m)^{1/2}$. These frequencies are called normal frequencies or eigen-frequencies.

Each coordinate of a normal mode is called a Normal Coordinate and each Normal Coordinate defines a degree of freedom, that is, an independent way in which a normal mode acquires energy.

The important property of normal modes of vibration is that they are entirely independent of each other. The energy associated with a normal mode is never exchanged with another mode; this is why we can add the energies of the separate modes to give the total energy of a system. If only one mode of our coupled pendulums is vibrating the other will always remain at rest.

Applying this principle to one simple harmonic oscillation; we associate one degree of freedom with potential energy, designated by the normal coordinate X and a second degree of freedom with kinetic energy designated by the velocity normal coordinate $\dot{X}$. The total energy of this normal mode may therefore be written

$$E_x = A\dot{X}^2 + BX^2 \tag{4.3a}$$

where A and B are constant coefficients, Similarly

$$E_y = C\dot{Y}^2 + DY^2 \tag{4.3b}$$

where C and D are constant coefficients.

This is consistent with the usual notation

$$E = \frac{1}{2}m\dot{x}^2 + sx^2$$

where m and s are constant.

Our system of two coupled pendulums has, then, four degrees of freedom and four normal coordinates.

Any configuration of our coupled system may be represented by the superposition of the two normal modes

$$X = x + y = X_0 \cos(\omega_1 t + \phi_1)$$

and

$$Y = x - y = Y_0 \cos(\omega_2 t + \phi_2)$$

where X_0 and Y_0 are the normal mode amplitudes, whilst $\omega_1^2 = g/l$ and $\omega_2^2 = (g/l + 2s/m)$ are the normal mode frequencies. To simplify the discussion let us choose

$$X_0 = Y_0 = 2a$$

and put

$$\phi_1 = \phi_2 = 0$$

The pendulum displacements are then given by

$$x = \frac{1}{2}(X + Y) = a \cos \omega_1 t + a \cos \omega_2 t$$

and

$$y = \frac{1}{2}(X - Y) = a \cos \omega_1 t - a \cos \omega_2 t$$

with velocities

$$\dot{x} = -a\omega_1 \sin \omega_1 t - a\omega_2 \sin \omega_2 t$$

and

$$\dot{y} = -a\omega_1 \sin \omega_1 t + a\omega_2 \sin \omega_2 t$$

Now let us set the system in motion by displacing the right hand mass a distance $x = 2a$ and releasing both masses from rest so that $\dot{x} = \dot{y} = 0$ at time $t = 0$.

Figure 4.3 shows that our initial displacement $x = 2a$, $y = 0$ at $t = 0$ may be seen as a combination of the 'in phase' mode ($x = y = a$ so that $x + y = X_0 = 2a$) and of the 'out of phase' mode ($x = -y = a$ so that $Y_0 = 2a$). After release, the motion of the right-hand pendulum is given by

$$x = a \cos \omega_1 t + a \cos \omega_2 t$$
$$= 2a \cos \frac{(\omega_2 - \omega_1)t}{2} \cos \frac{(\omega_1 + \omega_2)t}{2}$$

and that of the left-hand pendulum is given by

$$y = a \cos \omega_1 t - a \cos \omega_2 t$$
$$= -2a \sin \frac{(\omega_1 - \omega_2)t}{2} \sin \frac{(\omega_1 + \omega_2)t}{2}$$
$$= 2a \sin \frac{(\omega_2 - \omega_1)t}{2} \sin \frac{(\omega_1 + \omega_2)t}{2}$$

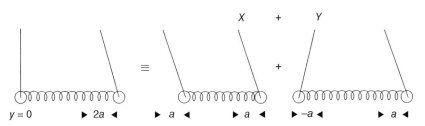

Figure 4.3 *The displacement of one pendulum by an amount 2a is shown as the combination of the two normal coordinates X + Y.*

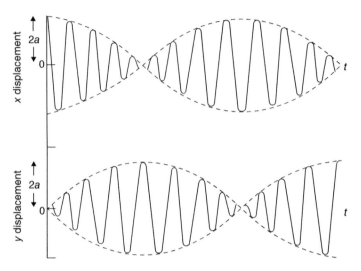

Figure 4.4   *Behaviour with time of individual pendulums, showing complete energy exchange between the pendulums as x decreases from 2a to zero whilst y grows from zero to 2a.*

If we plot the behaviour of the individual masses by showing how x and y change with time (Figure 4.4), we see that after drawing the first mass aside a distance $2a$ and releasing it x follows a cosinusoidal behaviour at a frequency which is the average of the two normal mode frequencies, but its amplitude varies cosinusoidally with a low frequency which is half the difference between the normal mode frequencies. On the other hand, y, which started at zero, vibrates sinusoidally with the average frequency but its amplitude builds up to $2a$ and then decays sinusoidally at the low frequency of half the difference between the normal mode frequencies. In short, the y displacement mass acquires all the energy of the x displacement mass which is stationary when y is vibrating with amplitude $2a$, but the energy is then returned to the mass originally displaced. This complete energy exchange is only possible when the masses are identical and the ratio $(\omega_1 + \omega_2)/(\omega_2 - \omega_1)$ is an integer, otherwise neither will ever be quite stationary. The slow variation of amplitude at half the normal mode frequency difference is the phenomenon of 'beats' which occurs between two oscillations of nearly equal frequencies. We shall discuss this further in the section on wave groups in Chapter 6.

The important point to recognize, however, is that although the individual pendulums may exchange energy, there is no energy exchange between the normal modes. Figure 4.3 showed the initial configuration $x = 2a$, $y = 0$, decomposed into the X and Y modes. The higher frequency of the Y mode ensures that after a number of oscillations the Y mode will have gained half a vibration (a phase of π rad) on the X mode; this is shown in Figure 4.5. The combination of the X and Y modes then gives y the value of $2a$ and $x = 0$, and the process is repeated. When Y gains another half vibration then x equals $2a$ again. The pendulums may exchange energy; the normal modes do not.

To reinforce the importance of normal modes and their coordinates let us return to equations 4.3(a) and 4.3(b). If we modify our normal coordinates to read

$$X_q = \left(\frac{m}{2}\right)^{1/2} (x + y) \quad \text{and} \quad Y_q = \left(\frac{m}{2}\right)^{1/2} (x - y)$$

then we find that the kinetic energy in those equations becomes, after dividing by m

$$E_k = T = A\dot{X}^2 + C\dot{Y}^2 = \frac{1}{2}\dot{X}_q^2 + \frac{1}{2}\dot{Y}_q^2 \tag{4.4a}$$

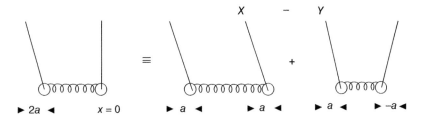

Figure 4.5 *The faster vibration of the Y mode results in a phase gain of π rad over the X mode of vibration, to give $y = 2a$, which is shown here as a combination of the normal modes X – Y.*

and the potential energy

$$V = BX^2 + DY^2 = \frac{1}{2}\left(\frac{g}{l}\right)X_q^2 + \frac{1}{2}\left(\frac{g}{l} + \frac{2s}{m}\right)Y_q^2$$

(4.4b)

$$= \frac{1}{2}\omega_0^2 X_q^2 + \frac{1}{2}\omega_s^2 Y_q^2,$$

where $\omega_0^2 = g/l$ and $\omega_s^2 = g/l + 2s/m$.

Note that the coefficients of X_q^2 and Y_q^2 depend only on the mode frequencies and that the properties of individual parts of the system are no longer explicit.

The total energy of the system is the sum of the energies of each separate excited mode for there are no cross products $X_q Y_q$ in the energy expression of our example, i.e.,

$$E = T + V = \left(\frac{1}{2}\dot{X}_q^2 + \frac{1}{2}\omega_0^2 X_q^2\right) + \left(\frac{1}{2}\dot{Y}_q^2 + \frac{1}{2}\omega_s^2 Y_q^2\right)$$

Worked Example

In the coupled pendulums of Figure 4.3 let us write the modulated frequency $\omega_m = (\omega_2 - \omega_1)/2$ and the average frequency $\omega_a = (\omega_2 + \omega_1)/2$ and assume that the spring is so weak that it stores a negligible amount of energy. Let the modulated amplitude

$$2a\cos\omega_m t \quad \text{or} \quad 2a\sin\omega_m t$$

be constant over one cycle at the average frequency ω_a to show that the energies of the masses may be written

$$E_x = 2ma^2\omega_a^2\cos^2\omega_m t$$

and

$$E_y = 2ma^2\omega_a^2\sin^2\omega_m t$$

Show that the total energy E remains constant and that the energy difference at any time is

$$E_x - E_y = E\cos(\omega_2 - \omega_1)t$$

Prove that

$$E_x = \frac{E}{2}[1 + \cos(\omega_2 - \omega_1)t]$$

and

$$E_y = \frac{E}{2}[1 - \cos(\omega_2 - \omega_1)t]$$

to show that the constant total energy is completely exchanged between the two pendulums at the beat frequency $(\omega_2 - \omega_1)$.

Solution

The pendulum motions in Figure 4.3 are given by

$$x = 2a\cos\frac{(\omega_2 - \omega_1)t}{2}\cos\frac{(\omega_1 + \omega_2)t}{2} = 2a\cos\omega_m t\cos\omega_a t$$

and

$$y = 2a\sin\frac{(\omega_2 - \omega_1)t}{2}\sin\frac{(\omega_1 + \omega_2)t}{2} = 2a\sin\omega_m t\sin\omega_a t$$

where the amplitudes of the masses $2a\cos\omega_m t$ and $2a\sin\omega_m t$ are constant over one cycle of the frequency ω_a. For small s, we have

$$\frac{g}{l} = \omega_1^2 \approx \omega_2^2 \approx \left(\frac{\omega_1 + \omega_2}{2}\right)^2 = \omega_a^2$$

so, for $s_x = m\omega_a^2 = mg/l$,

$$E_x = \frac{1}{2}s_x a_x^2 = \frac{1}{2}\frac{mg}{l}(2a\cos\omega_m t)^2 = 2ma^2\omega_a^2\cos^2\omega_m t$$

and

$$E_y = \frac{1}{2}s_y a_y^2 = \frac{1}{2}\frac{mg}{l}(2a\sin\omega_m t)^2 = 2ma^2\omega_a^2\sin^2\omega_m t$$

with total energy $= E_x + E_y = 2ma^2\omega_a^2$ because $\sin^2\omega_m t + \cos^2\omega_m t = 1$.
    Since $\omega_m = (\omega_2 - \omega_1)/2$ then

$$E_x = 2ma^2\omega_a^2\cos^2(\omega_2 - \omega_1)t = \frac{E}{2}[1 + \cos(\omega_2 - \omega_1)t]$$

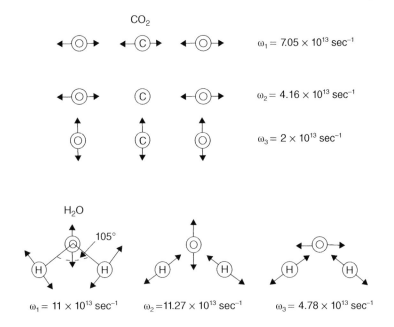

CO_2

$\omega_1 = 7.05 \times 10^{13}$ sec^{-1}

$\omega_2 = 4.16 \times 10^{13}$ sec^{-1}

$\omega_3 = 2 \times 10^{13}$ sec^{-1}

H_2O

105°

$\omega_1 = 11 \times 10^{13}$ sec^{-1} $\omega_2 = 11.27 \times 10^{13}$ sec^{-1} $\omega_3 = 4.78 \times 10^{13}$ sec^{-1}

Figure 4.6   *Normal modes of vibration for triatomic molecules CO_2 and H_2O.*

and

$$E_y = 2ma^2\omega_a^2 \sin^2(\omega_2 - \omega_1)t = \frac{E}{2}[1 - \cos(\omega_2 - \omega_1)t]$$

Note that the total energy E is completely exchanged at the beat frequency $(\omega_2 - \omega_1)$ and that $E_x - E_y = E\cos(\omega_2 - \omega_1)t$.

Atoms in polyatomic molecules behave as the masses of our pendulums; the normal modes of two tri-atomic molecules CO_2 and H_2O are shown with their frequencies in Figure 4.6. Normal modes and their vibrations will occur frequently throughout this book.

4.3   Mass or Inductance Coupling

In a later chapter we shall discuss in detail the propagation of voltage and current waves along a transmission line which may be considered as a series of coupled oscillators having identical values of inductance and of capacitance. For the moment we shall consider the energy transfer between two electrical circuits which are inductively coupled.

A mutual inductance (shared mass) exists between two electrical circuits when the magnetic flux from the current flowing in one circuit threads the second circuit. Any change of flux induces a voltage in both circuits.

A transformer depends upon mutual inductance for its operation. The power source is connected to the transformer primary coil of n_p turns, over which is wound in the same sense a secondary coil of n_s turns. If unit current flowing in a single turn of the primary coil produces a magnetic flux ϕ, then the flux threading each primary turn (assuming no flux leakage outside the coil) is $n_p\phi$ and the total flux threading all n_p turns of the primary is

$$L_p = n_p^2\phi$$

where L_p is the self inductance of the primary coil. If unit current in a single turn of the secondary coil produces a flux ϕ, then the flux threading each secondary turn is $n_s\phi$ and the total flux threading the secondary coil is

$$L_s = n_s^2\phi,$$

where L_s is the self inductance of the secondary coil.

If all the flux lines from unit current in the primary thread all the turns of the secondary, then the total flux lines threading the secondary defines the mutual inductance

$$M = n_s\,(n_p\phi) = \sqrt{L_pL_s}$$

In practice, because of flux leakage outside the coils, $M < \sqrt{L_pL_s}$ and the ratio

$$\frac{M}{\sqrt{L_pL_s}} = k, \text{ the } coefficient \text{ } of \text{ } coupling.$$

If the primary current I_p varies with $e^{i\omega t}$, a change of I_p gives an induced voltage $-L_p \mathrm{d}I_p/\mathrm{d}t = -i\omega L I_p$ in the primary and an induced voltage $-M\mathrm{d}I_p/\mathrm{d}t = -i\omega M I_p$ in the secondary.

If we consider now the two resistance-free circuits of Figure 4.7, where L_1 and L_2 are coupled by flux linkage and allowed to oscillate at some frequency ω (the voltage and current frequency of both circuits), then the voltage equations are

$$i\omega L_1 I_1 - i\frac{1}{\omega C_1}I_1 + i\omega M I_2 = 0 \tag{4.5}$$

and

$$i\omega L_2 I_2 - i\frac{1}{\omega C_2}I_2 + i\omega M I_1 = 0 \tag{4.6}$$

where M is the mutual inductance.

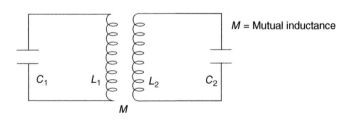

Figure 4.7   *Inductively (mass) coupled LC circuits with mutual inductance M.*

Multiplying (4.5) by ω/iL_1 gives

$$\omega^2 I_1 - \frac{I_1}{L_1 C_1} + \frac{M}{L_1}\omega^2 I_2 = 0$$

and multiplying (4.6) by ω/iL_2 gives

$$\omega^2 I_2 - \frac{I_2}{L_2 C_2} + \frac{M}{L_2}\omega^2 I_1 = 0,$$

where the natural frequencies of the circuit $\omega_1^2 = 1/L_1 C_1$ and $\omega_2^2 = 1/L_2 C_2$ give

$$\left(\omega_1^2 - \omega^2\right) I_1 = \frac{M}{L_1}\omega^2 I_2 \tag{4.7}$$

and

$$\left(\omega_2^2 - \omega^2\right) I_2 = \frac{M}{L_2}\omega^2 I_1 \tag{4.8}$$

The product of equations (4.7) and (4.8) gives

$$\left(\omega_1^2 - \omega^2\right)\left(\omega_2^2 - \omega^2\right) = \frac{M^2}{L_1 L_2}\omega^4 = k^2\omega^4, \tag{4.9}$$

where k is the coefficient of coupling.

Solving for ω gives the frequencies at which energy exchange between the circuits allows the circuits to resonate. If the circuits have equal natural frequencies $\omega_1 = \omega_2 = \omega_0$, say, then equation (4.9) becomes

$$\left(\omega_0^2 - \omega^2\right)^2 = k^2\omega^4$$

or

$$\left(\omega_0^2 - \omega^2\right) = \pm k\omega^2$$

that is

$$\omega = \pm\frac{\omega_0}{\sqrt{1 \pm k}}$$

The positive sign gives two frequencies

$$\omega' = \frac{\omega_0}{\sqrt{1 + k}} \quad \text{and} \quad \omega'' = \frac{\omega_0}{\sqrt{1 - k}}$$

at which, if we plot the current amplitude versus frequency, two maxima appear (Figure 4.8).

In loose coupling k and M are small, and $\omega' \approx \omega'' \approx \omega_0$, so that both systems behave almost independently. In tight coupling the frequency difference $\omega'' - \omega'$ increases, the peak values of current are displaced and the dip between the peaks is more pronounced. In this simple analysis the effect of resistance has been ignored. In practice some resistance is always present to limit the amplitude maximum.

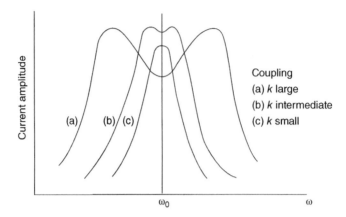

Figure 4.8   *Variation of the current amplitude in each circuit near the resonant frequency. A small resistance prevents the amplitude at resonance from reaching infinite values but this has been ignored in the simple analysis. Flattening of the response curve maximum gives 'frequency band pass' coupling.*

Worked Example

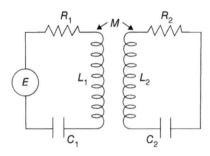

The two circuits in the diagram are coupled by a variable mutual inductance M and Kirchhoff's Law gives

$$Z_1I_1 + Z_MI_2 = E \tag{1}$$

and

$$Z_MI_1 + Z_2I_2 = 0, \quad \therefore \ I_1 = -\frac{Z_2}{Z_m}I_2 \tag{2}$$

where

$$Z_M = +i\omega M$$

M is varied at a frequency where the reactance $X_1 = X_2 = 0$ to give a maximum value of I_2. Show that the condition for this maximum is $\omega M = \sqrt{(R_1R_2)}$ and that this defines a 'critical coefficient of coupling' $k = (Q_1Q_2)^{-1/2}$, where the Q's are the quality factors of the circuits. $Q_1 = WL_1/R_1$.

Solution

Equation (2) into equation (1) gives

$$-\frac{Z_1 Z_2}{M} I_2 + Z_m I_2 = E$$

$$\therefore I_2 = \frac{E}{\left(Z_m - \frac{Z_1 Z_2}{M}\right)} I_1$$

Now $Z_m = i\omega M$ and I_2 is a maximum when $X_1 = X_2 = 0$, that is, when $Z_1 = R_1$ and $Z_2 = R_2$. Thus

$$|I_2| = \frac{E}{\left|i\omega M - \frac{R_1 R_2}{i\omega M}\right|} |I_1| = \frac{E}{\omega M + \frac{R_1 R_2}{\omega M}} |I_1| \leq \frac{E}{2\sqrt{\omega M \frac{R_1 R_2}{\omega M}}} = \frac{E}{2\sqrt{R_1 R_2}} I_1$$

so $|I_2|$ has a maximum value of

$$\frac{E}{2\sqrt{R_1 R_2}} |I_1|$$

when

$$\omega M = \frac{R_1 R_2}{\omega M} \quad \text{i.e.} \quad \omega M = \sqrt{R_1 R_2}.$$

$$k^2 = \frac{M^2}{L_1 L_2} \quad \therefore k^2_{\text{critical}} = \frac{R_1 R_2}{\omega^2 L_1 L_2} = \frac{1}{Q_1 Q_2}$$

4.4   Coupled Oscillations of a Loaded String

As a final example involving a large number of coupled oscillators we shall consider a light string supporting n equal masses m spaced at equal distance a along its length. The string is fixed at both ends; it has a length $(n + 1)a$ and a constant tension T exists at all points and all times in the string.

Small simple harmonic oscillations of the masses are allowed in only one plane and the problem is to find the frequencies of the normal modes and the displacement of each mass in a particular normal mode.

This problem was first treated by Lagrange, its particular interest being the use it makes of normal modes and the light it throws upon the wave motion and vibration of a continuous string to which it approximates as the linear separation and the magnitude of the masses are progressively reduced.

Figure 4.9 shows the displacement y_r of the rth mass together with those of its two neighbours. The equation of motion of this mass may be written by considering the components of the tension directed towards the equilibrium position. The rth mass is pulled downwards towards the equilibrium position by a force $T \sin \theta_1$ due to the tension on its left and a force $T \sin \theta_2$ due to the tension on its right where

$$\sin \theta_1 = \frac{y_r - y_{r-1}}{a}$$

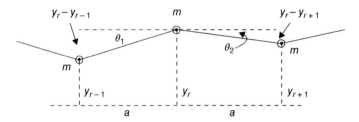

Figure 4.9 *Displacements of three masses on a loaded string under tension T giving equation of motion* $m\ddot{y}_r = T(y_{r+1} - 2y_r + y_{r-1})/a$.

and

$$\sin \theta_2 = \frac{y_r - y_{r+1}}{a}$$

Hence the equation of motion is given by

$$m\frac{\mathrm{d}^2 y_r}{\mathrm{d}t^2} = -T(\sin \theta_1 + \sin \theta_2)$$

$$= -T \left(\frac{y_r - y_{r-1}}{a} + \frac{y_r - y_{r+1}}{a} \right) \tag{4.10a}$$

so

$$\frac{\mathrm{d}^2 y_r}{\mathrm{d}t^2} = \ddot{y}_r = \frac{T}{ma} (y_{r-1} - 2y_r + y_{r+1}) \tag{4.10b}$$

If, in a normal mode of oscillation of frequency ω, the time variation of y_r is simple harmonic about the equilibrium axis, we may write the displacement of the rth mass in this mode as

$$y_r = A_r e^{i\omega t}$$

where A_r is the maximum displacement. Similarly $y_{r+1} = A_{r+1} e^{i\omega t}$ and $y_{r-1} = A_{r-1} e^{i\omega t}$. Using these values of y in the equation of motion gives

$$-\omega^2 A_r e^{i\omega t} = \frac{T}{ma} (A_{r-1} - 2A_r + A_{r+1}) e^{i\omega t}$$

or

$$\boxed{-A_{r-1} + \left(2 - \frac{ma\omega^2}{T} \right) A_r - A_{r+1} = 0} \tag{4.11}$$

This is the fundamental equation.

The procedure now is to start with the first mass $r = 1$ and move along the string, writing out the set of similar equations as r assumes the values $r = 1, 2, 3, \ldots, n$ remembering that, because the ends are fixed

$$y_0 = A_0 = 0 \quad \text{and} \quad y_{n+1} = A_{n+1} = 0$$

Thus, when $r = 1$ the equation becomes

$$\left(2 - \frac{ma\omega^2}{T}\right) A_1 - A_2 = 0 \quad (A_0 = 0)$$

When $r = 2$ we have

$$-A_1 + \left(2 - \frac{ma\omega^2}{T}\right) A_2 - A_3 = 0$$

and when $r = n$ we have

$$-A_{n-1} + \left(2 - \frac{ma\omega^2}{T}\right) A_n = 0 \quad (A_{n+1} = 0)$$

Thus, we have a set of n equations which, when solved, will yield n different values of ω^2, each value of ω being the frequency of a normal mode, the number of normal modes being equal to the number of masses.

The formal solution of this set of n equations involves the theory of matrices. However, we may easily solve the simple cases for one or two masses on the string ($n = 1$ or 2) and, in addition, it is possible to show what the complete solution for n masses must be without using sophisticated mathematics.

First, when $n = 1$, one mass on a string of length $2a$, we need only the equation for $r = 1$ where the fixed ends of the string give $A_0 = A_2 = 0$.

Hence we have

$$\left(2 - \frac{ma\omega^2}{T}\right) A_1 = 0$$

giving

$$\omega^2 = \frac{2T}{ma}$$

a single allowed frequency of vibration (Figure 4.10a).

When $n = 2$, string length $3a$ (Figure 4.10b) we need the equations for both $r = 1$ and $r = 2$; that is

$$\left(2 - \frac{ma\omega^2}{T}\right) A_1 - A_2 = 0$$

and

$$-A_1 + \left(2 - \frac{ma\omega^2}{T}\right) A_2 = 0 \quad (A_0 = A_3 = 0)$$

Eliminating A_1 or A_2 shows that these two equations may be solved (are consistent) when

$$\left(2 - \frac{ma\omega^2}{T}\right)^2 - 1 = 0$$

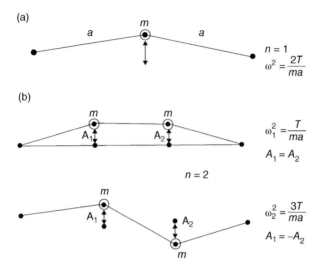

Figure 4.10 (a) Normal vibration of a single mass m on a string of length 2a at a frequency $\omega^2 = 2T/ma$. (b) Normal vibrations of two masses on a string of length 3a showing the loose coupled 'in phase' mode of frequency $\omega_1^2 = T/ma$ and the tighter coupled 'out of phase' mode of frequency $\omega_2^2 = 3T/ma$. The number of normal modes of vibration equals the number of masses.

that is

$$\left(2 - \frac{ma\omega^2}{T} - 1\right)\left(2 - \frac{ma\omega^2}{T} + 1\right) = 0$$

Thus, there are two normal mode frequencies

$$\omega_1^2 = \frac{T}{ma} \quad \text{and} \quad \omega_2^2 = \frac{3T}{ma}$$

Using the values of ω_1 in the equations for $r = 1$ and $r = 2$ gives $A_1 = A_2$ the slow 'in phase' oscillation of Figure 4.10b, whereas ω_2 gives $A_1 = -A_2$ the faster 'anti-phase' oscillation resulting from the increased coupling.

To find the general solution for any value of n let us rewrite the equation

$$\boxed{-A_{r-1} + \left(2 - \frac{ma\omega^2}{T}\right)A_r - A_{r+1} = 0} \tag{4.11}$$

in the form

$$\frac{A_{r-1} + A_{r+1}}{A_r} = \frac{2\omega_0^2 - \omega^2}{\omega_0^2} \quad \text{where} \quad \omega_0^2 = \frac{T}{ma}$$

We see that for any particular fixed value of the normal mode frequency ω (ω_j say) the right-hand side of this equation is constant, independent of r, so the equation holds for all values of r. What values can we give to A_r which will satisfy this equation, meeting the boundary conditions $A_0 = A_{n+1} = 0$ at the end of the string?

Let us assume that we may express the amplitude of the rth mass at the frequency ω_j as

$$A_r = Ce^{ir\theta}$$

where C is a constant and θ is some constant angle for a given value of ω_j. This θ has no relation to θ in Figure 4.9. The left-hand side of the equation then becomes

$$\frac{A_{r-1} + A_{r+1}}{A_r} = \frac{C(e^{i(r-1)\theta} + e^{i(r+1)\theta})}{Ce^{ir\theta}} = (e^{-i\theta} + e^{i\theta})$$
$$= 2\cos\theta$$

which is constant and independent of r.

The value of θ_j (constant at ω_j) is easily found from the boundary conditions

$$A_0 = A_{n+1} = 0 \quad \text{(fixed ends} \therefore \text{ no cosine terms)}$$

Using $\sin r\theta$ from $e^{ir\theta}$ gives

$$A_0 = C\sin r\theta = 0 \quad \text{(automatically at } r = 0)$$

and

$$A_{n+1} = C\sin(n+1)\theta = 0$$

when

$$(n+1)\theta_j = j\pi \quad \text{for} \quad j = 1, 2, \ldots, n$$

Hence

$$\theta_j = \frac{j\pi}{n+1}$$

and

$$A_r = C\sin r\theta_j = C\sin\frac{rj\pi}{n+1}$$

which is the amplitude of the rth mass at the fixed normal mode frequency ω_j.

To find the allowed values of ω_j we write

$$\frac{A_{r-1} + A_{r+1}}{A_r} = \frac{2\omega_0^2 - \omega_j^2}{\omega_0^2} = 2\cos\theta_j = 2\cos\frac{j\pi}{n+1}$$

giving

$$\omega_j^2 = 2\omega_0^2 \left[1 - \cos \frac{j\pi}{n+1} \right] \tag{4.12}$$

where j may take the values $j = 1, 2, \ldots, n$ and $\omega_0^2 = T/ma$.

Note that there is a maximum frequency of oscillation $\omega_j^2 = 2\omega_0^2$. This is called the 'cut off' frequency and such an upper frequency limit is characteristic of all oscillating systems composed of similar elements (the masses) repeated periodically throughout the structure of the system. We shall meet this in the Chapter 6 as a feature of wave propagation in crystals.

To summarize, we have found the normal modes of oscillation of n coupled masses on the string to have frequencies given by

$$\omega_j^2 = \frac{2T}{ma} \left[1 - \cos \frac{j\pi}{n+1} \right] \quad (j = 1, 2, 3 \ldots n)$$

At each frequency ω_j the rth mass has an amplitude

$$A_r = C \sin \frac{rj\pi}{n+1}$$

where C is a constant.

Worked Example

An electrical transmission line consists of equal inductances L and capacitances C arranged as shown. Using the equations

$$\frac{L dI_{r-1}}{dt} = V_{r-1} - V_r = \frac{q_{r-1} - q_r}{C}$$

and

$$I_{r-1} - I_r = \frac{dq_r}{dt},$$

show that an expression for I_r may be derived which is equivalent to that for y_r in the case of the mass-loaded string.

$$m \frac{d^2 y_r}{dt^2} = -T(\sin \theta_1 + \sin \theta_2)$$

$$= -T \left(\frac{y_r - y_{r-1}}{a} + \frac{y_r - y_{r+1}}{a} \right) \tag{4.10a}$$

(This acts as a low pass electric filter and has a cut-off frequency as in the case of the string. This cut-off frequency is a characteristic of wave propagation in periodic structures and electromagnetic wave guides.)

Solution

$$L\frac{dI_r}{dt} = \frac{1}{C}(q_r - q_{r+1}) \quad \text{where } L \text{ is the inductance.}$$

$$L\frac{d^2I_r}{dt^2} = \frac{1}{C}\left(\frac{dq_r}{dt} - \frac{dq_{r+1}}{dt}\right) = \frac{1}{C}[(I_{r-1} - I_r) - (I_r - I_{r+1})]$$

$$\therefore \frac{d^2I_r}{dt^2} = \frac{1}{LC}\left[\frac{(I_{r+1} - I_r)}{a} - \frac{(I_r - I_{r-1})}{a}\right]$$

for each unit a in length.

4.5 The Wave Equation

Finally, in this chapter, we show how the coupled vibrations in the periodic structure of our loaded string become waves in a continuous medium.

We found the equation of motion of the rth mass to be

$$\frac{d^2y_r}{dt^2} = \frac{T}{ma}(y_{r+1} - 2y_r + y_{r-1}) \tag{4.10b}$$

We know also that in a given normal mode all masses oscillate with the same mode frequency ω, so all y_r's have the same time dependence. However, as we see in Figure 4.10(b) where A_1 and A_2 are anti-phase, the transverse displacement y_r also depends upon the value of r; that is, the position of the rth mass on the string. In other words, y_r is a function of two independent variables, the time t and the location of r on the string.

If we use the separation $a \approx \delta x$ and let $\delta x \to 0$, the masses become closer and we can consider positions along the string in terms of a continuous variable x and any transverse displacement as $y(x, t)$, a function of both x and t.

The partial derivative notation $\partial y(x,t)/\partial t$ expresses the variation with time of $y(x,t)$ while x is kept constant.

The partial derivative $\partial y(x,t)/\partial x$ expresses the variation with x of $y(x,t)$ while the time t is kept constant. (Chapter 5 begins with an extended review of this process for students unfamiliar with this notation.)

In the same way, the second derivative $\partial^2 y(x,t)/\partial t^2$ continues to keep x constant and $\partial^2 y(x,t)/\partial x^2$ keeps t constant.

Recalling that in section 2.2.1 (3) we saw that

$$\frac{de^{\alpha x}}{dx} = \alpha e^{\alpha x}$$

and writing

$$y = e^{i(\omega t + kx)} = e^{i\omega t}e^{ikx}$$

where we have separated the function of t and x then we have

$$\frac{\partial y}{\partial t} = i\omega e^{i\omega t}e^{ikx} = i\omega y \quad \text{and} \quad \frac{\partial^2 y}{\partial t^2} = -\omega^2 e^{i\omega t}e^{ikx} = -\omega^2 y$$

while

$$\frac{\partial y}{\partial x} = ike^{i\omega t}e^{ikx} = iky \quad \text{and} \quad \frac{\partial^2 y}{\partial x^2} = -k^2 e^{i\omega t}e^{ikx} = -k^2 y$$

If we now locate the transverse displacement y_r at a position $x = x_r$ along the string, then the left-hand side of equation (4.10a) becomes

$$\frac{\partial^2 y_r}{\partial t^2} \rightarrow \frac{\partial^2 y}{\partial t^2},$$

where y is evaluated at $x = x_r$ and now, as $a = \delta x \rightarrow 0$, we may write $x_r = x$, $x_{r+1} = x + \delta x$ and $x_{r-1} = x - \delta x$ with $y_r(t) \rightarrow y(x, t)$, $y_{r+1}(t) \rightarrow y(x + \delta x, t)$ and $y_{r-1}(t) \rightarrow y(x - \delta x, t)$.

Here we need to explain Taylor's series which is formally derived in Appendix 2. The definition of the first differential coefficient of a function f(x) is written

$$\frac{f(x + dx) - f(x)}{dx} = \frac{df}{dx}$$

We can rearrange this to read

$$f(x + dx) = f(x) + \frac{df}{dx}dx$$

which is a first approximation to expressing f(x+dx) in terms of f(x). Taylor's series improves the approximation by a series of terms each of which is a higher derivative of f(x). That is, from Appendix 2,

$$f(x + dx) = f(x)_0 + \left(\frac{df}{dx}\right)_0 dx + \frac{1}{2!}\left(\frac{d^2 f}{dx^2}\right)_0 dx^2 + \frac{1}{3!}\left(\frac{d^3 f}{dx^3}\right)_0 dx^3$$
$$\cdots \frac{1}{n!}\left(\frac{d^n f}{dx^n}\right)_0 dx^n$$

where each of the right-hand terms is evaluated at x_0. Each term is smaller than its predecessor so that the series more accurately represents f(x+dx) in terms of the nearby $f(x)_0$. Here we need only the first two derivatives. The sign of dx (or δx) may be positive or negative and, in the Taylor series expression, to express $y(x \pm \delta x, t)$ in terms of partial derivatives of y with respect to x (keeping t constant) we have

$$y(x \pm \delta x, t) = y(x) \pm \delta x \frac{\partial y}{\partial x} + \frac{1}{2}(\pm\delta x)^2 \frac{\partial^2 y}{\partial x^2}$$

and equation (4.10a) becomes after substitution

$$\frac{\partial^2 y}{\partial t^2} = \frac{T}{m}\left(\frac{y_{r+1}-y_r}{a} - \frac{y_r - y_{r-1}}{a}\right)$$
$$= \frac{T}{m}\left(\frac{y(x+\delta x)-y(x)}{a} - \frac{y(x)-y(x-\delta x)}{a}\right)$$
$$= \frac{T}{m}\left(\frac{\delta x\frac{\partial y}{\partial x}+\frac{1}{2}(\delta x)^2\frac{\partial^2 y}{\partial x^2}}{\delta x} - \frac{\delta x\frac{\partial y}{\partial x}-\frac{1}{2}(\delta x)^2\frac{\partial^2 y}{\partial x^2}}{\delta x}\right)$$

so

$$\frac{\partial^2 y}{\partial t^2} = \frac{T}{m}\frac{(\delta x)^2}{\delta x}\frac{\partial^2 y}{\partial x^2} = \frac{T}{m}\delta x\frac{\partial^2 y}{\partial x^2}$$

If we now write $m = \rho\delta x$ where ρ is the linear density (mass per unit length) of the string, the masses must $\rightarrow 0$ as $\delta x \rightarrow 0$ to avoid infinite mass density. Thus, we have

$$\frac{\partial^2 y}{\partial t^2} = \frac{T}{\rho}\frac{\partial^2 y}{\partial x^2}$$

This is the Wave Equation.

T/ρ has the dimensions of the square of a velocity, the velocity with which the wave, that is, the phase of oscillation, is propagated. The solution for y at any particular point along the string is always that of a harmonic oscillation.

Problem 4.1.   Show that the choice of new normal coordinates X_q and Y_q expresses equations (4.3a) and (4.3b) as equations (4.4a) and (4.4b).

Problem 4.2.   The central vibration mode of the CO_2 molecule in Figure 4.6 has a stationary carbon molecule and an angular frequency of $\omega = 4.16 \times 10^{13}\,\text{s}^{-1}$. The mass of an oxygen atom is 26.56×10^{-27} kg. Show that the strength of the chemical bond for this mode is $46\,\text{N}\,\text{m}^{-1}$.

Problem 4.3.   Figures 4.3 and 4.5 show how the pendulum configurations $x = 2a$, $y = 0$ and $x = 0$, $y = 2a$ result from the superposition of the normal modes X and Y. Using the same initial conditions $(x = 2a, y = 0, \dot{x} = \dot{y} = 0)$ draw similar sketches to show how X and Y superpose to produce $x = -2a$, $y = 0$ and $x = 0$, $y = -2a$.

Problem 4.4.

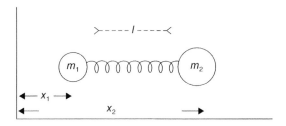

In the figure two masses m_1 and m_2 are coupled by a spring of stiffness s and natural length 1. If x is the extension of the spring show that equations of motion along the x axis are

$$m_1 \ddot{x}_1 = +sx$$

and

$$m_2 \ddot{x}_2 = -sx$$

where

$$x = x_2 - x_1$$

and combine these to show that the system oscillates with a frequency

$$\omega^2 = \frac{s}{\mu},$$

where

$$\mu = \frac{m_1 m_2}{m_1 + m_2}$$

is called the reduced mass.

The figure now represents a diatomic molecule as a harmonic oscillator with an effective mass equal to its reduced mass. If a sodium chloride molecule has a natural vibration frequency $= 1.14 \times 10^{13}$ Hz (in the infrared region of the electromagnetic spectrum) show that the interatomic force constant $s = 120 \, \mathrm{N\,m^{-1}}$ (this simple model gives a higher value for s than more refined methods which account for other interactions within the salt crystal lattice)

$$\text{Mass of Na atom} = 23 \text{ a.m.u.}$$
$$\text{Mass of Cl atom} = 35 \text{ a.m.u.}$$
$$1 \text{ a.m.u.} = 1.67 \times 10^{-27} \text{ kg}$$

Problem 4.5.   The equal masses in the figure oscillate in the vertical direction. Show that the frequencies of the normal modes of oscillation are given by

$$\omega^2 = (3 \pm \sqrt{5}) \frac{s}{2m}$$

and that in the slower mode the ratio of the amplitude of the upper mass to that of the lower mass is $\frac{1}{2}(\sqrt{5} - 1)$ whilst in the faster mode this ratio is $-\frac{1}{2}(\sqrt{5} + 1)$.

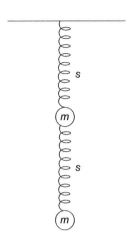

In the calculations it is not necessary to consider gravitational forces because they play no part in the forces responsible for the oscillation.

Problem 4.6. When the masses of the coupled pendulums of Figure 4.1 are no longer equal the equations of motion become

$$m_1\ddot{x} = -m_1(g/l)x - s(x - y)$$

and

$$m_2\ddot{y} = -m_2(g/l)y + s(x - y)$$

Show that we may choose the normal coordinates

$$X = \frac{m_1 x + m_2 y}{m_1 + m_2}$$

with a normal mode frequency $\omega_1^2 = g/l$ and $Y = x - y$ with a normal mode frequency $\omega_2^2 = g/l + s(1/m_1 + 1/m_2)$.

Note that X is the coordinate of the centre of mass of the system whilst the effective mass in the Y mode is the reduced mass μ of the system where $1/\mu = 1/m_1 + 1/m_2$.

Problem 4.7. The diagram shows an oscillatory force $F_o \cos \omega t$ acting on a mass M which is part of a simple harmonic system of stiffness k and is connected to a mass m by a spring of stiffness s. If all oscillations are along the x axis show that the condition for M to remain stationary is $\omega^2 = s/m$. (This is a simple version of small mass loading in engineering to quench undesirable oscillations.)

$F_0 \cos \omega t$

$\longleftrightarrow$ M ⟡⟡⟡⟡⟡ m

Problem 4.8. The figure below shows two identical LC circuits coupled by a common capacitance C with the directions of current flow indicated by arrows. The voltage equations are

$$V_1 - V_2 = L\frac{dI_a}{dt}$$

and

$$V_2 - V_3 = L\frac{\mathrm{d}I_b}{\mathrm{d}t}$$

whilst the currents are given by

$$\frac{\mathrm{d}q_1}{\mathrm{d}t} = -I_a \quad \frac{\mathrm{d}q_2}{\mathrm{d}t} = I_a - I_b$$

and

$$\frac{\mathrm{d}q_3}{\mathrm{d}t} = I_b$$

Solve the voltage equations for the normal coordinates $(I_a + I_b)$ and $(I_a - I_b)$ to show that the normal modes of oscillation are given by

$$I_a = I_b \quad \text{at} \quad \omega_1^2 = \frac{1}{LC}$$

and

$$I_a = -I_b \quad \text{at} \quad \omega_2^2 = \frac{3}{LC}$$

Note that when $I_a = I_b$ the coupling capacitance may be removed and $q_1 = -q_2$. When $I_a = -I_b$, $q_2 = -2q_1 = -2q_3$.

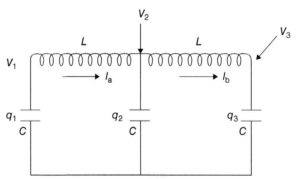

Problem 4.9.   A generator of e.m.f. E is coupled to a load Z by means of an ideal transformer. From the diagram, Kirchhoff's Law gives

$$E = -e_1 = \mathrm{i}\omega L_p I_1 - \mathrm{i}\omega M I_2$$

and

$$I_2 Z_2 = e_2 = \mathrm{i}\omega M I_1 - \mathrm{i}\omega L_s I_2.$$

Show that E/I_1, the impedance of the whole system seen by the generator, is the sum of the primary impedance and a 'reflected impedance' from the secondary circuit of $\omega^2 M^2/Z_s$ where $Z_s = Z_2 + \mathrm{i}\omega L_s$.

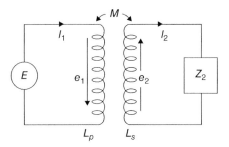

Problem 4.10. Show, for the perfect transformer of Problem 4.9, that the impedance seen by the generator consists of the primary impedance in parallel with an impedance $(n_p/n_s)^2 Z_2$, where n_p and n_s are the number of primary and secondary transformer coil turns respectively.

Problem 4.11. If the generator delivers maximum power when its load equals its own internal impedance show how an ideal transformer may be used as a device to match a load to a generator, e.g. a loudspeaker of a few ohms impedance to an amplifier output of $10^3\ \Omega$ impedance.

Problem 4.12. Consider the case when the number of masses on the loaded string of this chapter is $n = 3$. Use equation (4.12) to show that the normal mode frequencies are given by

$$\omega_1^2 = (2 - \sqrt{2})\omega_0^2; \quad \omega_2^2 = 2\omega_0^2$$

and

$$\omega_3^2 = (2 + \sqrt{2})\omega_0^2$$

Problem 4.13. Show that the relative displacements of the masses in the modes of Problem 4.12 are $1 : \sqrt{2} : 1$, $1 : 0 : -1$, and $1 : -\sqrt{2} : 1$. Show by sketching these relative displacements that tighter coupling increases the mode frequency.

Problem 4.14. Expand the value of

$$\omega_j^2 = \frac{2T}{ma}\left(1 - \cos\frac{j\pi}{n+1}\right)$$

when $j \ll n$ in powers of $(j/n + 1)$ to show that in the limit of very large values of n, a low frequency

$$\omega_J = \frac{j\pi}{l}\sqrt{\frac{T}{\rho}},$$

where $\rho = m/a$ and $l = (n+1)a$.

Problem 4.15.

An electrical transmission line consists of equal inductances L and capacitances C arranged as shown. Using the equations

$$\frac{L dI_{r-1}}{dt} = V_{r-1} - V_r = \frac{q_{r-1} - q_r}{C}$$

and

$$I_{r-1} - I_r = \frac{dq_r}{dt},$$

show that an expression for V_r may be derived which is equivalent to that for y_r in the case of the mass-loaded string, equation 4.10a and b. (This acts as a low pass electric filter and has a cut-off frequency as in the case of the string. This cut-off frequency is a characteristic of wave propagation in periodic structures and electromagnetic wave guides.) See worked example at the end of section 4.4 Coupled Oscillations of a Loaded String.

Problem 4.16.   Show that

$$y = e^{i\omega t} e^{ikx}$$

(a) satisfies the wave equation

$$\frac{\partial^2 y}{\partial t^2} = c^2 \frac{\partial^2 y}{\partial x^2}, \quad \text{if} \quad \omega = ck$$

and (b) that

$$\frac{\partial}{\partial x}\left(\frac{\partial y}{\partial t}\right)_x = \frac{\partial}{\partial t}\left(\frac{\partial y}{\partial x}\right)_t = -\omega k y$$

5

Transverse Wave Motion (1)

Introduction

We started this book with simple harmonic oscillators and ended the last chapter by deriving the wave equation. These are the tools which we now use in discussing waves. We have seen that the energy of a simple harmonic oscillator can be transferred by coupling to a neighbour and as we increase the number of oscillators we end up with a medium through which a wave propagates. In particular, the oscillators or particles in a medium do not move through the medium but only vibrate about their equilibrium positions so that what we observe as waves is the changing relative displacements of neighbouring oscillators.

We shall show by treating the string as a forced oscillator how it behaves as a medium with an impedance which stores wave energy, how power fed into one end of the string propagates and maintains waves along the string and how the wave energy is distributed along the string.

When the wave meets a boundary between two different impedances some energy is reflected and some is transmitted. We begin by extending our familiarity with partial differentiation using a range of different examples.

5.1 Partial Differentiation

From this chapter onwards we shall often need to use the notation of partial differentiation.

When we are dealing with a function of only one variable, $y = f(x)$ say, we write the differential coefficient

$$\frac{dy}{dx} = \lim_{\delta x \to 0} \frac{f(x + \delta x) - f(x)}{\delta x}$$

but if we consider a function of two or more variables, the value of this function will vary with a change in any or all of the variables. For instance, the value of the coordinate z on the surface of a sphere whose

Introduction to Vibrations and Waves, First Edition. H. J. Pain and Patricia Rankin.
© 2015 John Wiley & Sons, Ltd. Published 2015 by John Wiley & Sons, Ltd.
Companion website: http://booksupport.wiley.com

equation is $x^2 + y^2 + z^2 = a^2$, where a is the radius of the sphere, will depend on x and y so that z is a function of x and y written $z = z(x, y)$. The differential change of z which follows from a change of x and y may be written

$$dz = \left(\frac{\partial z}{\partial x}\right)_y dx + \left(\frac{\partial z}{\partial y}\right)_x dy$$

where $(\partial z/\partial x)_y$ means differentiating z with respect to x whilst y is kept constant, so that

$$\left(\frac{\partial z}{\partial x}\right)_y = \lim_{\delta x \to 0} \frac{z(x + \delta x, y) - z(x, y)}{\delta x}$$

The total change dz is found by adding the separate increments due to the change of each variable in turn whilst the others are kept constant. In Figure 5.1 we can see that keeping y constant isolates a plane which cuts the spherical surface in a curved line, and the incremental contribution to dz along this line is exactly as though z were a function of x only. Now by keeping x constant we turn the plane through $90°$ and repeat the process with y as a variable so that the total increment of dz is the sum of these two processes.

If only two independent variables are involved, the subscript showing which variable is kept constant is omitted without ambiguity.

In wave motion our functions will be those of variables of distance and time, and we shall write $\partial/\partial x$ and $\partial^2/\partial x^2$ for the first or second derivatives with respect to x, whilst the time t remains constant. Again, $\partial/\partial t$ and $\partial^2/\partial t^2$ will denote first and second derivatives with respect to time, implying that x is kept constant.

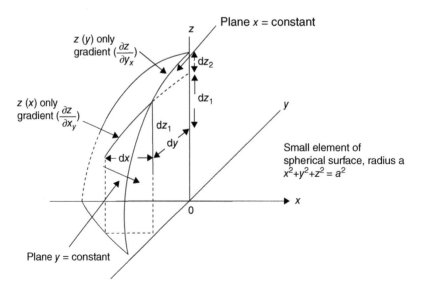

Figure 5.1   *Small element of a spherical surface showing $dz = dz_1 + dz_2 = (\partial z/\partial x)_y dx + (\partial z/\partial y)_x dy$ where each gradient is calculated with one variable remaining constant.*

Examples of Partial Differentiation

We now consider the partial differentiation of functions of z where z is itself a function of two variables, e.g. x and y or x and t, that is

$$f(z) \text{ where } z = z(x, y) \text{ or } z = z(x, t).$$

The rate of change of $f(z)$ with x if y remains constant is

$$\left(\frac{\partial f(z)}{\partial x}\right)_y = \frac{df(z)}{dz}\left(\frac{\partial z}{\partial x}\right)_y,$$

that is the change of $f(z)$ with z times the change of z with x with y constant.
   Similarly

$$\left(\frac{\partial f(z)}{\partial y}\right)_x = \frac{df(z)}{dz}\left(\frac{\partial z}{\partial y}\right)_x$$

(a) $f(z) = z$ where $z = (3x - 2y)$

$$\left(\frac{\partial f(z)}{\partial x}\right)_y = \frac{df(z)}{dz}\left(\frac{\partial z}{\partial x}\right)_y = 1 \cdot 3 = 3$$

$$\left(\frac{\partial f(z)}{\partial y}\right)_x = \frac{df(z)}{dz}\left(\frac{\partial z}{\partial y}\right)_x = 1 \cdot -2 = -2$$

(b) $f(z) = z^2$ where $z = (3x - 2y)$

$$\left(\frac{\partial f(z)}{\partial x}\right)_y = \frac{df(z)}{dz}\left(\frac{\partial z}{\partial x}\right)_y = 2z \cdot 3 = 6(3x - 2y)$$

$$\left(\frac{\partial f(z)}{\partial y}\right)_x = \frac{df(z)}{dz}\left(\frac{\partial z}{\partial y}\right)_x = 2z \cdot -2 = -4(3x - 2y)$$

(c) $f(z) = e^z$ where $z = x + iy$ so $e^z = e^{x+iy}$

$$\left(\frac{\partial f(z)}{\partial x}\right)_y = \frac{df(z)}{dz}\left(\frac{\partial z}{\partial x}\right)_y = e^z \cdot 1 = e^z = e^{x+iy}$$

$$\left(\frac{\partial f(z)}{\partial y}\right)_x = \frac{df(z)}{dz}\left(\frac{\partial z}{\partial y}\right)_x = e^z \cdot i = i e^z = i e^{x+iy}$$

The following function is very important in wave motion

(d) $f(z) = e^z$ where $z = i(\omega t - kx)$

$$\left(\frac{\partial f(z)}{\partial t}\right)_x = \frac{df(z)}{dz}\left(\frac{\partial z}{\partial t}\right)_x = i\omega e^z = i\omega e^{(\omega t - kx)}$$

$$\left(\frac{\partial f(z)}{\partial x}\right)_t = \frac{df(z)}{dz}\left(\frac{\partial z}{\partial x}\right)_t = -ike^z = -ike^{(\omega t - kx)}$$

5.2   Waves

One of the simplest ways to demonstrate wave motion is to take the loose end of a long rope which is fixed at the other end and to move the loose end quickly up and down. Crests and troughs of the waves move down the rope, and if the rope were infinitely long such waves would be called progressive waves – these are waves travelling in an unbounded medium free from possible reflection (Figure 5.2).

If the medium is limited in extent, for example, if the rope were reduced to a violin string, fixed at both ends, the progressive waves travelling on the string would be reflected at both ends; the vibration of the string would then be the combination of such waves moving to and fro along the string and standing waves would be formed.

Waves on strings are transverse waves where the displacements or oscillations in the medium are transverse to the direction of wave propagation. When the oscillations are parallel to the direction of wave propagation the waves are longitudinal. Sound waves are longitudinal waves; a gas can sustain only longitudinal waves because transverse waves require a shear force to maintain them. Both transverse and longitudinal waves can travel in a solid.

In this book we are going to discuss plane waves only. When we see wave motion as a series of crests and troughs we are in fact observing the vibrational motion of the individual oscillators in the medium, and in particular all of those oscillators in a plane of the medium which, at the instant of observation, have the same phase in their vibrations. When all the vibrations are restricted to one plane the wave is said to be *plane polarized*.

If we take a plane perpendicular to the direction of wave propagation and all oscillators lying within that plane have a common phase, we shall observe with time how that plane of common phase progresses through the medium. Over such a plane, all parameters describing the wave motion remain constant. The crests and troughs are planes of maximum amplitude of oscillation which are π rad out of phase; a crest is a plane of maximum positive amplitude, while a trough is a plane of maximum negative amplitude. In formulating such wave motion in mathematical terms we shall have to relate the phase difference between any two planes to their physical separation in space. We have, in principle, already done this in our discussion on oscillators.

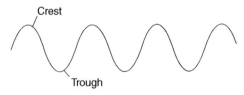

Progressive waves on infinitely long string

Figure 5.2   *Progressive transverse waves moving along a string.*

Spherical waves are waves in which the surfaces of common phase are spheres and the source of waves is a central point, e.g. an explosion; each spherical surface defines a set of oscillators over which the radiating disturbance has imposed a common phase in vibration. In practice, spherical waves become plane waves after travelling a very short distance. A small section of a spherical surface is a very close approximation to a plane.

5.3 Velocities in Wave Motion

At the outset we must be very clear about one point. The individual oscillators which make up the medium do not progress through the medium with the waves. Their motion is simple harmonic, limited to oscillations, transverse or longitudinal, about their equilibrium positions. It is their phase relationships we observe as waves, not their progressive motion through the medium.

There are three velocities in wave motion which are quite distinct although they are connected mathematically. They are

(1)   The particle velocity, which is the simple harmonic velocity of the oscillator about its equilibrium position.
(2)   The wave or phase velocity, the velocity with which planes of equal phase, crests or troughs, progress through the medium.
(3)   The group velocity. A number of waves of different frequencies, wavelengths and velocities may be superposed to form a group. Waves rarely occur as single monochromatic components; a white light pulse consists of an infinitely fine spectrum of frequencies and the motion of such a pulse would be described by its group velocity. Such a group would, of course, 'disperse' with time because the wave velocity of each component would be different in all media except free space. Only in free space would it remain as white light. We shall discuss group velocity as a separate topic in Chapter 6. Its importance is that it is the velocity with which the energy in the wave group is transmitted. For a monochromatic wave the group velocity and the wave velocity are identical. Here we shall concentrate on particle and wave velocities.

5.4 The Wave Equation

This equation will dominate the rest of this text and we shall derive it, first of all, by considering the motion of transverse waves on a string.

We shall consider the vertical displacement y of a very short section of a uniform string. This section will perform vertical simple harmonic motions; it is our simple oscillator. The displacement y will, of course, vary with the time and also with x, the position along the string at which we choose to observe the oscillation.

The wave equation therefore will relate the displacement y of a single oscillator to distance x and time t. We shall consider oscillations only in the plane of the paper, so that our transverse waves on the string are *plane polarized.*

The mass of the uniform string per unit length or its linear density is ρ, and a constant tension T exists throughout the string although it is slightly extensible.

This requires us to consider such a short length and such small oscillations that we may linearize our equations. The effect of gravity is neglected.

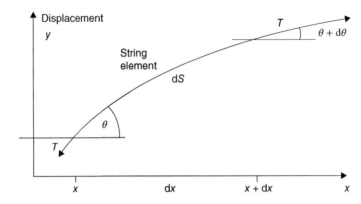

Figure 5.3   *Displaced element of string of length ds ≈ dx with tension T acting at an angle θ at x and at θ + dθ at x + dx.*

Thus in Figure 5.3 the forces acting on the curved element of length ds are T at an angle θ to the axis at one end of the element, and T at an angle $\theta + \mathrm{d}\theta$ at the other end. The length of the curved element is

$$\mathrm{d}s = \left[1 + \left(\frac{\partial y}{\partial x}\right)^2\right]^{1/2} \mathrm{d}x$$

but within the limitations imposed $\partial y/\partial x$ is so small that we ignore its square and take $\mathrm{d}s = \mathrm{d}x$. The mass of the element of string is therefore $\rho \mathrm{d}s = \rho \mathrm{d}x$. Its equation of motion is found from Newton's Law, force equals mass times acceleration.

The perpendicular force on the element dx is $T \sin(\theta + \mathrm{d}\theta) - T \sin \theta$ in the positive y direction, which equals the product of $\rho \mathrm{d}x$ (mass) and $\partial^2 y/\partial t^2$ (acceleration).

Since θ is very small $\sin \theta \approx \tan \theta = \partial y/\partial x$, so that the force is given by

$$T\left[\left(\frac{\partial y}{\partial x}\right)_{x+\mathrm{d}x} - \left(\frac{\partial y}{\partial x}\right)_{x}\right]$$

where the subscripts refer to the point at which the partial derivative is evaluated. The difference between the two terms in the bracket defines the differential coefficient of the partial derivative $\partial y/\partial x$ times the space interval dx, so that the force is

$$T\frac{\partial^2 y}{\partial x^2}\mathrm{d}x$$

The equation of motion of the small element dx then becomes

$$T\frac{\partial^2 y}{\partial x^2}\mathrm{d}x = \rho \mathrm{d}x \frac{\partial^2 y}{\partial t^2}$$

or

$$\frac{\partial^2 y}{\partial x^2} = \frac{\rho}{T}\frac{\partial^2 y}{\partial t^2}$$

giving

$$\frac{\partial^2 y}{\partial x^2} = \frac{1}{c^2}\frac{\partial^2 y}{\partial t^2}$$

where T/ρ has the dimensions of a velocity squared, so c in the preceding equation is a velocity. **This is the wave equation**.

It relates the acceleration of a simple harmonic oscillator in a medium to the second derivative of its displacement with respect to its position, x, in the medium. The position of the term c^2 in the equation is always shown by a rapid dimensional analysis.

So far we have not explicitly stated which velocity c represents. We shall see that it is the wave or phase velocity, the velocity with which planes of common phase are propagated. In the string the velocity arises as the ratio of the tension to the inertial density of the string. We shall see, whatever the waves, that the wave velocity can always be expressed as a function of the elasticity or potential energy storing mechanism in the medium and the inertia of the medium through which its kinetic or inductive energy is stored. For longitudinal waves in a solid the elasticity is measured by Young's modulus, in a gas by γP, where γ is the specific heat ratio and P is the gas pressure.

5.5 Solution of the Wave Equation

The solution of the wave equation

$$\frac{\partial^2 y}{\partial x^2} = \frac{1}{c^2}\frac{\partial^2 y}{\partial t^2}$$

will, of course, be a function of the variables x and t. We are going to show that any function of the form $y = f_1(ct - x)$ is a solution. Moreover, any function $y = f_2(ct + x)$ will be a solution so that, generally, their superposition $y = f_1(ct - x) + f_2(ct + x)$ is the complete solution.

If f_1' represents the differentiation of the function with respect to the bracket $(ct - x)$, then using the chain rule which also applies to partial differentiation

$$\frac{\partial y}{\partial x} = -f_1'(ct - x)$$

and

$$\frac{\partial^2 y}{\partial x^2} = f_1''(ct - x)$$

also

$$\frac{\partial y}{\partial t} = cf_1'(ct - x)$$

and

$$\frac{\partial^2 y}{\partial t^2} = c^2 f_1''(ct - x)$$

so that

$$\frac{\partial^2 y}{\partial x^2} = \frac{1}{c^2}\frac{\partial^2 y}{\partial t^2}$$

for $y = f_1(ct - x)$. When $y = f_2(ct + x)$ a similar result holds.

Worked Example

$$y = f_2(ct + x)$$

$$\frac{\partial y}{\partial x} = f_2'(ct + x) \qquad \frac{\partial^2 y}{\partial x^2} = f_2''(ct + x)$$

$$\frac{\partial y}{\partial t} = cf_2'(ct + x) \qquad \frac{\partial^2 y}{\partial t^2} = c^2 f_2''(ct + x)$$

$$\therefore \frac{\partial^2 y}{\partial x^2} = \frac{1}{c^2}\frac{\partial^2 y}{\partial t^2}$$

If y is the simple harmonic displacement of an oscillator at position x and time t we would expect, from Chapter 1, to be able to express it in the form $y = a\sin(\omega t - \phi)$, and in fact all of the waves we discuss in this book will be described by sine or cosine functions.

The bracket $(ct - x)$ in the expression $y = f(ct-x)$ has the dimensions of a length and, for the function to be a sine or cosine, its argument must have the dimensions of radians so that $(ct - x)$ must be multiplied by a factor $2\pi/\lambda$, where λ is a length to be defined.

We can now write

$$y = a\sin(\omega t - \phi) = a\sin\frac{2\pi}{\lambda}(ct - x)$$

as a solution to the wave equation if $2\pi c/\lambda = \omega = 2\pi\nu$, where ν is the oscillation frequency and $\phi = 2\pi x/\lambda$.

This means that if a wave, moving to the right, passes over the oscillators in a medium and a photograph is taken at time $t = 0$, the locus of the oscillator displacements (Figure 5.4) will be given by the expression

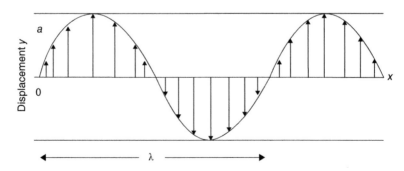

Figure 5.4 *Locus of oscillator displacements in a continuous medium as a wave passes over them travelling in the positive x direction. The wavelength λ is defined as the distance between any two oscillators having a phase difference of 2π rad.*

$y = a\sin(\omega t - \phi) = a\sin 2\pi(ct - x)/\lambda$. If we now observe the motion of the oscillator at the position $x = 0$ it will be given by $y = a\sin \omega t$.

Any oscillator to its right at some position x will be set in motion at some later time by the wave moving to the right; this motion will be given by

$$y = a\sin(\omega t - \phi) = a\sin \frac{2\pi}{\lambda}(ct - x)$$

having a phase lag of ϕ with respect to the oscillator at $x = 0$. This phase lag $\phi = 2\pi x/\lambda$, so that if $x = \lambda$ the phase lag is 2π rad that is, equivalent to exactly one complete vibration of an oscillator.

This defines λ as the wavelength, the separation in space between any two oscillators with a phase difference of 2π rad. The expression $2\pi c/\lambda = \omega = 2\pi\nu$ gives $c = \nu\lambda$, where c, the wave or phase velocity, is the product of the frequency and the wavelength. Thus, $\lambda/c = 1/\nu = \tau$, the period of oscillation, showing that the wave travels one wavelength in this time. An observer at any point would be passed by ν wavelengths per second, a distance per unit time equal to the velocity c of the wave.

If the wave is moving to the left the sign of ϕ is changed because the oscillation at x begins before that at $x = 0$. Thus, the bracket

$$(ct - x) \text{ denotes a wave moving to the right}$$

and

$$(ct + x) \text{ gives a wave moving in the direction of negative } x.$$

There are several equivalent expressions for $y = f(ct - x)$ which we list here as sine functions, although cosine functions are equally valid.
   They are:

$$y = a\sin \frac{2\pi}{\lambda}(ct - x)$$
$$y = a\sin 2\pi\left(\nu t - \frac{x}{\lambda}\right)$$
$$y = a\sin \omega\left(t - \frac{x}{c}\right)$$
$$y = a\sin(\omega t - kx)$$

where $k = 2\pi/\lambda = \omega/c$ is called the wave number; also $y = ae^{i(\omega t - kx)}$, the exponential representation of both sine and cosine.

Each of the expressions above is a solution to the wave equation giving the displacement of an oscillator and its phase with respect to some reference oscillator. The changes of the displacements of the oscillators and the propagation of their phases are what we observe as wave motion.

The wave or phase velocity is, of course, $\partial x/\partial t$, the rate at which the disturbance moves across the oscillators; the oscillator or particle velocity is the simple harmonic velocity $\partial y/\partial t$.

Choosing any one of the expressions above for a right-going wave, e.g.

$$y = a\sin(\omega t - kx)$$

we have

$$\frac{\partial y}{\partial t} = wa\cos(\omega t - kx)$$

and

$$\frac{\partial y}{\partial x} = -ka\cos(\omega t - kx)$$

so that

$$\frac{\partial y}{\partial t} = -\frac{\omega}{k}\frac{\partial y}{\partial x} = -c\frac{\partial y}{\partial x}\left(= -\frac{\partial x}{\partial t}\frac{\partial y}{\partial x}\right)$$

The particle velocity $\partial y/\partial t$ is therefore given as the product of the wave velocity

$$c = \frac{\partial x}{\partial t}$$

and the gradient of the wave profile preceded by a negative sign for a right-going wave

$$y = f(ct - x)$$

In Figure 5.5 the arrows show the direction of the particle velocity at various points of the right-going wave. It is evident that the particle velocity increases in the same direction as the transverse force in the wave and we shall see in the next section that this force is given by

$$-T\partial y/\partial x$$

where T is the tension in the string.

Worked Example

Show that, for a left-going wave

$$\frac{\partial y}{\partial t} = c\frac{\partial y}{\partial x}$$

Solution

A left-going wave is

$$y = a\sin(\omega t + kx)$$

with

$$\frac{\partial y}{\partial x} = +ka\cos(\omega t + kx)$$

and

$$\frac{\partial y}{\partial t} = +wa\cos(\omega t + kx)$$

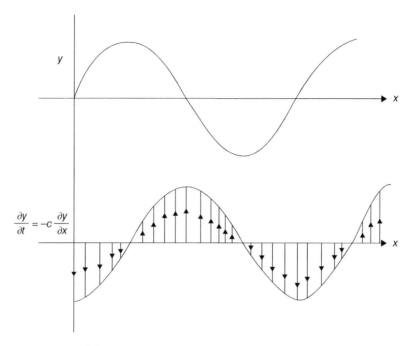

Figure 5.5 *The magnitude and direction of the particle velocity $\partial y/\partial t = -c(\partial y/\partial x)$ at any point x is shown by an arrow in the right-going sine wave above.*

so

$$\frac{\partial y}{\partial t} = \frac{\omega}{k}\frac{\partial y}{\partial x} = c\frac{\partial y}{\partial x}$$

c is a magnitude with no sign.

5.6 Characteristic Impedance of a String (the String as a Forced Oscillator)

Any medium through which waves propagate will present an impedance to those waves. If the medium is lossless, and possesses no resistive or dissipation mechanism, this impedance will be determined by the two energy storing parameters, inertia and elasticity, and it will be real. The presence of a loss mechanism will introduce a complex term into the impedance.

A string presents such an impedance to progressive waves and this is defined, because of the nature of the waves, as the transverse impedance

$$Z = \frac{\text{transverse force}}{\text{transverse velocity}} = \frac{F}{v}$$

The following analysis will emphasize the dual role of the string as a medium and as a forced oscillator.

In Figure 5.6 we consider progressive waves on the string which are generated at one end by an oscillating force, $F_0 e^{i\omega t}$, which is restricted to the direction transverse to the string and operates only in the

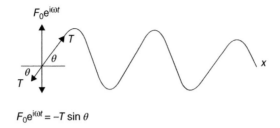

$$F_0 e^{i\omega t} = -T \sin \theta$$

Figure 5.6   *The string as a forced oscillator with a vertical force $F_0 e^{i\omega t}$ driving it at one end.*

plane of the paper. The tension in the string has a constant value, T, and at the end of the string the balance of forces shows that the applied force is equal and opposite to $T \sin \theta$ at all time, so that

$$F_0 e^{i\omega t} = -T \sin \theta \approx -T \tan \theta = -T \left(\frac{\partial y}{\partial x} \right)$$

where θ is small.

The displacement of the progressive waves may be represented exponentially by

$$y = A \mathrm{e}^{\mathrm{i}(\omega t - kx)}$$

where the amplitude A may be complex because of its phase relation with F. At the end of the string, where $x = 0$,

$$F_0 e^{i\omega t} = -T \left(\frac{\partial y}{\partial x} \right)_{x=0} = \mathrm{i} k T A \mathrm{e}^{\mathrm{i}(\omega t - k \cdot 0)}$$

giving

$$A = \frac{F_0}{\mathrm{i} k T} = \frac{F_0}{\mathrm{i} \omega} \left(\frac{c}{T} \right)$$

and

$$y = \frac{F_0}{\mathrm{i} \omega} \left(\frac{c}{T} \right) \mathrm{e}^{\mathrm{i}(\omega t - kx)}$$

(since $c = \omega / k$).

The transverse velocity

$$v = \dot{y} = F_0 \left(\frac{c}{T} \right) \mathrm{e}^{\mathrm{i}(\omega t - kx)}$$

where the velocity amplitude $v = F_0 / Z$, gives a transverse impedance

$$Z = \frac{T}{c} = \rho c \, (\text{since } T = \rho c^2)$$

or characteristic impedance of the string.

Since the velocity c is determined by the inertia and the elasticity, the impedance is also governed by these properties.

(We can see that the amplitude of displacement $y = F_0/\omega Z$, with the phase relationship $-\mathrm{i}$ with respect to the force, is in complete accord with our discussion in Chapter 3.)

Rate of Wave Energy Transmission along the String

In moving the end of the string vertically up and down to sustain the wave motion along the string, the power, that is the *work rate* by the force is $F_0\, \mathrm{e}^{\mathrm{i}\omega t}v$ where v is the transverse simple harmonic velocity $\partial y/\partial t$, so

$$F_0\, \mathrm{e}^{\mathrm{i}\omega t}v = -T\frac{\partial y}{\partial x}\frac{\partial y}{\partial t}$$

From the worked example at the end of section 5.5, for a right-going wave we have

$$\frac{\partial y}{\partial t} = -c\frac{\partial y}{\partial x}$$

so the rate of working $=$

$$-T\frac{\partial y}{\partial x}\frac{\partial y}{\partial t} = -\rho c^2\frac{\partial y}{\partial x}\frac{\partial y}{\partial t} = \rho c\left(\frac{\partial y}{\partial t}\right)^2$$

where c is the phase velocity of the wave.

But $\rho \mathrm{d}x(\partial y/\partial t)^2_{\max}$ is the total harmonic energy of an elemental length $\mathrm{d}x$ of the oscillating string so

$$F_0\, \mathrm{e}^{\mathrm{i}\omega t}v = \rho c\left(\frac{\partial y}{\partial t}\right)^2_{\max}$$

equals the amount of wave energy travelling down the string per second which is stored and maintained in the string via its impedance as a medium.

Distribution of Wave Energy along a Vibrating String

A vibrating string possesses both kinetic and potential energy. The kinetic energy of an element of length $\mathrm{d}x$ and linear density ρ is given by

$$E_{\mathrm{kin}} = \frac{1}{2}\rho \mathrm{d}x\left(\frac{\partial y}{\partial t}\right)^2.$$

The potential energy is the work done by the tension T in extending an element $\mathrm{d}x$ to a new length $\mathrm{d}s$ when the string is vibrating. Thus

$$E_{\mathrm{pot}} = T(\mathrm{d}s - \mathrm{d}x) = T\left\{\left[1 + \left(\frac{\partial y}{\partial x}\right)^2\right]^{\frac{1}{2}} - 1\right\}\mathrm{d}x = \frac{1}{2}T\left(\frac{\partial y}{\partial x}\right)^2 \mathrm{d}x$$

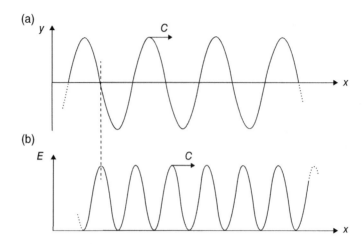

Figure 5.7   *Distribution of total energy E in a wave (b) versus wavelengths (a). The wave velocity is c. The peaks of E coincide with the wave amplitude zeros and the zeros of E coincide with the crests and troughs of the waves.*

Provided $\left(\frac{\partial y}{\partial x}\right)$ in the wave is of the first order of small quantities, the change in T is of the second order and T may be considered constant. But

$$\frac{1}{2}T\left(\frac{\partial y}{\partial x}\right)^2 dx = \frac{1}{2}\rho c^2 \left(\frac{\partial y}{\partial x}\right)^2 dx = \frac{1}{2}\rho dx \left(\frac{\partial y}{\partial t}\right)^2$$

so the instantaneous values of the kinetic and potential energies in the wave are equal at all points.

In particular their maximum values occur at $x = 0$ where $\partial y/\partial t$ is a maximum and $\partial y/\partial x$ has a maximum and minimum value of ± 1.

Note that both $\partial y/\partial t$ and $\partial y/\partial x$ are zero at the crests and troughs of the waves.

Figure 5.7 shows the total wave energy distribution along wavelengths of the string.

Treating the string as a forced oscillator has allowed us to demonstrate (a) its function as a medium with an impedance capable of storing wave energy, (b) the rate at which the wave energy propagates in the medium and (c) the distribution of that energy within the medium.

5.7   Reflection and Transmission of Waves on a String at a Boundary

We have seen that a string presents a characteristic impedance ρc to waves travelling along it, and we ask how the waves will respond to a sudden change of impedance; that is, of the value ρc. We shall ask this question of all the waves we discuss, acoustic waves, voltage and current waves and electromagnetic waves, and we shall find a remarkably consistent pattern in their behaviour.

We suppose that a string consists of two sections smoothly joined at a point $x = 0$ with a constant tension T along the whole string. The two sections have different linear densities ρ_1 and ρ_2, and therefore different wave velocities $T/\rho_1 = c_1^2$ and $T/\rho_2 = c_2^2$. The specific impedances are $\rho_1 c_1$ and $\rho_2 c_2$, respectively.

An incident wave travelling along the string meets the discontinuity in impedance at the position $x = 0$ in Figure 5.8. At this position, $x = 0$, a part of the incident wave will be reflected and part of it will be transmitted into the region of impedance $\rho_2 c_2$.

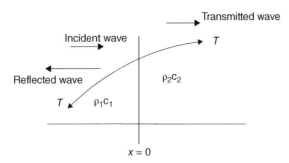

Figure 5.8   *Waves on a string of impedance $\rho_1 c_1$ reflected and transmitted at the boundary $x = 0$ where the string changes to impedance $\rho_2 c_2$.*

We shall denote the impedance $\rho_1 c_1$ by Z_1 and the impedance $\rho_2 c_2$ by Z_2. We write the displacement of the incident wave as $y_i = A_1 e^{i(\omega t - kx)}$, a wave of real (not complex) amplitude A_1 travelling in the positive x direction with velocity c_1. The displacement of the reflected wave is $y_r = B_1 e^{i(\omega t + k_1 x)}$, of amplitude B_1 and travelling in the negative x direction with velocity c_1.

The transmitted wave displacement is given by $y_t = A_2 e^{i(\omega t - k_2 x)}$, of amplitude A_2 and travelling in the positive x direction with velocity c_2.

We wish to find the reflection and transmission amplitude coefficients; that is, the relative values of B_1 and A_2 with respect to A_1. We find these via two boundary conditions which must be satisfied at the impedance discontinuity at $x = 0$.

The boundary conditions which apply at $x = 0$ are:

(1)   A geometrical condition that the displacement is the same immediately to the left and right of $x = 0$ for all time, so that there is no discontinuity of displacement.
(2)   A dynamical condition that there is a continuity of the transverse force $T(\partial y / \partial x)$ at $x = 0$, and therefore a continuous slope. This must hold, otherwise a finite difference in the force acts on an infinitesimally small mass of the string giving an infinite acceleration; this is not permitted.

Condition (1) at $x = 0$ gives

$$y_i + y_r = y_t$$

or

$$A_1 e^{i(\omega t - k_1 x)} + B_1 e^{i(\omega t + k_1 x)} = A_2 e^{i(\omega t - k_2 x)}$$

At $x = 0$ we may cancel the exponential terms giving

$$A_1 + B_1 = A_2 \tag{5.1}$$

Condition (2) gives

$$T \frac{\partial}{\partial x}(y_i + y_r) = T \frac{\partial}{\partial x} y_t$$

at $x = 0$ for all t, so that

$$-k_1 TA_1 + k_1 TB_1 = -k_2 TA_2 \tag{5.1a}$$

or

$$-\omega \frac{T}{c_1} A_1 + \omega \frac{T}{c_1} B_1 = -\omega \frac{T}{c_2} A_2 \tag{5.1b}$$

after cancelling exponentials at $x = 0$. But $T/c_1 = \rho_1 c_1 = Z_1$ and $T/c_2 = \rho_2 c_2 = Z_2$, so that

$$Z_1(A_1 - B_1) = Z_2 A_2 \tag{5.2}$$

Equations (5.1) and (5.2) give the

$$\textit{Reflection coefficient of amplitude, } \frac{B_1}{A_1} = \frac{Z_1 - Z_2}{Z_1 + Z_2}$$

and the

$$\textit{Transmission coefficient of amplitude, } \frac{A_2}{A_1} = \frac{2Z_1}{Z_1 + Z_2}$$

We see immediately that these coefficients are independent of ω and hold for waves of all frequencies; they are real and therefore free from phase changes other than that of π rad which will change the sign of a term. Moreover, these ratios depend entirely upon the ratios of the impedances. (See summary in Appendix 8). If $Z_2 = \infty$, this is equivalent to $x = 0$ being a fixed end to the string because no transmitted wave exists. This gives $B_1/A_1 = -1$, so that the incident wave is completely reflected (as we expect) with a phase change of π (phase reversal) – conditions we shall find to be necessary for standing waves to exist. A group of waves having many component frequencies will retain its shape upon reflection at $Z_2 = \infty$, but will suffer reversal (Figure 5.9). If $Z_2 = 0$, so that $x = 0$ is a free end of the string, then $B_1/A_1 = 1$ and $A_2/A_1 = 2$. This explains the 'flick' at the end of a whip or free ended string when a wave reaches it.

The use of a pulse is a convenient (but artificial) way of showing that the same phase change of all its component frequencies inverts the shape of the pulse. Without energy input this does not happen in practice and the pulse changes shape as it travels.

Replacement of Z by k and $\sqrt{\rho}$ in Transmission and Reflection Coefficients

Particular wave properties may replace the symbol Z where more convenient in a problem, e.g. equations 5.1 and 5.1a may be used to show that the reflection coefficient of amplitude

$$\frac{B_1}{A_1} = \frac{k_1 - k_2}{k_1 + k_2}$$

and the transmission coefficient of amplitude

$$\frac{A_2}{A_1} = \frac{2k_1}{k_1 + k_2}$$

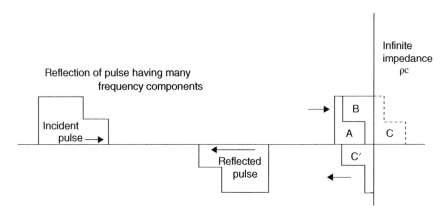

Figure 5.9   *A pulse of arbitrary shape is reflected at an infinite impedance with a phase change of π rad, so that the reflected pulse is the inverted and reversed shape of the initial waveform. The pulse at reflection is divided in the figure into three sections A, B, and C. At the moment of observation section C has already been reflected and suffered inversion and reversal to become C'. The actual shape of the pulse observed at this instant is A being A + B − C' where B = C'. The displacement at the point of reflection must be zero.*

Moreover on a string

$$T = \rho c^2 = \rho \frac{\omega^2}{k^2}$$

so $k \propto \sqrt{\rho}$ gives

$$\frac{B_1}{A_1} = \frac{\sqrt{\rho_1} - \sqrt{\rho_2}}{\sqrt{\rho_1} + \sqrt{\rho_2}}$$

and

$$\frac{A_2}{A_1} = \frac{2\sqrt{\rho_1}}{\sqrt{\rho_1} + \sqrt{\rho_2}}$$

In electromagnetic waves we shall find

$$Z = \frac{1}{n} = \frac{v}{c} \quad \text{where} \quad n = \frac{c}{v}$$

is the refractive index of the material.

Worked Example

A transverse sinusodial wave of amplitude 3.0 cm and wavelength 25 cm travels along a light string of 1 gram · cm^{-1} mass, which is joined to a heavier string of 4.0 gram · cm^{-1} mass. The joined strings are held under constant tension. (a) What is the wavelength and amplitude of the wave as it travels along the heavier string and (b) what fraction of wave power is reflected at the boundary of the two strings?

Solution

$$\frac{\lambda_2}{\lambda_1} = \sqrt{\frac{\rho_1}{\rho_2}} \quad \therefore \lambda_2 = 12.5 \, \text{cm}.$$

$$\frac{A_2}{A_1} = \frac{2\sqrt{\rho_1}}{\sqrt{\rho_1} + \sqrt{\rho_2}} \quad \therefore A_2 = 2 \, \text{cm}.$$

$$\frac{B_1}{A_1} = \frac{\sqrt{\rho_1} - \sqrt{\rho_2}}{\sqrt{\rho_1} + \sqrt{\rho_2}} = -\frac{1}{3} \quad \therefore \left(\frac{B_1}{A_1}\right)^2 = \frac{1}{9}.$$

5.8 Reflection and Transmission of Energy

Our interest in waves, however, is chiefly concerned with their function of transferring energy throughout a medium, and we shall now consider what happens to the energy in a wave when it meets a boundary between two media of different impedance values.

If we consider each unit length, mass ρ, of the string as a simple harmonic oscillator of maximum amplitude A, we know that its total energy will be $E = \frac{1}{2}\rho\omega^2 A^2$, where ω is the wave frequency.

The wave is travelling at a velocity c so that as each unit length of string takes up its oscillation with the passage of the wave the rate at which energy is being carried along the string is

$$(\text{energy} \times \text{velocity}) = \frac{1}{2}\rho\omega^2 A^2 c$$

Thus, the rate of energy arriving at the boundary $x = 0$ is the energy arriving with the incident wave; that is

$$\frac{1}{2}\rho_1 c_1 \omega^2 A_1^2 = \frac{1}{2}Z_1 \omega^2 A_1^2$$

The rate at which energy leaves the boundary, via the reflected and transmitted waves, is

$$\frac{1}{2}\rho_1 c_1 \omega^2 B_1^2 + \frac{1}{2}\rho_2 c_2 \omega^2 A_2^2 = \frac{1}{2}Z_1 \omega^2 B_1^2 + \frac{1}{2}Z_2 \omega^2 A_2^2$$

which, from the ratio B_1/A_1 and A_2/A_1,

$$= \frac{1}{2}\omega^2 A_1^2 \frac{Z_1(Z_1 - Z_2)^2 + 4Z_1^2 Z_2}{(Z_1 + Z_2)^2} = \frac{1}{2}Z_1 \omega^2 A_1^2$$

Thus, energy is conserved, and all energy arriving at the boundary in the incident wave leaves the boundary in the reflected and transmitted waves.

5.9   The Reflected and Transmitted Intensity Coefficients

These are given by

$$\frac{\text{Reflected Energy}}{\text{Incident Energy}} = \frac{Z_1 B_1^2}{Z_1 A_1^2} = \left(\frac{B_1}{A_1}\right)^2 = \left(\frac{Z_1 - Z_2}{Z_1 + Z_2}\right)^2$$

$$\frac{\text{Transmitted Energy}}{\text{Incident Energy}} = \frac{Z_2 A_2^2}{Z_1 A_1^2} = \frac{4 Z_1 Z_2}{(Z_1 + Z_2)^2}$$

We see that if $Z_1 = Z_2$ no energy is reflected and the impedances are said to be matched.

5.10   Matching of Impedances

We have just seen that at the boundary between two unequal impedances wave energy transport will be lost due to reflection and only a fraction of the energy will be transmitted. This happens in all media, on strings, acoustically, in optics, in electrical cables and when light waves enter a dielectric. The solution is common to all media. It is the insertion of a layer of a medium with an impedance equal to the harmonic mean of the unmatched impedances having a thickness of $\lambda/4$ of a wavelength measured in the intermediate impedance. Two unmatched impedances Z_1 and Z_3 are matched when a medium Z_2 is inserted between them, where $Z_2^2 = Z_1 Z_3$ of thickness $\lambda/4$ measured in Z_2.

We shall prove this statement in section 8.9, Matching Impedances, for the very common example of electrical cables.

Worked Example

For an electromagnetic wave travelling in a dielectric the impedance equals $1/n$ where n is the refractive index

$$\frac{c}{v} = \frac{\nu \lambda_0}{\nu \lambda_{\text{dielectric}}}$$

where λ_0 is the wavelength in free space.

To avoid reflection a camera lens ($n = 1.9$) is coated with a $\lambda/4$ thickness of a dielectric with refractive index n_2. Calculate the value of n_2 and the thickness of the layer if the wavelength in air is 550 nm.

Solution

$$n_2^2 = n_1 n_3 \; (n_1 = 1, n_3 = 1.9) \quad \therefore n_2 = 1.38$$

$$\text{Thickness} = \frac{1}{4}\frac{\lambda_{\text{air}}}{n_2} = \frac{550}{4 \times 1.38} = 99 \, \text{nm}$$

5.11   Standing Waves on a String of Fixed Length

We have already seen that a progressive wave is completely reflected at an infinite impedance with a π phase change in amplitude. A string of fixed length l with both ends rigidly clamped presents an infinite impedance at each end; we now investigate the behaviour of waves on such a string. Let us consider the simplest case of a monochromatic wave of one frequency ω with an amplitude a travelling in the positive

x direction and an amplitude b travelling in the negative x direction. The displacement on the string at any point would then be given by

$$y = a\,e^{i(\omega t - kx)} + b\,e^{i(\omega t + kx)}$$

with the boundary condition that $y = 0$ at $x = 0$ and $x = l$ at all times.

The condition $y = 0$ at $x = 0$ gives $0 = (a + b)\,e^{i\omega t}$ for all t, so that $a = -b$. This expresses physically the fact that a wave in either direction meeting the infinite impedance at either end is completely reflected with a π phase change in amplitude. This is a general result for all wave shapes and frequencies.

Thus

$$y = a\,e^{i\omega t}(e^{-ikx} - e^{ikx}) = (-2i)a\,e^{i\omega t} \sin kx \tag{5.3}$$

an expression for y which satisfies the standing wave time-independent form of the wave equation

$$\partial^2 y / \partial x^2 + k^2 y = 0$$

because $(1/c^2)(\partial^2 y / \partial t^2) = (-\omega^2/c^2)y = -k^2 y$. The condition that $y = 0$ at $x = l$ for all t requires

$$\sin kl = \sin \frac{\omega l}{c} = 0 \ \text{ or } \ \frac{\omega l}{c} = n\pi$$

limiting the values of allowed frequencies to

$$\omega_n = \frac{n\pi c}{l}$$

or

$$\nu_n = \frac{nc}{2l} = \frac{c}{\lambda_n}$$

that is

$$l = \frac{n\lambda_n}{2}$$

giving

$$\sin \frac{\omega_n x}{c} = \sin \frac{n\pi x}{l}$$

These frequencies are the normal frequencies or modes of vibration we first met in Chapter 4. They are often called eigenfrequencies, particularly in wave mechanics.

Such allowed frequencies define the length of the string as an exact number of half wavelengths, and Figure 5.10 shows the string displacement for the first four harmonics ($n = 1, 2, 3, 4$). The value for $n = 1$ is called the fundamental.

As with the loaded string of Chapter 4, all normal modes may be present at the same time and the general displacement is the superposition of the displacements at each frequency. This is a more complicated problem which we discuss in Chapter 11 (Fourier Methods).

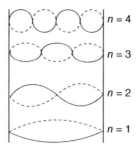

Figure 5.10 *The first four harmonics, n = 1, 2, 3, 4 of the standing waves allowed between the two fixed ends of a string.*

For the moment we see that for each single harmonic $n > 1$ there will be a number of positions along the string which are always at rest. These points occur where

$$\sin \frac{\omega_n x}{c} = \sin \frac{n\pi x}{l} = 0$$

or

$$\frac{n\pi x}{l} = r\pi \ (r = 0, 1, 2, 3, \ldots n)$$

The values $r = 0$ and $r = n$ give $x = 0$ and $x = l$, the ends of the string, but between the ends there are $n - 1$ positions equally spaced along the string in the nth harmonic where the displacement is always zero. These positions are called nodes or nodal points, being the positions of zero motion in a system of standing waves. Standing waves arise when a single mode is excited and the incident and reflected waves are superposed. If the amplitudes of these progressive waves are equal and opposite (resulting from complete reflection), nodal points will exist. Often, however, the reflection is not quite complete and the waves in the opposite direction do not cancel each other to give complete nodal points. In this case we speak of a standing wave ratio which we shall discuss in the next section.

Whenever nodal points exist, however, we know that the waves travelling in opposite directions are exactly equal in all respects so that the energy carried in one direction is exactly equal to that carried in the other. This means that the total energy flux, that is, the energy carried across unit area per second in a standing wave system, is zero.

Returning to equation 5.3, we see that the complete expression for the displacement of the nth harmonic is given by

$$y_n = 2a(-\mathrm{i})(\cos \omega_n t + \mathrm{i} \sin \omega_n t) \sin \frac{\omega_n x}{c}$$

We can express this in the form

$$y_n = (A_n \cos \omega_n t + B_n \sin \omega_n t) \sin \frac{\omega_n x}{c} \tag{5.4}$$

where the amplitude of the nth mode is given by $(A_n^2 + B_n^2)^{1/2} = 2a$.

5.12   Standing Wave Ratio

When a wave is completely reflected the superposition of the incident and reflected amplitudes will give nodal points (zero amplitude) where the incident and reflected amplitudes cancel each other, and points of maximum displacement equal to twice the incident amplitude where they reinforce.

If a progressive wave system is partially reflected from a boundary let the amplitude reflection coefficient B_1/A_1 of the earlier section be written as r, where $r < 1$.

The maximum amplitude at reinforcement is then A_1+B_1; the minimum amplitude is given by A_1-B_1. In this case the ratio of maximum to minimum amplitudes in the standing wave system is called the

$$Standing\ Wave\ Ratio\ = \frac{A_1+B_1}{A_1-B_1} = \frac{1+r}{1-r}$$

where $r = B_1/A_1$.

Measuring the values of the maximum and minimum amplitudes gives the value of the reflection coefficient for

$$r = B_1/A_1 = \frac{\text{SWR}-1}{\text{SWR}+1}$$

where SWR refers to the Standing Wave Ratio.

Worked Example

A travelling wave $y_1 = A\cos(\omega t - kx)$ combines with the reflected wave $y_2 = rA\cos(\omega t + kx)$ to produce a standing wave. Show that the standing wave can be represented by $y = 2rA\cos \omega t \cos kx + A(1-r)\cos(\omega t - kx)$. Show that SWR$= \frac{1+r}{1-r}$.

Solution

At reflection incident wave amplitude is reduced to $A(1-r)$ and reflected amplitude is rA. At reflection phase of reflected wave is $\cos(\omega t + kx) + \cos(\omega t - kx) = 2\cos \omega t \cos kx$ so reflected wave is $2rA\cos \omega t \cos kx$ and incident wave is $A(1-r)\cos(\omega t - kx)$. Max. amplitude $= 2rA + A(1-r)$ at antinode of the reflected wave. Max. amplitude $= A(1-r)$ at node of reflected wave.

$$\text{SWR} = \frac{2rA + A(1-r)}{A(1-r)} = \frac{1+r}{1-r}$$

5.13   Energy in Each Normal Mode of a Vibrating String

The total displacement y in the string is the superposition of the displacements y_n of the individual harmonics and we can find the energy in each harmonic by replacing y_n for y in the results on the last page of section 5.11, Standing Waves on a String of Fixed Length. Thus, the kinetic energy in the nth harmonic is

$$E_n(\text{kinetic}) = \frac{1}{2}\int_0^l \rho\dot{y}_n^2 \mathrm{d}x$$

for a string of length l and the potential energy is

$$E_n(\text{potential}) = \frac{1}{2}T \int_0^l \left(\frac{\partial y_n}{\partial x}\right)^2 dx$$

Since we have already shown for standing waves at the end of section 5.11, Standing Waves on a String of Fixed Length, that

$$y_n = (A_n \cos \omega_n t + B_n \sin \omega_n t) \sin \frac{\omega_n x}{c}$$

then

$$\dot{y}_n = (-A_n \omega_n \sin \omega_n t + B_n \omega_n \cos \omega_n t) \sin \frac{\omega_n x}{c}$$

and

$$\frac{\partial y_n}{\partial x} = \frac{\omega_n}{c}(A_n \cos \omega_n t + B_n \sin \omega_n t) \cos \frac{\omega_n x}{c}$$

Thus

$$E_n(\text{kinetic}) = \frac{1}{2}\rho \omega_n^2[-A_n \sin \omega_n t + B_n \cos \omega_n t]^2 \int_0^l \sin^2 \frac{\omega_n x}{c} dx$$

and

$$E_n(\text{potential}) = \frac{1}{2}T\frac{\omega_n^2}{c^2}[A_n \cos \omega_n t + B_n \sin \omega_n t]^2 \int_0^l \cos^2 \frac{\omega_n x}{c} dx$$

Remembering that $T = \rho c^2$ we have

$$E_n(\text{kinetic + potential}) = \frac{1}{4}\rho l \omega_n^2 (A_n^2 + B_n^2)$$
$$= \frac{1}{4}m \omega_n^2 (A_n^2 + B_n^2)$$

where m is the mass of the string and $(A_n^2 + B_n^2)$ is the square of the maximum displacement (amplitude) of the mode. To find the exact value of the total energy E_n of the mode we would need to know the precise value of A_n and B_n and we shall evaluate these in Chapter 11 on Fourier Methods. The total energy of the vibrating string is the sum of all the E_n's of the normal modes.

Note that the distribution of energy along the normal mode of a vibration is the same as that of a travelling wave.

Problem 5.1.   Show that the wave profile, that is,

$$y = f_1(ct - x)$$

remains unchanged with time when c is the wave velocity. To do this consider the expression for y at a time $t + \Delta t$ where $\Delta t = \Delta x/c$.
   Repeat the problem for $y = f_2(ct + x)$.

Problem 5.2.   A triangular shaped pulse of length l is reflected at the fixed end of the string on which it travels ($Z_2 = \infty$). Sketch the shape of the pulse (see Figure 5.9) after a length (a) $l/4$, (b) $l/2$, (c) $3l/4$ and (d) l of the pulse has been reflected.

Problem 5.3.   An electrically driven oscillator at the end of string propagates a sinusoidal wave along the string which has a linear density $\rho = 30$ g·m^{-1} and is under a constant tension $T = 12$ N. What power is required to sustain a frequency of $\nu = 300$ Hz and an amplitude of 1.5 cm? What power is required if (a) the frequency is doubled and (b) the amplitude is halved?

Problem 5.4.   A cello string has a linear density of $\rho = 1.7$ g m^{-1} and a length $L = 0.7$ m. A tension T in the string times it to 220 Hz. What is T?

Problem 5.5.   A point mass M is concentrated at a point on a string of characteristic impedance ρc. A transverse wave of frequency ω moves in the positive x direction and is partially reflected and transmitted at the mass. The boundary conditions are that the string displacements just to the left and right of the mass are equal ($y_i + y_r = y_t$) and that the difference between the transverse forces just to the left and right of the mass equal the mass times its acceleration. If A_1, B_1 and A_2 are respectively the incident, reflected and transmitted wave amplitudes the values

$$\frac{B_1}{A_1} = \frac{-iq}{1 + iq} \text{ and } \frac{A_2}{A_1} = \frac{1}{1 + iq}$$

where $q = \omega M/2\rho c$ and $i^2 = -1$. Writing $q = \tan\theta$, show that A_2 lags A_1 by θ and that B_1 lags A_1 by $(\pi/2 + \theta)$ for $0 < \theta < \pi/2$.
   Show also that the reflected and transmitted energy coefficients are represented by $\sin^2\theta$ and $\cos^2\theta$, respectively.

Problem 5.6.   A transverse harmonic force of peak value 0.3 N and frequency 5 Hz initiates waves of amplitude 0.1 m at one end of a very long string of linear density 0.01 kg/m. Show that the rate of energy transfer along the string is $3\pi/20$ W and that the wave velocity is $30/\pi$ m s^{-1}.

Problem 5.7.   The tension in a string produces a fundamental frequency of 440 Hz. (a) What are the frequencies of the 2nd and 3rd harmonics? (b) The average human ear can register 16,000 kHz. How many harmonics does this represent? (c) If the violin string is 32 cm long how far from its end should the string be pressed to shorten its length and produce a fundamental of 523 Hz?

Problem 5.8.   The relation between the impedance Z and the refractive index n of a dielectric is given by $Z = 1/n$. Light travelling in free space enters a glass lens which has a refractive index of 1.5 for a free space wavelength of 5.5×10^{-7} m. Show that reflections at this wavelength are avoided by a coating of refractive index 1.22 and thickness 1.12×10^{-7} m.

Problem 5.9.   Prove that the displacement y_n of the standing wave expression in equation (5.4) satisfies the time-independent form of the wave equation

$$\frac{\partial^2 y}{\partial x^2} + k^2 y = 0.$$

Problem 5.10.   The total energy E_n of a normal mode may be found by an alternative method. Each section dx of the string is a simple harmonic oscillator with total energy equal to the maximum kinetic energy of oscillation

$$k.e._{\text{max}} = \frac{1}{2}\rho dx(\dot{y}_n^2)_{\text{max}} = \frac{1}{2}\rho dx \omega_n^2 (y_n^2)_{\text{max}}$$

Now the value of $(y_n^2)_{\text{max}}$ at a point x on the string is given by

$$(y_n^2)_{\text{max}} = (A_n^2 + B_n^2) \sin^2 \frac{\omega_n x}{c}$$

Show that the sum of the energies of the oscillators along the string, that is, the integral

$$\frac{1}{2}\rho \omega_n^2 \int_0^l (y_n^2)_{\text{max}} dx$$

gives the expected result.

Problem 5.11.   The displacement of a wave on a string which is fixed at both ends is given by

$$y(x,\ t) = A \cos(\omega t - kx) + rA \cos(\omega t + kx)$$

where r is the coefficient of amplitude reflection. Show that this may be expressed as the superposition of standing waves

$$y(x,\ t) = A(1 + r) \cos \omega t \, \cos kx + A(1 - r) \sin \omega t \sin kx.$$

6

Transverse Wave Motion (2)

Introduction

Waves are rarely monochromatic, that is, limited to a single frequency, but are usually made of a mixture of frequencies. First of all we consider the superposition of two waves of equal amplitudes and phase velocities but with slightly different frequencies. Then the two waves are allowed different phase velocities and finally multiple waves over a narrow frequency range are superposed to form a pulse. This leads to the concepts of group velocity, beats, and dispersion. The Bandwidth Theorem is derived and its connection to Heisenberg's Uncertainty Principle is explored. The propagation of transverse waves in a periodic structure such as an ionic crystal explains how infrared radiation is absorbed. The Diffusion Equation, applied to the periodic structure of a transmission line, is used to account for energy loss in wave propagation.

6.1 Wave Groups, Group Velocity and Dispersion

Our discussion so far has been limited to monochromatic waves – waves of a single frequency and wavelength. It is much more common for waves to occur as a mixture of a number or group of component frequencies; white light, for instance, is composed of a continuous visible wavelength spectrum extending from about 3000 Å in the blue to 7000 Å in the red. Examining the behaviour of such a group leads to the third kind of velocity mentioned early in the last chapter, that is, the group velocity.

6.1.1 Superposition of Two Waves of Almost Equal Frequencies

We begin by considering a group which consists of two components of equal amplitude a but frequencies ω_1 and ω_2 which differ by a small amount.

Their separate displacements are given by

$$y_1 = a \cos \left(\omega_1 t - k_1 x \right)$$

Introduction to Vibrations and Waves, First Edition. H. J. Pain and Patricia Rankin.
© 2015 John Wiley & Sons, Ltd. Published 2015 by John Wiley & Sons, Ltd.
Companion website: http://booksupport.wiley.com

and

$$y_2 = a \cos(\omega_2 t - k_2 x)$$

Superposition of amplitude and phase gives

$$y = y_1 + y_2 = 2a \cos \left[\frac{(\omega_1 - \omega_2)t}{2} - \frac{(k_1 - k_2)x}{2} \right] \cos \left[\frac{(\omega_1 + \omega_2)t}{2} - \frac{(k_1 + k_2)x}{2} \right]$$

a wave system with a frequency $(\omega_1 + \omega_2)/2$ which is very close to the frequency of either component but with a maximum amplitude of $2a$, modulated in space and time by a very slowly varying envelope of frequency $(\omega_1 - \omega_2)/2$ and wave number $(k_1 - k_2)/2$.

This system is shown in Figure 6.1 and shows a behaviour similar to that of the equivalent coupled oscillators in Chapter 4. The velocity of the new wave is $(\omega_1 - \omega_2)/(k_1 - k_2)$ which, if the phase velocities $\omega_1/k_1 = \omega_2/k_2 = c$, gives

$$\frac{\omega_1 - \omega_2}{k_1 - k_2} = c\frac{(k_1 - k_2)}{k_1 - k_2} = c$$

so that the component frequencies and their superposition, or group will travel with the same velocity, the profile of their combination in Figure 6.1 remaining constant.

If the waves are sound waves the intensity is a maximum whenever the amplitude is a maximum of $2a$; this occurs twice for every period of the modulating frequency; that is, at a frequency $\nu_1 - \nu_2$.

The beats of maximum intensity fluctuations thus have a frequency equal to the difference $\nu_1 - \nu_2$ of the components. In the example here where the components have equal amplitudes a, superposition will produce an amplitude which varies between $2a$ and 0; this is called complete or 100% modulation.

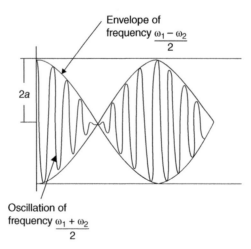

Figure 6.1 *The superposition of two waves of slightly different frequency ω_1 and ω_2 forms a group. The faster oscillation occurs at the average frequency of the two components $(\omega_1 + \omega_2)/2$ and the slowly varying group envelope has a frequency $(\omega_1 - \omega_2)/2$, half the frequency difference between the components.*

More generally an amplitude modulated wave may be represented by

$$y = A \cos(\omega t - kx)$$

where the modulated amplitude

$$A = a + b \cos \omega' t$$

This gives

$$y = a \cos(\omega t - kx) + \frac{b}{2}\{\cos[(\omega + \omega')t - kx] + \cos[(\omega - \omega')t - kx]\}$$

so that here amplitude modulation has introduced two new frequencies $\omega \pm \omega'$, known as combination tones or sidebands. Amplitude modulation of a carrier frequency is a common form of radio transmission, but its generation of sidebands has led to the crowding of radio frequencies and interference between stations.

6.1.2 Wave Groups, Group Velocity and Dispersion

Suppose now that the two frequency components of the last section have different phase velocities so that $\omega_1/k_1 \neq \omega_2/k_2$. The velocity of the maximum amplitude of the group, that is, the group velocity

$$\frac{\omega_1 - \omega_2}{k_1 - k_2} = \frac{\Delta\omega}{\Delta k}$$

is now different from each of these velocities; the superposition of the two waves will no longer remain constant and the group profile will change with time.

A medium in which the phase velocity is frequency dependent (ω/k not constant) is known as a dispersive medium and a dispersion relation expresses the variation of ω as a function of k. If a group contains a number of components of frequencies which are nearly equal the original expression for the group velocity is written

$$\frac{\Delta\omega}{\Delta k} = \frac{d\omega}{dk}$$

The group velocity is that of the maximum amplitude of the group so that it is the velocity with which the energy in the group is transmitted. Since $\omega = kv$, where v is the phase velocity, the group velocity

$$v_g = \frac{d\omega}{dk} = \frac{d}{dk}(kv) = v + k\frac{dv}{dk}$$
$$= v - \lambda\frac{dv}{d\lambda}$$

where $k = 2\pi/\lambda$. Usually $dv/d\lambda$ is positive, so that $v_g < v$. This is called normal dispersion, but anomalous dispersion can arise when $dv/d\lambda$ is negative, so that $v_g > v$.

We shall see when we discuss electromagnetic waves that an electrical conductor is anomalously dispersive to these waves whilst a dielectric is normally dispersive except at the natural resonant frequencies of its atoms. In the chapter on forced oscillations we saw that the wave then acted as a driving force upon

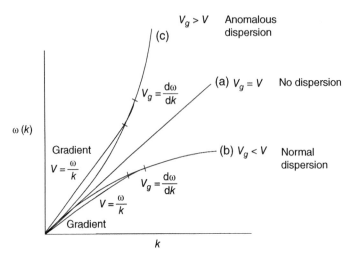

Figure 6.2   *Curves illustrating dispersion relations: (a) a straight line representing a non-dispersive medium,* $v = v_g$; *(b) a normal dispersion relation where the gradient* $v = \omega/k > v_g = d\omega/dk$; *(c) an anomalous dispersion relation where* $v < v_g$.

the atomic oscillators and that strong absorption of the wave energy was represented by the dissipation fraction of the oscillator impedance, whilst the anomalous dispersion curve followed the value of the reactive part of the impedance.

The three curves of Figure 6.2 represent

- A non-dispersive medium where ω/k is constant, so that $v_g = v$, for instance free space behaviour towards light waves.
- A normal dispersion relation $v_g < v$.
- An anomalous dispersion relation $v_g > v$.

Worked Example

The electric vector of an electromagnetic wave propagates in a dielectric with a velocity $v = (\mu\varepsilon)^{-1/2}$ where μ is the permeability and ε is the permittivity. In free space the velocity is that of light, $c = (\mu_0\varepsilon_0)^{-1/2}$. The refractive index $n = c/v = \sqrt{\mu\varepsilon/\mu_0\varepsilon_0} = \sqrt{\mu_r\varepsilon_r}$ where $\mu_r = \mu/\mu_0$ and $\varepsilon_r = \varepsilon/\varepsilon_0$. For many substances μ_r is constant and $\sim$1, but ε_r is frequency dependent, so that v depends on λ.

The group velocity

$$v_g = v - \lambda dv/d\lambda = v\left(1 + \frac{\lambda}{2\varepsilon_r}\frac{\partial\varepsilon_r}{\partial\lambda}\right)$$

so that $v_g > v$ (anomalous dispersion) when $\partial\varepsilon_r/\partial\lambda$ is +ve. Figure 6.3 shows the behaviour of the refractive index $n = \sqrt{\varepsilon_r}$ versus ω, the frequency, and λ, the wavelength, in the region of anomalous dispersion associated with a resonant frequency. The dotted curve shows the energy absorption (compare this with Figure 3.11).

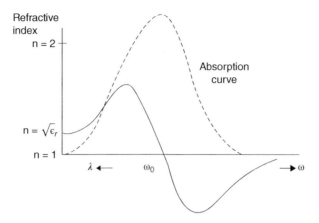

Figure 6.3   *Anomalous dispersion showing the behaviour of the refractive index $n = \sqrt{\varepsilon_r}$ versus ω and λ, where ω_0 is a resonant frequency of the atoms of the medium. The absorption in such a region is shown by the dotted line (see Figure 3.11).*

6.2   Wave Group of Many Components. The Bandwidth Theorem

We have so far considered wave groups having only two frequency components. We may easily extend this to the case of a group of many frequency components, each of amplitude a, lying within the narrow frequency range $\Delta\omega$.

The essential physics of this problem is shown in Appendix 3, where we find the sum of the series, with δ as the constant phase difference between n successive equal components to be

$$R = \sum_{0}^{n-1} a \, \cos\left(\omega t + n\delta\right)$$

Here we are concerned with the constant phase difference $(\delta\omega)t$ which results from a constant frequency difference $\delta\omega$ between successive components. The spectrum or range of frequencies of this group is shown in Figure 6.4a and we wish to follow its behaviour with time.

We seek the amplitude which results from the superposition of the frequency components and write it

$$R = a \cos \omega_1 t + a \cos\left(\omega_1 + \delta\omega\right)t + a \cos\left(\omega_1 + 2\delta\omega\right)t + \cdots$$
$$+ \, a \cos\left[\omega_1 + (n-1)(\delta\omega)\right]t$$

The result is given in Appendix 3 as

$$R = a \frac{\sin[n(\delta\omega)t/2]}{\sin[(\delta\omega)t/2]} \cos \overline{\omega} t$$

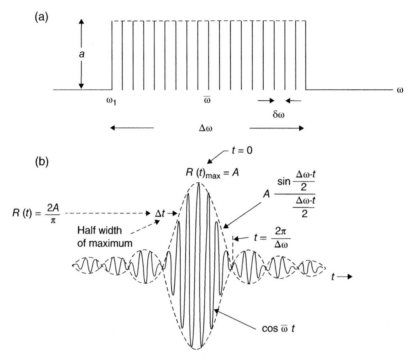

Figure 6.4   *A rectangular wave band of width $\Delta\omega$ having n frequency components of amplitude a with a common frequency difference $\delta\omega$. (b) Representation of the frequency band on a time axis is a cosine curve at the average frequency $\bar{\omega}$, amplitude modulated by a $\sin\alpha/\alpha$ curve where $\alpha = \Delta\omega \cdot t/2$. After a time $t = 2\pi/\Delta\omega$ the superposition of the components gives a zero amplitude.*

where the average frequency in the group or band is

$$\bar{\omega} = \omega_1 + \frac{1}{2}(n-1)(\delta\omega)$$

Now $n(\delta\omega) = \Delta\omega$, the bandwidth, so the behaviour of the resultant R with time may be written

$$R(t) = a\frac{\sin(\Delta\omega \cdot t/2)}{\sin(\Delta\omega \cdot t/n2)}\cos\bar{\omega}t = na\frac{\sin(\Delta\omega \cdot t/2)}{\Delta\omega \cdot t/2}\cos\bar{\omega}t$$

when n is large, and

$$\sin(\Delta\omega \cdot t/n2) \rightarrow \frac{\Delta\omega \cdot t}{n2}$$

or

$$R(t) = A\frac{\sin\alpha}{\alpha}\cos\bar{\omega}t$$

where $A = na$ and $\alpha = \Delta\omega \cdot t/2$ is half the phase difference between the first and last components at time t.

This expression gives us the time behaviour of the band and is displayed on a time axis in Figure 6.4b. We see that the amplitude $R(t)$ is given by the cosine curve of the average frequency $\bar{\omega}$ modified by the $A \sin \alpha / \alpha$ term.

At $t = 0$, $\sin \alpha / \alpha \rightarrow 1$ and all the components superpose with zero phase difference to give the maximum amplitude $R(t) = A = na$. After some time interval Δt when

$$\alpha = \frac{\Delta \omega \Delta t}{2} = \pi$$

the phases between the frequency components are such that the resulting amplitude $R(t)$ is zero.

The time Δt which is a measure of the width of the central pulse of Figure 6.4b is therefore given by

$$\frac{\Delta \omega \Delta t}{2} = \pi$$

or $\Delta \nu \, \Delta t = 1$ where $\Delta \omega = 2\pi \Delta \nu$.

The true width of the base of the central pulse is $2\Delta t$ but the interval Δt is taken as an arbitrary measure of time, centred about $t = 0$, during which the amplitude $R(t)$ remains significantly large $(>A/2)$. With this arbitrary definition the exact expression

$$\Delta \nu \, \Delta t = 1$$

becomes the approximation

$$\Delta \nu \, \Delta t \approx 1 \quad \text{or} \quad (\Delta \omega \, \Delta t \approx 2\pi)$$

and this approximation is known as the Bandwidth Theorem.

It states that the components of a band of width $\Delta \omega$ in the frequency range will superpose to produce a significant amplitude $R(t)$ only for a time Δt before the band decays from random phase differences. The greater the range $\Delta \omega$ the shorter the period Δt.

Alternatively, the theorem states that a single pulse of time duration Δt is the result of the superposition of frequency components over the range $\Delta \omega$; the shorter the period Δt of the pulse the wider the range $\Delta \omega$ of the frequencies required to represent it.

When $\Delta \omega$ is zero we have a single frequency, the monochromatic wave which is therefore required (in theory) to have an infinitely long time span.

We have chosen to express our wave group in the two parameters of frequency and time (having a product of zero dimensions), but we may just as easily work in the other pair of parameters wave number k and distance x.

Replacing ω by k and t by x would define the length of the wave group as Δx in terms of the range of component wavelengths $\Delta(1/\lambda)$.

The Bandwidth Theorem then becomes

$$\Delta x \, \Delta k \approx 2\pi$$

or

$$\Delta x \Delta(1/\lambda) \approx 1 \quad \text{i.e.} \quad \Delta x \approx \lambda^2 / \Delta \lambda$$

Note again that a monochromatic wave with $\Delta k = 0$ requires $\Delta x \rightarrow \infty$; that is, an infinitely long wavetrain.

In the wave group we have just considered the problem has been simplified by assuming all frequency components to have the same amplitude a. When this is not the case, the different values $a(\omega)$ are treated by Fourier methods as we shall see in Chapter 11.

We shall meet the ideas of this section several times in the course of this text, noting particularly that in modern physics the Bandwidth Theorem becomes Heisenberg's Uncertainty Principle.

Worked Example

A pulse of white light has a frequency range $\Delta\nu$ between 769 and 384 times 10^{12} Hz, i.e. $\Delta\nu \approx 385 \times 10^{12}$ Hz. The Bandwidth Theorem gives $\Delta\nu\Delta t \approx 1$ $\therefore$ $\Delta\nu = 1/\Delta t$ and the coherent length of the wavetrain of such a pulse is $c\Delta t = c/\Delta\nu = 779 \times 10^{-9}$ m, that is, one wavelength at the red end of the visible spectrum.

6.3   Heisenberg's Uncertainty Principle

Compton (in 1922–23) fired X-rays of a known frequency at thin foils of different materials and found that the scattered radiation was independent of the foil material and that his results were consistent only if momentum and energy were conserved in an elastic collision between two 'particles', an electron and an X-ray of energy $h\nu$, rest mass m_0 and (from Einstein's relativistic energy equation) a momentum $p = E/c = h\nu/c = h/\lambda$ where $c = \nu\lambda$ and h is Planck's constant.

In 1924 de Broglie proposed that if the dual wave particle nature of electromagnetic fields (X-rays) required a particle momentum of $p = h/\lambda$ it was possible that a wavelength λ of a 'matter' field could be associated with *any* particle $p = mv$ to give the relation $p = h/\lambda$. He showed that the velocity v in mv was the *group velocity* of a pulse (not a single frequency) so

$$p = \frac{h}{\lambda} = \frac{hk}{2\pi}$$

and

$$\Delta p = \frac{h}{2\pi}\Delta k$$

But the Bandwidth Theorem shows that a group in the wave number range Δk superposed in space over a distance Δx obeys the relation

$$\Delta x \Delta k \approx 2\pi$$

so

$$\Delta x \Delta p \approx h$$

This is Heisenberg's Uncertainty Principle.

This relation sets a fundamental limit on the ultimate precision with which we can know the position x of a particle and the x component of its momentum simultaneously (Figure 6.5). More advanced mathematics shows that a 'wave packet' of typical shape (Gaussian in Figure 6.5), representing an electron localized at time $t = 0$ to within a distance of $\Delta x = 10^{-10}$ m (atomic dimensions) with $\Delta p_x = h/\Delta x \approx 10^{-24}$ kg $\cdot$ m $\cdot$ s^{-1} will spread to twice its length in time $t = 10^{-16}$ sec.

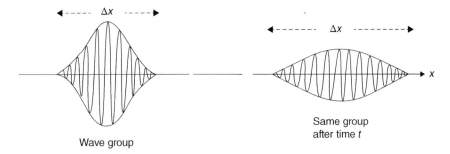

Wave group

Same group
after time t

Figure 6.5 *A wave group representing a particle showing dispersion after time t. The square of the wave amplitude at any point represents the probability of the particle being in that position, and the dispersion represents the increasing uncertainty of the particle position with time (Heisenberg's Uncertainty Principle).*

6.4 Transverse Waves in Periodic Structures (1) Waves in a Crystal

At the end of the chapter on coupled oscillations we discussed the normal transverse vibrations of n equal masses of separation a along a light string of length $(n+1)a$ under a tension T with both ends fixed. The equation of motion of the rth particle was found to be

$$m\ddot{y}_r = \frac{T}{a}(y_{r+1} + y_{r-1} - 2y_r) \tag{6.1}$$

and for n masses the frequencies of the normal modes of vibration were given by

$$\omega_j^2 = \frac{2T}{ma}\left(1 - \cos\frac{j\pi}{n+1}\right) \tag{6.2}$$

where $j = 1, 2, 3, \dots, n$. When the separation a becomes infinitesimally small ($= \delta x$, say) the term in the equation of motion

$$\frac{1}{a}(y_{r+1} + y_{r-1} - 2y_r) \rightarrow \frac{1}{\delta x}(y_{r+1} + y_{r-1} - 2y_r)$$
$$= \frac{(y_{r+1} - y_r)}{\delta x} - \frac{(y_r - y_{r-1})}{\delta x} = \left(\frac{\partial y}{\partial x}\right)_{r+1/2} - \left(\frac{\partial y}{\partial x}\right)_{r-1/2} = \left(\frac{\partial^2 y}{\partial x^2}\right)_r dx$$

so that the equation of motion becomes

$$\frac{\partial^2 y}{\partial t^2} = \frac{T}{\rho}\frac{\partial^2 y}{\partial x^2},$$

the wave equation, where $\rho = m/\delta x$, the linear density and

$$y \propto e^{i(\omega t - kx)}$$

We are now going to consider the propagation of transverse waves along a linear array of atoms, mass m, in a crystal lattice where the tension T now represents the elastic force between the atoms (so that T/a is the stiffness) and a, the separation between the atoms, is about 1 Å or 10^{-10} m. When the clamped ends of the string are replaced by the ends of the crystal we can express the displacement of the rth particle due to the transverse waves as

$$y_r = A_r \, e^{i(\omega t - kx)} = A_r \, e^{i(\omega t - kra)},$$

since $x = ra$. The equation of motion then becomes

$$-\omega^2 m = \frac{T}{a}(e^{ika} + e^{-ika} - 2)$$

$$= \frac{T}{a}(e^{ika/2} - e^{-ika/2})^2 = -\frac{4T}{a}\sin^2\frac{ka}{2}$$

giving the permitted frequencies

$$\omega^2 = \frac{4T}{ma}\sin^2\frac{ka}{2} \tag{6.3}$$

This expression for ω^2 is equivalent to our earlier value at the end of Chapter 4:

$$\omega_j^2 = \frac{2T}{ma}\left(1 - \cos\frac{j\pi}{n+1}\right) = \frac{4T}{ma}\sin^2\frac{j\pi}{2(n+1)} \tag{6.4}$$

if

$$\frac{ka}{2} = \frac{j\pi}{2(n+1)}$$

where $j = 1, 2, 3, \ldots, n$.

But $(n+1)a = l$, the length of the string or crystal, and we have seen that wavelengths λ are allowed where $p\lambda/2 = l = (n+1)a$.

Thus

$$\frac{ka}{2} = \frac{2\pi}{\lambda}\cdot\frac{a}{2} = \frac{\pi a}{\lambda} = \frac{ja\pi}{2(n+1)a} = \frac{j}{p}\cdot\frac{\pi a}{\lambda}$$

if $j = p$. When $j = p$, a unit change in j corresponds to a change from one allowed number of half wavelengths to the next so that the minimum wavelength is $\lambda = 2a$, giving a maximum frequency $\omega_m^2 = 4T/ma$. Thus, both expressions may be considered equivalent.

When $\lambda = 2a$, $\sin ka/2 = 1$ because $ka = \pi$, and neighbouring atoms are exactly π rad out of phase because

$$\frac{y_r}{y_{r+1}} \propto e^{ika} = e^{i\pi} = -1$$

The highest frequency is thus associated with maximum coupling, as we expect.

If in equation 6.1 we plot $|\sin ka/2|$ against k (Figure 6.6) we find that when ka is increased beyond π the phase relationship is the same as for a negative value of ka beyond $-\pi$. It is, therefore, sufficient to restrict the values of k to the region

$$\frac{-\pi}{a} \le k \le \frac{\pi}{a}$$

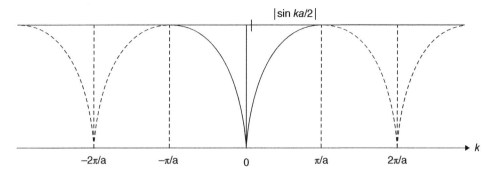

Figure 6.6 $|\sin\frac{ka}{2}|$ *versus k from equation (6.3) shows the repetition of values beyond the region $\frac{-\pi}{a} \leq k \leq \frac{\pi}{a}$; this region defines a Brillouin zone.*

which is known as the first Brillouin zone. We shall use this concept in the section on electron waves in solids in Chapter 13.

For long wavelengths or low values of the wave number k, $\sin ka/2 \rightarrow ka/2$ so that

$$\omega^2 = \frac{4T}{ma}\frac{k^2a^2}{4}$$

and the velocity of the wave is given by

$$c^2 = \frac{\omega^2}{k^2} = \frac{Ta}{m} = \frac{T}{\rho}$$

as before, where $\rho = m/a$.

In general the phase velocity is given by

$$v = \frac{\omega}{k} = c\left[\frac{\sin ka/2}{ka/2}\right] \tag{6.5}$$

a dispersion relation which is shown in Figure 6.7. Only at very short wavelengths does the atomic spacing of the crystal structure affect the wave propagation, and here the limiting or maximum value of the wave number $k_m = \pi/a \approx 10^{10}$ m^{-1}.

The elastic force constant T/a for a crystal is about 15 N m^{-1}; a typical 'reduced' atomic mass is about 60×10^{-27} kg. These values give a maximum frequency

$$\omega^2 = \frac{4T}{ma} \approx \frac{60}{60 \times 10^{-27}} = 10^{27} \text{ rad s}^{-1}$$

that is, a frequency $\nu \approx 5 \times 10^{12}$ Hz.

(Note that the value of T/a used here for the crystal is a factor of 8 lower than that found in Problem 4.4 for a single molecule. This is due to the interaction between neighbouring ions and the change in their equilibrium separation.)

The frequency $\nu = 5 \times 10^{12}$ Hz is in the infrared region of the electromagnetic spectrum. We shall see later that electromagnetic waves of frequency ω have a transverse electric field vector $E = E_0\,e^{i\omega t}$,

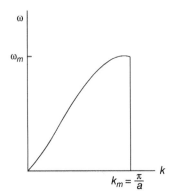

Figure 6.7   *The dispersion relation ω(k) versus k for waves travelling along a linear one-dimensional array of atoms in a periodic structure.*

where E_0 is the maximum amplitude, so that charged atoms or ions in a crystal lattice could respond as forced oscillators to radiation falling upon the crystal, which would absorb any radiation at the resonant frequency of its oscillating atoms.

6.5   Linear Array of Two Kinds of Atoms in an Ionic Crystal

We continue the discussion of this problem using a one-dimensional line which contains two kinds of atoms with separation a as before, those atoms of mass M occupying the odd numbered positions, $2r-1, 2r+1$, etc. and those of mass m occupying the even numbered positions, $2r, 2r+2$, etc. The equations of motion for each type are

$$m\ddot{y}_{2r} = \frac{T}{a}(y_{2r+1} + y_{2r-1} - 2y_{2r})$$

and

$$M\ddot{y}_{2r+1} = \frac{T}{a}(y_{2r+2} + y_{2r} - 2y_{2r+1})$$

with solutions

$$y_{2r} = A_m\,e^{i(\omega t - 2rka)}$$
$$y_{2r+1} = A_M\,e^{i(\omega t - (2r+1)ka)}$$

where A_m and A_M are the amplitudes of the respective masses.
   The equations of motion thus become

$$-\omega^2 m A_m = \frac{TA_M}{a}\left(e^{-ika} + e^{ika}\right) - \frac{2TA_m}{a}$$

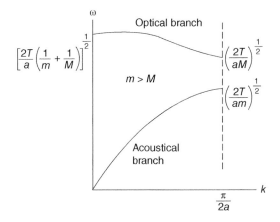

Figure 6.8 *Dispersion relations for the two modes of transverse oscillation in a crystal structure.*

and

$$-\omega^2 M A_M = \frac{TA_m}{a}(e^{-ika} + e^{ika}) - \frac{2TA_M}{a}$$

equations which are consistent when

$$\omega^2 = \frac{T}{a}\left(\frac{1}{m} + \frac{1}{M}\right) \pm \frac{T}{a}\left[\left(\frac{1}{m} + \frac{1}{M}\right)^2 - \frac{4\sin^2 ka}{mM}\right]^{1/2} \qquad (6.6)$$

Plotting the dispersion relation ω versus k for the positive sign and $m > M$ gives the upper curve of Figure 6.8 with

$$\omega^2 = \frac{2T}{a}\left(\frac{1}{m} + \frac{1}{M}\right) \quad \text{for} \quad k = 0 \qquad (6.7)$$

and

$$\omega^2 = \frac{2T}{aM} \quad \text{for} \quad k_m = \frac{\pi}{2a} \text{ (minimum } \lambda = 4a)$$

The negative sign in equation 6.6 gives the lower curve of Figure 6.8 with

$$\omega^2 = \frac{2Tk^2a^2}{a(M+m)} \quad \text{for very small } k$$

and

$$\omega^2 = \frac{2T}{am} \quad \text{for} \quad k = \frac{\pi}{2a}$$

The upper curve is called the 'optical' branch and the lower curve is known as the 'acoustical' branch. The motions of the two types of atom for each branch are shown in Figure 6.9.

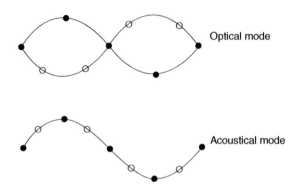

Figure 6.9   *The displacements of the different atomic species in the two modes of transverse oscillations in a crystal structure: (a) the optical mode and (b) the acoustic mode. The black and clear circles represent ions of opposite charge.*

In the optical branch for long wavelengths and small k, $A_m/A_M = -M/m$, and the atoms vibrate against each other, so that the centre of mass of the unit cell in the crystal remains fixed. This motion can be generated by the action of an electromagnetic wave when alternate atoms are ions of opposite charge; hence the name 'optical branch'. In the acoustic branch, long wavelengths and small k give $A_m = A_M$, and the atoms and their centre of mass move together (as in longitudinal sound waves). We shall see in the next chapter that the atoms may also vibrate in a longitudinal wave.

The transverse waves we have just discussed are polarized in one plane; they may also vibrate in a plane perpendicular to the plane considered here. The vibrational energy of these two transverse waves, together with that of the longitudinal wave to be discussed in the next chapter, form the basis of the theory of the specific heats of solids.

Absorption of Infrared Radiation by Ionic Crystals

Radiation of frequency 3×10^{12} Hz. gives an infrared wavelength of 100 µm $(10^{-4}$ m$)$ and a wave number $k = 2\pi/\lambda \approx 6.10^4$ m^{-1}. We found the cut-off frequency in the crystal lattice to give a wave number $k_m \approx 10^{10}$ m^{-1}, so that the k value of infrared radiation is a negligible quantity relative to k_m and may be taken as zero. When the ions of opposite charge $\pm e$ move under the influence of the electric field vector $E = E_0\,e^{i\omega t}$ of electromagnetic radiation, with $k = 0$ the appropriate frequency of their vibration becomes the low k limit of the optical branch.

$$\omega_0^2 = \frac{2T}{a}\left(\frac{1}{m} + \frac{1}{M}\right) \tag{6.8}$$

Worked Example

A sodium chloride crystal has a sodium ion Na of mass $23 \times 1.66 \times 10^{-27}$ kg and a chloride ion of $35 \times 1.66 \times 10^{-27}$ kg. The value of $T/a = 15$ N $\cdot$ m^{-1}. At what frequency will it absorb electromagnetic radiation?

Solution

Using these values in

$$\omega_0^2 = \frac{2T}{a}\left(\frac{1}{m} + \frac{1}{M}\right)$$

gives a value of ν which when converted to $\lambda = c/\nu$ gives a wavelength $\lambda = 66 \times 10^{-6}$ m. Experimentally sodium chloride is found to absorb strongly at $\lambda = 61 \times 10^{-6}$ m.

6.6 Transverse Waves in Periodic Structures (2) The Diffusion Equation, Energy Loss from Wave Systems

Energy loss from waves is not limited to periodic structures but the transmission line units of the worked example at the end of section 4.4 Coupled Oscillations of a Loaded String, and Problem 4.15 form a particularly revealing model. There we saw that in the voltage wave equation

$$\frac{\partial^2 V}{\partial x^2} = \frac{1}{c^2}\frac{\partial^2 V}{\partial t^2}$$

the coefficient $1/c^2$, where c is the wave velocity, depends only on L_0 and C_0, the energy storing parameters. The equation has constant sine or cosine solutions and the wave energy suffers no loss. This ideal situation changes when loss mechanisms are involved. These arise from particle collisions in the medium causing loss of mass (diffusion), momentum (friction or viscosity) and energy (thermal conductivity). All three are non-equilibrium irreversible thermodynamic processes which are unidirectional in the sense that the equation which describes them has a solution with a decaying exponential which is always directed to the equilibrium position. This equation, the *diffusion equation*, is written

$$\frac{\partial^2 V}{\partial x^2} = \frac{1}{d}\frac{\partial V}{\partial t}$$

where d the diffusivity is always the denominator of the right-hand coefficient (like c^2 in the wave equation). Where the right-hand term $\partial^2 V/\partial t^2$ in the wave equation is an acceleration, the $\partial V/\partial t$ in the diffusion equation is the velocity with which the energy V is moving towards equilibrium, often towards zero. This velocity is the gradient of the exponential so the greater the distance from equilibrium the greater the rate of change.

We see from the dimensions of the diffusion equation

$$\frac{V}{L^2} = \frac{1}{d}\frac{V}{t}$$

that the diffusivity d has the dimension L^2/t and this is interpreted as a velocity $L/t^{1/2}$ (not L/t) or as a measure of the distance L travelled in time $t^{1/2}$ when particles collide with each other. A well-known example is Einstein's relation for Brownian motion. This incoherent behaviour is known as Random or Drunk Man's Walk. The details of this are outlined in Appendix 4 but the principle is as follows. A very drunk man clings to a lamp post before setting out for home. He lurches 100 steps each of length 1 metre with no relation between the directions of consecutive steps. The question is, after 100 steps how far is he from the lamp post? The answer is, statistically, somewhere near or on the circumference of a circle of radius 10 metres centred on the lamp post. In random processes after n steps (or events) the distance from the origin is $\sqrt{n}$ and not n times the length of each step. This is a well-established rule for non-coherent processes.

Worked Example

A localized magnetic field H in an electrically conductive medium of permeability μ and conductivity σ will diffuse through the medium at a rate given by

$$\frac{\partial^2 H}{\partial x^2} = \mu\sigma \frac{\partial H}{\partial t} = \frac{1}{d}\frac{\partial H}{\partial t}$$

where d is the magnetic diffusivity $(\mu\sigma)^{-1}$. Show that the time of decay of the field in the x direction is approximately $L^2\mu\sigma$ where L is the extent of the medium and show that for a copper sphere of radius 1 metre this time $t < 100$ sec. $\mu(\text{copper}) = 1.26 \times 10^{-6}\,\mathrm{H \cdot m^{-1}}$; $\sigma(\text{copper}) = 5.8 \times 10^{7}\,\mathrm{s\,m^{-1}}$.

Solution

$$\mu\sigma \text{ has dimensions } \frac{t}{L^2} \text{ from } \frac{\partial^2 H}{\partial x^2} = \mu\sigma \frac{\partial H}{\partial t}$$
$$\therefore\ t = L^2\mu\sigma = 1.26 \times 5.8 \times 10 \approx 73\,\text{seconds}$$

If the earth's core were molten iron its field would freely decay in about 15×10^3 years. In the sun the local field would take 10^{10} years to decay.

When σ is very high the local field will change only by the movement of the medium – the field lines are 'frozen' into the medium and they stretch to oppose the motion.

The Wave Equation with Diffusion Effects

We can rarely find waves which propagate free from the energy-loss mechanisms we have been discussing – the exception being electromagnetic waves in regions of free space. If we try to solve the equation combining wave and diffusion effects

$$\frac{\partial^2 \phi}{\partial x^2} = \frac{1}{c^2}\frac{\partial^2 \phi}{\partial t^2} + \frac{1}{d}\frac{\partial \phi}{\partial t}$$

we shall not obtain a pure sine or cosine solution. Let us try the solution

$$\phi = \phi_m\, \mathrm{e}^{\mathrm{i}(\omega t - \gamma x)}$$

where ϕ_m is the maximum amplitude. This gives

$$\mathrm{i}^2\gamma^2 = \mathrm{i}^2\frac{\omega^2}{c^2} + \mathrm{i}\frac{\omega}{d}$$

or

$$\gamma^2 = \frac{\omega^2}{c^2} - \mathrm{i}\frac{\omega}{d}$$

giving a complex value for γ. But $\omega^2/c^2 = k^2$ where k is the wave number and if we put $\gamma = k - \mathrm{i}\alpha$ we obtain

$$\gamma^2 = k^2 - 2\mathrm{i}k\alpha - \alpha^2.$$

There are two possibilities: $k \gg \alpha$ or $\alpha \gg k$.

First of all we consider $k \gg \alpha$. The solution for ϕ then becomes

$$\phi = \phi_m \, \mathrm{e}^{\mathrm{i}(\omega t - \gamma x)} = \phi_m \, \mathrm{e}^{-\alpha x} \mathrm{e}^{\mathrm{i}(\omega t - kx)}$$

i.e. a sine or cosine solution of maximum amplitude ϕ_m which decays exponentially with distance. The physical significance of $k = 2\pi/\lambda \gg \alpha$ is that many wavelengths λ are contained in the distance $x = 1/\alpha$ before the amplitude decays to $\phi_m \, \mathrm{e}^{-1}$. Energy decays as the square of the amplitude, that is, as $\mathrm{e}^{-2\alpha x}$. This expression is familiar to us from from Chapter 2 and suggests that an attenuating wave is a travelling damped simple harmonic oscillator.

When $\alpha \gg k$,

$$\phi = \phi_m \, \mathrm{e}^{-\alpha x} \mathrm{e}^{\mathrm{i}(\omega t - kx)}$$

where the wave term is quickly extinguished by a rapidly decaying exponential term. In fact, there are no oscillations. We have a dead beat condition. So we can identify lightly damped wave attenuation with k and heavily damped diffusion behaviour with α. We shall discuss the reasons for this in the next section.

Energy Loss on a Transmission Line

Let us redraw a unit of the transmission line, Figure 6.10, to include a small resistance R_0 per unit length. Using the worked example at the end of section 4.4 Coupled Oscillations of a Loaded String, and Problem 4.15, we have

$$\frac{\partial V}{\partial x} = -L_0 \frac{\partial I}{\partial t} \gg R_0 I \mathrm{d}x \tag{6.9a}$$

$$\frac{\partial I}{\partial x} = -C_0 \frac{\partial V}{\partial t} - \frac{V}{R_0} \tag{6.9b}$$

Now

$$\frac{\partial^2 I}{\partial x \partial t} = \frac{\partial^2 I}{\partial t \partial x}$$

so applying this to equations 6.9 a and b

$$\frac{\partial^2 V}{\partial x^2} = -L_0 \frac{\partial^2 I}{\partial x \partial t} \tag{6.10a}$$

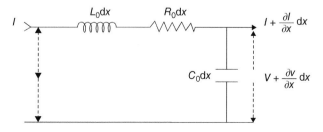

Figure 6.10 *Real transmission line unit with small R_0.*

and

$$\frac{\partial^2 I}{\partial t \partial x} = -C_0 \frac{\partial^2 V}{\partial t^2} - \frac{1}{R_0} \frac{\partial V}{\partial t} \qquad (6.10b)$$

then equations 6.10a and 6.10b yield

$$\boxed{\frac{\partial^2 V}{\partial x^2} = L_0 C_0 \frac{\partial^2 V}{\partial t^2} + \frac{L_0}{R_0} \frac{\partial V}{\partial t}} \qquad (6.11)$$

This is a most important equation. When V is replaced by E or H it describes the behaviour of electromagnetic waves in a medium having permeability μ, permittivity ϵ and conductivity σ replacing L_0, C_0 and $1/R_0$ respectively. It displays the important characteristic of electromagnetic waves, proposed by Maxwell, that the change of voltage (or E or H) is due to the change of *two* different currents, one related to $L_0 C_0 (\mu \epsilon)$ which conserves energy and one involving $R_0 (1/\sigma)$ which dissipates energy, the ohmic current.

The diffusion coefficient

$$\frac{L_0}{R_0} \equiv \mu \sigma \equiv \frac{t}{L^2}$$

so

$$\frac{1}{L^2} = \frac{L_0}{R_0 t} \approx \frac{\omega L_0}{R_0}$$

At very high electromagnetic frequencies $\omega L_0 \gg R_0$ so L^2, the region of heavy damping, is very short, much less than one wavelength (see Chapter 9). At the voltage and current wave frequencies on transmission lines R_0 and L_0 are precisely manufactured to produce light damping (see Chapter 8).

Problem 6.1.   A wave group consists of two wavelengths λ and $\lambda + \Delta\lambda$ where $\Delta\lambda/\lambda$ is very small.
    Show that the number of wavelengths λ contained between two successive zeros of the modulating envelope is $\approx \lambda/\Delta\lambda$.

Problem 6.2.   The phase velocity v of transverse waves in a crystal of atomic separation a is given by

$$v = c \left(\frac{\sin(ka/2)}{(ka/2)} \right)$$

where k is the wave number and c is constant. Show that the value of the group velocity is

$$c \cos \frac{ka}{2}$$

What is the limiting value of the group velocity for long wavelengths?

Problem 6.3.   The dielectric constant of a gas at a wavelength λ is given by

$$\varepsilon_r = \frac{c^2}{v^2} = A + \frac{B}{\lambda^2} - D\lambda^2$$

where A, B and D are constants, c is the velocity of light in free space and v is its phase velocity. If the group velocity is V_g show that

$$V_g \varepsilon_r = v(A - 2D\lambda^2)$$

Problem 6.4. Problem 3.10 shows that the relative permittivity of an ionized gas is given by

$$\varepsilon_r = \frac{c^2}{v^2} = 1 - \left(\frac{\omega_e}{\omega}\right)^2$$

where v is the phase velocity, c is the velocity of light and ω_e is the constant value of the electron plasma frequency. Show that this yields the dispersion relation $\omega^2 = \omega_e^2 + c^2 k^2$, and that as $\omega \rightarrow \omega_e$ the phase velocity exceeds that of light, c, but that the group velocity (the velocity of energy transmission) is always less than c.

Problem 6.5. The electron plasma frequency of Problem 2.8 is given by

$$\omega_e^2 = \frac{n_e e^2}{m_e \varepsilon_0}.$$

Show that for an electron number density $n_e \sim 10^{20}$ (10^{-5} of an atmosphere), electromagnetic waves must have wavelengths $\lambda < 3 \times 10^{-3}$m (in the microwave region) to propagate. These are typical wavelengths for probing thermonuclear plasmas at high temperatures.

$$\varepsilon_0 = 8.8 \times 10^{-12} \mathrm{F\ m^{-1}}$$
$$m_e = 9.1 \times 10^{-31} \mathrm{kg}$$
$$e = 1.6 \times 10^{-19} \mathrm{C}$$

Problem 6.6. In relativistic wave mechanics the dispersion relation for an electron of velocity $v = \hbar k/m$ is given by $\omega^2/c^2 = k^2 + m^2 c^2/\hbar^2$, where c is the velocity of light, m is the electron mass (considered constant at a given velocity) $\hbar = h/2\pi$ and h is Planck's constant. Show that the product of the group and particle velocities is c^2.

Problem 6.7. An electron (mass 9.1×10^{-31} kg) accelerated through 1 volt has a kinetic energy of 1.6×10^{19} J and a momentum of 5.4×10^{-25} kg$\cdot$m$\cdot$s^{-1}. Show that its de Broglie wavelength is 1.2 nm.

Problem 6.8. The figure shows a pulse of length Δt given by $y = A \cos \omega_0 t$.
  Show that the frequency representation

$$y(\omega) = a \cos \omega_1 t + a \cos (\omega_1 + \delta\omega)t \cdots + a \cos [\omega_1 + (n-1)(\delta\omega)]t$$

is centred on the average frequency ω_0 and that the range of frequencies making significant contributions to the pulse satisfy the criterion

$$\Delta\omega \, \Delta t \approx 2\pi$$

Repeat this process for a pulse of length Δx with $y = A \cos k_0 x$ to show that in k space the pulse is centred at k_0 with the significant range of wave numbers Δk satisfying the criterion $\Delta x \, \Delta k \approx 2\pi$.

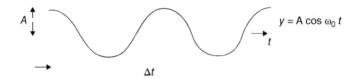

Problem 6.9.   The elastic force constant for an ionic crystal is $\sim$15 N m^{-1}. Show that the experimental value of 71×10^{-6} m for the wavelength of infrared absorption in KCl is in reasonable agreement with the calculated value.

$$1 \text{ a.m.u.} = 1.66 \times 10^{-27} \text{kg}$$
$$\text{K mass} = 39 \text{ a.m.u.}$$
$$\text{Cl mass} = 35 \text{ a.m.u.}$$

Problem 6.10.   Light near the blue end of the visible spectrum has a wavelength $\lambda \approx 4 \times 10^{-10}$ m. The last section of this chapter showed that the diffusion length in copper is given by

$$\frac{1}{L^2} = \omega\mu\sigma$$

If the light strikes a copper sheet, show that the wave is extinguished in a distance $L = 1/10$ of λ (blue). μ(copper) $= 1.26 \times 10^{-6}$ H $\cdot$ m^{-1}. σ(copper) $= 5.8 \times 10^{7}$ s $\cdot$ m^{-1}.

7

Longitudinal Waves

Introduction

In transverse waves the particles in the medium are displaced in a direction perpendicular to the wave direction. In longitudinal waves the particle displacements and wave directions are parallel. Sound waves in gases are the most common longitudinal waves. Fluids cannot sustain the transverse shear necessary for transverse waves. Solids can sustain both longitudinal and transverse waves because compressions can generate lateral distortions which may maintain transverse forces. One result of this is that earthquakes produce both transverse and longitudinal waves which radiate from the epicentre in three dimensions. The longitudinal waves are deeper and faster than the transverse waves which follow as aftershocks. Waves in deep and shallow water are compared and this chapter ends with the Doppler effect.

7.1 Sound Waves in Gases

Let us consider a fixed mass of gas, which at a pressure P_0 occupies a volume V_0 with a density ρ_0. These values define the equilibrium state of the gas which is disturbed, or deformed, by the compressions and rarefactions of the sound waves. Under the influence of the sound waves

$$\text{the pressure } P_0 \text{ becomes } P = P_0 + p$$
$$\text{the volume } V_0 \text{ becomes } V = V_0 + v$$

and

$$\text{the density } \rho_0 \text{ becomes } \rho = \rho_0 + \rho_d.$$

The excess pressure p_m is the maximum pressure amplitude of the sound wave and p is an alternating component superimposed on the equilibrium gas pressure P_0.

Introduction to Vibrations and Waves, First Edition. H. J. Pain and Patricia Rankin.
© 2015 John Wiley & Sons, Ltd. Published 2015 by John Wiley & Sons, Ltd.
Companion website: http://booksupport.wiley.com

The fractional change in volume is called the dilatation, written $v/V_0 = \delta$, and the fractional change of density is called the condensation, written $\rho_d/\rho_0 = s$. The values of δ and s are $\approx 10^{-3}$ for ordinary sound waves, and a value of $p_m = 2 \times 10^{-5} \mathrm{N \, m}^{-2}$ (about 10^{-10} of an atmosphere) gives a sound wave which is still audible at 1000 Hz. Thus, the changes in the medium due to sound waves are of an extremely small order and define limitations within which the wave equation is appropriate.

The fixed mass of gas is equal to

$$\rho_0 V_0 = \rho V = \rho_0 V_0 (1 + \delta)(1 + s)$$

so that $(1 + \delta)(1 + s) = 1$, giving $s = -\delta$ to a very close approximation. The elastic property of the gas, a measure of its compressibility, is defined in terms of its bulk modulus

$$B = -\frac{\mathrm{d}P}{\mathrm{d}V/V} = -V\frac{\mathrm{d}P}{\mathrm{d}V}$$

the difference in pressure for a fractional change in volume, a volume increase with fall in pressure giving the negative sign. The value of B depends on whether the changes in the gas arising from the wave motion are adiabatic or isothermal. They must be thermodynamically reversible in order to avoid the energy loss mechanisms of diffusion, viscosity and thermal conductivity. The complete absence of these random, entropy-generating processes defines an adiabatic process, a thermodynamic cycle with a 100% efficiency in the sense that none of the energy in the wave, potential or kinetic, is lost. In a sound wave such thermodynamic concepts restrict the excess pressure amplitude; too great an amplitude raises the local temperature in the gas at the amplitude peaks and thermal conductivity removes energy from the wave system. Local particle velocity gradients will also develop, leading to diffusion and viscosity.

Using a constant value of the adiabatic bulk modulus limits sound waves to small oscillations since the total pressure $P = P_0 + p$ is taken as constant; larger amplitudes lead to non-linear effects and shock waves, which we shall discuss separately in Chapter 14.

All adiabatic changes in the gas obey the relation $PV^\gamma = \mathrm{constant}$, where γ is the ratio of the specific heats at constant pressure and volume, respectively.

Differentiation gives

$$V^\gamma \mathrm{d}P + \gamma PV^{\gamma-1}\mathrm{d}V = 0$$

or

$$-V\frac{\mathrm{d}P}{\mathrm{d}V} = \gamma P = B_a \text{ (where the subscript } a \text{ denotes adiabatic)}$$

so that the elastic property of the gas is γP, considered to be constant. Since $P = P_0 + p$, then $\mathrm{d}P = p$, the excess pressure, giving

$$B_a = -\frac{p}{v/V_0} \quad \text{or} \quad p = -B_a\delta = B_a s$$

In a sound wave the particle displacements and velocities are along the x axis and we choose the coordinate η to define the displacement where $\eta(x, t)$.

In obtaining the wave equation we consider the motion of an element of the gas of thickness Δx and unit cross section. Under the influence of the sound wave the behaviour of this element is shown in Figure 7.1. The particles in the layer x are displaced a distance η and those at $x + \Delta x$ are displaced a

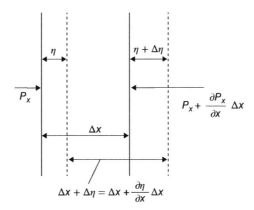

Figure 7.1 *Thin element of gas of unit cross-section and thickness Δx displaced an amount η and expanded by an amount $(\delta\eta/\partial x)\Delta x$ under the influence of a pressure difference $-(\partial P_x/\partial x)\Delta x$.*

distance $\eta + \Delta\eta$, so that the increase in the thickness Δx of the element of unit cross section (which therefore measures the increase in volume) is

$$\Delta\eta = \frac{\partial\eta}{\partial x}\Delta x$$

and

$$\delta = \frac{v}{V_0} = \left(\frac{\partial\eta}{\partial x}\right)\Delta x/\Delta x = \frac{\partial\eta}{\partial x} = -s$$

where $\partial\eta/\delta x$ is called the strain.

The medium is deformed because the pressures along the x axis on either side of the thin element are not in balance (Figure 7.1). The net force acting on the element is given by

$$P_x - P_{x+\Delta x} = \left[P_x - \left(P_x + \frac{\partial P_x}{\partial x}\Delta x\right)\right]$$

$$= -\frac{\partial P_x}{\partial x}\Delta x = -\frac{\partial}{\partial x}(P_0 + p)\Delta x = -\frac{\partial p}{\partial x}\Delta x$$

The mass of the element is $\rho_0\Delta x$ and its acceleration is given, to a close approxmation, by $\partial^2\eta/dt^2$.

From Newton's Law we have

$$-\frac{\partial p}{\partial x}\Delta x = \rho_0\Delta x\frac{\partial^2\eta}{\partial t^2}$$

where

$$p = -B_a\delta = -B_a\frac{\partial\eta}{\partial x}$$

so that

$$-\frac{\partial p}{\partial x} = B_a \frac{\partial^2 \eta}{\partial x^2}, \quad \text{giving} \quad B_a \frac{\partial^2 \eta}{\partial x^2} = \rho_0 \frac{\partial^2 \eta}{\partial t^2}$$

But $B_a/\rho_0 = \gamma P/\rho_0$ is the ratio of the elasticity to the inertia or density of the gas, and this ratio has the dimensions

$$\frac{\text{force}}{\text{area}} \cdot \frac{\text{velocity}}{\text{mass}} = (\text{velocity})^2, \quad \text{so} \quad \frac{\gamma P}{\rho_0} = c^2$$

where c is the sound wave velocity.
   Thus

$$\frac{\partial^2 \eta}{\partial x^2} = \frac{1}{c^2} \frac{\partial^2 \eta}{\partial t^2}$$

is the wave equation. Writing η_m as the maximum amplitude of displacement we have the following expressions for a wave in the positive x direction:

$$\eta = \eta_m e^{i(\omega t - kx)} \quad \dot{\eta} = \frac{\partial \eta}{\partial t} = i\omega\eta$$

$$\delta = \frac{\partial \eta}{\partial x} = -ik\eta = -s \quad (\text{so } s = ik\eta)$$

$$p = B_a s = iB_a k\eta$$

The phase relationships between these parameters (Figure 7.2a) show that when the wave is in the positive x direction, the excess pressure p, the fractional density increase s and the particle velocity $\dot{\eta}$ are all $\pi/2$ rad in phase ahead of the displacement η, whilst the volume change (π rad out of phase with the density

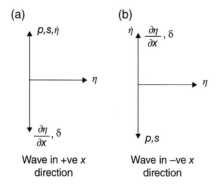

(a)   (b)

$p,s,\dot{\eta}$ $\dot{\eta}$ $\frac{\partial \eta}{\partial x}, \delta$

η η

$\frac{\partial \eta}{\partial x}, \delta$ p,s

Wave in +ve x Wave in −ve x
direction   direction

Figure 7.2   *Phase relationships between the particle displacement η, particle velocity $\dot{\eta}$, excess pressure p and condensation $s = -\delta$ (the dilatation) for waves travelling in the positive and negative x directions. The displacement η is taken in the positive x direction for both waves.*

change) is $\pi/2$ rad behind the displacement. These relationships no longer hold when the wave direction is reversed (Figure 7.2b); for a wave in the negative x direction

$$\eta = \eta_m\, e^{i(\omega t + kx)} \quad \dot{\eta} = \frac{\partial \eta}{\partial t} = i\omega\eta$$

$$\delta = \frac{\partial \eta}{\partial x} = +ik\eta = -s \quad (\text{so } s = -ik\eta)$$

$$p = B_a s = -iB_a k\eta$$

In both waves the particle displacement η is measured in the positive x direction and the thin element Δx of the gas oscillates about the value $\eta = 0$, which defines its central position. For a wave in the positive x direction the value $\eta = 0$, with $\dot{\eta}$ a maximum in the positive x direction, gives a maximum positive excess pressure (compression) with a maximum condensation s_m (maximum density) and a minimum volume. For a wave in the negative x direction, the same value $\eta = 0$, with $\dot{\eta}$ a maximum in the positive x direction, gives a maximum negative excess pressure (rarefaction), a maximum volume and a minimum density. To produce a compression in a wave moving in the negative x direction the particle velocity $\dot{\eta}$ must be a maximum in the negative x direction at $\eta = 0$. This distinction is significant when we are defining the impedance of the medium to the waves. A change of sign is involved with a change of direction – a convention we shall also have to follow when discussing the waves of Chapters 8 and 9.

7.2 Energy Distribution in Sound Waves

The kinetic energy in the sound wave is found by considering the motion of the individual gas elements of thickness Δx.

Each element will have a kinetic energy per unit cross section

$$\Delta E_{\text{kin}} = \frac{1}{2}\rho_0 \Delta x \dot{\eta}^2$$

where $\dot{\eta}$ will depend upon the position x of the element. The average value of the kinetic energy density is found by taking the value of $\dot{\eta}^2$ averaged over a region of n wavelengths.

Now

$$\dot{\eta} = \dot{\eta}_m \sin \frac{2\pi}{\lambda}(ct - x)$$

so that

$$\overline{\dot{\eta}^2} = \frac{\dot{\eta}_m^2 \int_0^{n\lambda} \sin^2 2\pi(ct-x)/\lambda \Delta x}{n\lambda} = \frac{1}{2}\dot{\eta}_m^2$$

so that the average kinetic energy density in the medium is

$$\overline{\Delta E}_{\text{kin}} = \frac{1}{4}\rho_0 \dot{\eta}_m^2 = \frac{1}{4}\rho_0 \omega^2 \eta_m^2$$

(a simple harmonic oscillator of maximum amplitude a has an average kinetic energy over one cycle of $\frac{1}{4}m\omega^2 a^2$).

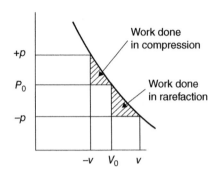

Figure 7.3 *Shaded triangles show that potential energy $\frac{pv}{2} = \frac{p_m v_m}{4}$ gained by gas in compression equals that gained in rarefaction when both p and v change sign.*

The potential energy density is found by considering the work $P\,dV$ done on the fixed mass of gas of volume V_0 during the adiabatic changes in the sound wave. This work is expressed for the complete cycle as

$$\Delta E_{\text{pot}} = -\int P\,dV = -\frac{-1}{2\pi}\int_0^{2\pi} pv\,d(\omega t) = \frac{p_m v_m}{2} : \left[\frac{p}{p_m} = \frac{-v}{v_m} = \sin(\omega t - kx)\right]$$

The negative sign shows that the potential energy change is positive in both a compression (p positive, dV negative) and a rarefaction (p negative, dV positive) (Figure 7.3).

The condensation

$$s = \frac{-\int dv}{V_0} = \frac{-v}{V_0} = -\delta$$

we write

$$\frac{s}{s_m} = \frac{-\delta}{\delta_m} = \sin(\omega t - kx) \quad \text{and} \quad -v = V_0 s$$

which, with

$$p = B_a s$$

gives

$$\Delta E_{\text{pot}} = \frac{-1}{2\pi}\int_0^{2\pi} pv\,d(\omega t) = \frac{B_a V_0}{2\pi}\int_0^{2\pi} s^2 d(\omega t)$$

where $s = -\delta$ and the thickness Δx of the element of unit cross section represents its volume V_0.

Now

$$\eta = \eta_m e^{i(\omega t \pm kx)}$$

so that

$$\delta = \frac{\partial \eta}{\partial x} = \pm \frac{1}{c}\frac{\partial \eta}{\partial t}, \quad \text{where} \quad c = \frac{\omega}{k}$$

Thus

$$\Delta E_{\text{pot}} = \frac{1}{2}\frac{B_a}{c^2}\dot{\eta}^2 \Delta x = \frac{1}{2}\rho_0 \dot{\eta}^2 \Delta x$$

and its average value over $n\lambda$ gives the potential energy density

$$\overline{\Delta E}_{\text{pot}} = \frac{1}{4}\rho_0 \dot{\eta}_m^2 = \overline{\Delta E}_{\text{kin}}$$

We see that the average values of the kinetic and potential energy density in the sound wave are equal, but more important, since the value of each for the element Δx is $\frac{1}{2}\rho_0 \dot{\eta}^2 \Delta x$, we observe that the element possesses maximum (or minimum) potential and kinetic energy at the same time. A compression or rarefaction produces a maximum in the energy of the element since the value $\dot{\eta}$ governs the energy content. Thus, the energy in the wave is distributed in the wave system with distance as shown in Figure 7.4. Note that this distribution is non-uniform with distance similar to that for a transverse wave.

Worked Example

Show that in a gas at temperature T the average molecular thermal velocity is approximately equal to the velocity of sound.

Molecular energy:

$$\frac{1}{2}mv^2 = \frac{3}{2}kT \quad (\frac{1}{2}kT \text{ for each dimension}).$$

For a mole volume V, N (Avogadro number) and mass M we have

$$c^2 = \frac{\gamma P}{\rho} \quad \text{i.e.} \quad \frac{\gamma PV}{M}(\text{per mole}) = \frac{\gamma RT}{M} = \gamma NkT = Mc^2$$

$$\therefore c^2 \text{ per particle} = \gamma kT = \frac{5}{3}kT$$

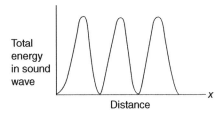

Total energy in sound wave

Distance

Figure 7.4   *Energy distribution in space for a sound wave in a gas. Both potential and kinetic energies are at a maximum when the particle velocity $\dot{\eta}$ is a maximum and zero at $\dot{\eta} = 0$.*

7.3   Intensity of Sound Waves

This is a measure of the energy flux, the rate at which energy crosses unit area, so that it is the product of the energy density (kinetic plus potential) and the wave velocity c. Normal sound waves range in intensity between 10^{-12} and 1 W m^{-2}, extremely low levels which testify to the sensitivity of the ear. The roar of a large football crowd greeting a goal will just about heat a cup of coffee.

The intensity may be written

$$I = \frac{1}{2}\rho_0 c \dot{\eta}_m^2 = \frac{1}{2}\rho_0 c \omega^2 \eta_m^2 = \rho_0 c \dot{\eta}_{\mathrm{rms}}^2 = p_{\mathrm{rms}}^2 / \rho_0 c = p_{\mathrm{rms}} \dot{\eta}_{\mathrm{rms}}$$

A commonly used standard of sound intensity is given by

$$I_0 = 10^{-2} \text{W m}^{-2}$$

which is about the level of the average conversational tone between two people standing next to each other. Shouting at this range raises the intensity by a factor of 100 and in the range $100\, I_0$ to $1000\, I_0$ (10 W m^{-2}) the sound is painful.

Whenever the sound intensity increases by a factor of 10 it is said to have increased by 1 B so the dynamic range of the ear is about 12 B. An intensity increase by a factor of

$$10^{0.1} = 1 \cdot 26$$

increases the intensity by 1 dB, a change of loudness which is just detected by a person with good hearing. dB is a decibel.

We see that the product $\rho_0 c$ appears in most of the expressions for the intensity; its significance becomes apparent when we define the impedance of the medium to the waves as the

$$\textit{Specific Acoustic Impedance} = \frac{\text{excess pressure}}{\text{particle velocity}} = \frac{p}{\dot{\eta}} = Z$$

(the ratio of a force per unit area to a velocity).

Now, for a wave in the positive x direction.

$$p = B_a s = iB_a k\eta \quad \text{and} \quad \dot{\eta} = i\omega\eta$$

so that,

$$\frac{p}{\dot{\eta}} = \frac{B_a k}{\omega} = \frac{B_a}{c} = \rho_0 c = Z_+$$

Thus, the acoustic impedance presented by the medium to these waves, as in the case of the transverse waves on the string, is given by the product of the density and the wave velocity and is governed by the elasticity and inertia of the medium. For a wave in the negative x direction, the specific acoustic impedance

$$\frac{p}{\dot{\eta}} = -\frac{iB_a k\eta}{i\omega\eta} = -\rho_0 c = Z_-$$

with a change of sign because of the changed phase relationship.

The units of $\rho_0 c$ are normally stated as kg m^{-2}s^{-1} in books on practical acoustics; in these units air has a specific acoustic impedance value of 400, water a value of 1.45×10^6 and steel a value of 3.9×10^7. These values will become more significant when we use them later in examples on the reflection and transmission of sound waves.

Although the specific acoustic impedance $\rho_0 c$ is a real quantity for plane sound waves, it has an added reactive component ik/r for spherical waves, where r is the distance travelled by the wavefront. This component tends to zero with increasing r as the spherical wave becomes effectively plane.

Worked Example

The velocity of sound in air of density 1.29 kg $\cdot$ m^{-3} may be taken as 330 m $\cdot$ s^{-1}. Show that the acoustic pressure for sound of an intensity of 1 W $\cdot$ m$^{-2} \approx 3 \times 10^{-4}$ of an atmosphere.

Solution

$$I = \frac{p_{\text{rms}}^2}{\rho_0 c} = \frac{p^2}{2\rho_0 c}$$
$$\therefore p = (1.29 \times 330 \times 2)^{\frac{1}{2}} \approx 30\,\text{N} \cdot \text{m}^{-2}$$

1 Atmosphere $\approx 10^5$ N $\cdot$ m^{-2} $\therefore p \approx 3 \times 10^{-4}$ atmospheres.

7.4   Longitudinal Waves in a Solid

The velocity of longitudinal waves in a solid depends upon the dimensions of the specimen in which the waves are travelling. If the solid is a thin bar of finite cross section the analysis for longitudinal waves in a gas is equally valid, except that the bulk modulus B_a is replaced by Young's modulus Y, the ratio of the longitudinal stress in the bar to its longitudinal strain.

The wave equation is then

$$\frac{\partial^2 \eta}{\partial x^2} = \frac{1}{c^2}\frac{\partial^2 \eta}{\partial t^2}, \quad \text{with} \quad c^2 = \frac{Y}{\rho}$$

A longitudinal wave in a bulk medium compresses the medium and distorts it laterally. Because a solid can develop a shear force in any direction, such a lateral distortion is accompanied by a transverse shear. The effect of this upon the wave motion in solids of finite cross section is quite complicated and has been ignored in the very thin specimen above. In bulk solids, however, the longitudinal and transverse modes may be considered separately.

We have seen that the longitudinal compression produces a strain $\partial\eta/\partial x$; the accompanying lateral distortion produces a strain $\partial\beta/\partial y$ (of opposite sign to $\partial\eta/\partial x$ and perpendicular to the x direction).

Here β is the displacement in the y direction and is a function of both x and y. The ratio of these strains

$$-\frac{\partial\beta}{\partial y} \bigg/ \frac{\partial\eta}{\partial x} = \sigma$$

is known as Poisson's ratio and is expressed in terms of Lamé's elastic constants λ and μ for a solid as

$$\sigma = \frac{\lambda}{2(\lambda + \mu)} \quad \text{where} \quad \lambda = \frac{\sigma Y}{(1 + \sigma)(1 - 2\sigma)}$$

These constants are always positive, so that $\sigma < \frac{1}{2}$, and is commonly $\approx \frac{1}{3}$. In terms of these constants Young's modulus becomes

$$Y = (\lambda + 2\mu - 2\lambda\sigma)$$

The constant μ is the transverse coefficient of rigidity; that is, the ratio of the transverse stress to the transverse strain. It plays the role of the elasticity in the propagation of pure transverse waves in a bulk solid which Young's modulus plays for longitudinal waves in a thin specimen. Figure 7.5 illustrates the shear in a transverse plane wave, where the transverse strain is defined by $\partial\beta/\partial x$. The transverse stress at x is therefore $T_x = \mu\partial\beta/\partial x$. The equation of transverse motion of the thin element dx is then given by

$$T_{x+dx} - T_{dx} = \rho \, dx \ddot{y}$$

Where ρ is the density, or

$$\frac{\partial}{\partial x}\left(\mu\frac{\partial\beta}{\partial x}\right) = \rho\ddot{y}$$

but $\ddot{y} = \partial^2\beta/\partial t^2$, hence

$$\frac{\partial^2\beta}{\partial x^2} = \frac{\rho}{\mu}\frac{\partial^2\beta}{\partial t^2}$$

the wave equation with a velocity given by $c^2 = \mu/\rho$.

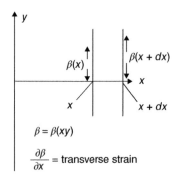

Figure 7.5   *Shear in a bulk solid producing a transverse wave. The transverse shear strain is $\partial\beta/\partial x$ and the transverse shear stress is $\mu\partial\beta/\partial x$, where μ is the shear modulus of rigidity.*

The effect of the transverse rigidity μ is to stiffen the solid and increase the elastic constant governing the propagation of longitudinal waves. In a bulk solid the velocity of these waves is no longer given by $c^2 = Y/\rho$, but becomes

$$c^2 = \frac{\lambda + 2\mu}{\rho}$$

Since Young's modulus $Y = \lambda + 2\mu - 2\lambda\sigma$, the elasticity is increased by the amount $2\lambda\sigma \approx \lambda$, so that longitudinal waves in a bulk solid have a higher velocity than the same waves along a thin specimen.

In an isotropic solid, where the velocity of propagation is the same in all directions, the concept of a bulk modulus, used in the discussion on waves in gases, holds equally well. Expressed in terms of Lamé's elastic constants the bulk modulus for a solid is written

$$B = \lambda + \frac{2}{3}\mu = Y[3(1 - 2\sigma)]^{-1}$$

the longitudinal wave velocity for a bulk solid becomes

$$c_L = \left(\frac{B + (4/3)\mu}{\rho} \right)^{1/2}$$

whilst the transverse velocity remains as

$$c_T = \left(\frac{\mu}{\rho} \right)^{1/2}$$

7.5 Application to Earthquakes

The values of these velocities are well known for seismic waves generated by earthquakes. Near the surface of the earth the longitudinal waves have a velocity of 8 km s^{-1} and the transverse waves travel at 4.45 km s^{-1}. The velocity of the longitudinal waves increases with depth until, at a depth of about 1800 miles, no waves are transmitted because of a discontinuity and severe mismatch of impedances associated with the fluid core.

At the surface of the earth the transverse wave velocity is affected by the fact that stress components directed through the surface are zero there and these waves, known as Rayleigh Waves, travel with a velocity given by

$$c = f(\sigma) \left(\frac{\mu}{\rho} \right)^{1/2}$$

where

$$f(\sigma) = 0.9194 \quad \text{when} \quad \sigma = 0{\cdot}25$$

and

$$f(\sigma) = 0.9553 \quad \text{when} \quad \sigma = 0{\cdot}5$$

The energy of the Rayleigh Waves is confined to two dimensions; their amplitude is often much higher than that of the three-dimensional longitudinal waves and therefore they are potentially more damaging.

In an earthquake the arrival of the fast longitudinal waves is followed by the Rayleigh Waves and then by a complicated pattern of reflected waves including those affected by the stratification of the earth's structure, known as Love Waves.

Worked Example

An earthquake is felt 6000 kilometres from its epicentre. The first shock is caused by a three-dimensional wave governed by Young's modulus with a velocity of 7.5 km·s^{-1}. It is followed by a two-dimensional transverse Rayleigh wave. If Poisson's ratio of the earth is 0.3 how much later does the aftershock arrive?

Solution

First shock arrives after $6000/7.5 \times 60 = 13.33$ minutes from epicentre.

$$\text{Poisson's ratio: } \sigma = 0.3 = \frac{\lambda}{2(\lambda + \mu)} \quad \therefore \frac{\lambda}{\mu} = \frac{3}{2}$$

$$\frac{Y}{\mu} = \frac{\lambda + 2\mu}{\mu} = \frac{7}{2} \quad \therefore \frac{c_Y}{c_\mu} = \left(\frac{7}{2}\right)^{\frac{1}{2}} = 1.88$$

time of arrival of aftershock given by

$$\frac{t_\mu}{t_Y} = 1.88 \times 13.33 = 25 \text{ mins}$$

$$\therefore \text{delay of aftershock} = 11 \text{ mins } 27 \text{ seconds}$$

7.6   Reflection and Transmission of Sound Waves at Boundaries

When a sound wave meets a boundary separating two media of different acoustic impedances two boundary conditions must be met in considering the reflection and transmission of the wave. They are that

(i) the particle velocity $\dot{\eta}$

and

(ii) the acoustic excess pressure p

are both continuous across the boundary. Physically this ensures that the two media are in complete contact everywhere across the boundary.

Figure 7.6 shows that we are considering a plane sound wave travelling in a medium of specific acoustic impedance $Z_1 = \rho_1 c_1$ and meeting, at normal incidence, an infinite plane boundary separating the first medium from another of specific acoustic impedance $Z_2 = \rho_2 c_2$. If the subscripts i, r and t denote incident, reflected and transmitted respectively, then the boundary conditions give

$$\dot{\eta}_i + \dot{\eta}_r = \dot{\eta}_t \tag{7.1}$$

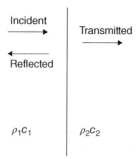

Figure 7.6   *Incident, reflected and transmitted sound waves at a plane boundary between media of specific acoustic impedances $\rho_1 c_1$ and $\rho_2 c_2$.*

and

$$p_i + p_r = p_t \tag{7.2}$$

For the incident wave $p_i = \rho_1 c_1 \dot{\eta}_i$ and for the reflected wave $p_r = -\rho_1 c_1 \dot{\eta}_r$, so equation 7.2 becomes

$$\rho_1 c_1 \dot{\eta}_i - \rho_1 c_1 \dot{\eta}_r = \rho_2 c_2 \dot{\eta}_t$$

or

$$Z_1 \dot{\eta}_i - Z_1 \dot{\eta}_r = Z_2 \dot{\eta}_t \tag{7.3}$$

Eliminating $\dot{\eta}_t$ from (7.1) and (7.3) gives

$$\frac{\dot{\eta}_r}{\dot{\eta}_i} = \frac{\omega \eta_r}{\omega \eta_i} = \frac{\eta_r}{\eta_i} = \frac{Z_1 - Z_2}{Z_1 + Z_2}$$

Eliminating $\dot{\eta}_r$ from (7.1) and (7.3) gives

$$\frac{\dot{\eta}_t}{\dot{\eta}_i} = \frac{\eta_t}{\eta_i} = \frac{2Z_1}{Z_1 + Z_2}$$

Now

$$\frac{p_r}{p_i} = -\frac{Z_1 \dot{\eta}_r}{Z_1 \dot{\eta}_i} = \frac{Z_2 - Z_1}{Z_1 + Z_2} = -\frac{\dot{\eta}_r}{\dot{\eta}_i}$$

and

$$\frac{p_t}{p_i} = \frac{Z_2 \dot{\eta}_t}{Z_1 \dot{\eta}_i} = \frac{2Z_2}{Z_1 + Z_2}$$

We see that if $Z_1 > Z_2$ the incident and reflected particle velocities are in phase, whilst the incident and reflected acoustic pressures are out of phase. The superposition of incident and reflected velocities which are in phase leads to a cancellation of pressure (a pressure node in a standing wave system). If $Z_1 < Z_2$ the pressures are in phase and the velocities are out of phase.

The transmitted particle velocity and acoustic pressure are always in phase with their incident counterparts.

At a rigid wall, where Z_2 is infinite, the velocity $\dot{\eta}_t = 0 = \dot{\eta}_i + \dot{\eta}_r$, which leads to a doubling of pressure at the boundary. (See Summary in Appendix 8.)

7.7   Reflection and Transmission of Sound Intensity

The intensity coefficients of reflection and transmission are given by

$$\frac{I_r}{I_i} = \frac{Z_1(\dot{\eta}_r^2)_{\text{rms}}}{Z_1(\dot{\eta}_i^2)_{\text{rms}}} = \left(\frac{Z_1 - Z_2}{Z_1 + Z_2}\right)^2$$

and

$$\frac{I_t}{I_i} = \frac{Z_2(\dot{\eta}_t^2)_{\text{rms}}}{Z_1(\dot{\eta}_i^2)_{\text{rms}}} = \frac{Z_2}{Z_1}\left(\frac{2Z_1}{Z_1 + Z_2}\right)^2 = \frac{4Z_1 Z_2}{(Z_1 + Z_2)^2}$$

The conservation of energy gives

$$\frac{I_r}{I_i} + \frac{I_t}{I_i} = 1 \quad \text{or} \quad I_i = I_t + I_r$$

Worked Example

Show that if waves travelling in water are normally incident on a plane water–ice interface 82.3% of the energy is transmitted.

(ρc values in $\text{kg·m}^{-2}\text{·s}^{-1}$) water $= 1.43 \times 10^6$ ice $= 3.49 \times 10^6$

Solution

$$\frac{I_t}{I_i} = \frac{4Z_1 Z_2}{(Z_1 + Z_2)^2} = \frac{4 \times 1.43 \times 3.49}{(1.43 + 3.49)^2} = 82.3\%$$

The great disparity between the specific acoustic impedance of air on the one hand and water or steel on the other leads to an extreme mismatch of impedances when the transmission of acoustic energy between these media is attempted.

There is an almost total reflection of sound wave energy at an air–water interface, independent of the side from which the wave approaches the boundary. Only 14% of acoustic energy can be transmitted at a steel–water interface, a limitation which has severe implications for underwater transmission and detection devices which rely on acoustics.

7.8   Water Waves

In deep water $hk \gg 1$ where h is the depth $\gg \lambda$ and k is the wave number the particle motion is circular in the vertical plane. This motion is forwards in the direction of the wave, i.e. clockwise for a right-going wave just below the crest, and backwards just beneath a trough. With increasing depth the circle diameter reduces and the particles become effectively stationary. There are no deep water waves. In shallow water $hk \ll 1$, the circular motion becomes elliptical, retaining a constant horizontal diameter but the vertical minor axis decreases with depth and flattens near the water bed. Waves near the surface of a non-viscous incompressible liquid of density ρ have a phase velocity given by

$$v^2(k) = \left(\frac{g}{k} + \frac{Tk}{\rho}\right) \tanh kh \qquad (7.5)$$

where T is the surface tension and g is the acceleration due to gravity. **For deep water, $hk \gg 1$, $\tanh kh = 1$. For shallow water $hk \ll 1$, $\tanh kh = k$.**

Worked Example

Show that when gravity and surface tension are equally important in deep water, the wave velocity is a minimum at $v^4 = 4gT/\rho$ at a 'critical' wavelength $\lambda_c = 2\pi(T/\rho g)^{\frac{1}{2}}$.

Solution

For deep water we have

$$v = \left(\frac{g}{k} + \frac{Tk}{\rho}\right)^{\frac{1}{2}}$$

and

$$\frac{dv}{dk} = \frac{1}{2}\left(\frac{g}{k} + \frac{Tk}{\rho}\right)^{-\frac{1}{2}}\left(-\frac{g}{k^2} + \frac{T}{\rho}\right) = 0$$

when $k^2 = g\rho/T = 4\pi^2/\lambda^2$

$$\therefore \lambda_c = 2\pi\sqrt{\frac{T}{\rho g}}$$

writing $g/k = a = Tk/\rho$ we have

$$v^2 = 2a = 2\sqrt{a^2} = 2\sqrt{\frac{g}{k}\frac{Tk}{\rho}}$$

$$\therefore v^2 = 2\sqrt{\frac{gT}{\rho}} \text{ giving } v^4 = \frac{4Tg}{\rho} \text{ with } k^2 = \frac{g\rho}{T} \text{ giving } \lambda_c.$$

Putting it in

$$v^2 = \left(g\sqrt{\frac{T}{\rho g}} + \frac{T}{\rho}\sqrt{\frac{\rho g}{T}}\right) = 2\sqrt{\frac{gT}{\rho}}$$

confirming

$$v_{min}^4 = \frac{4gT}{\rho}.$$

7.9 Doppler Effect

In the absence of dispersion the velocity of waves sent out by a moving source is constant but the wavelength and frequency noted by a stationary observer are altered.

In Figure 7.7 a stationary source S emits a signal of frequency ν and wavelength λ for a period t so the distance to a stationary observer O is $\nu\lambda t$. If the source S′ moves towards O at a velocity u during the period t then O registers a new frequency ν'.

We see that

$$\nu\lambda t = ut + \nu\lambda' t$$

which, for

$$c = \nu\lambda = \nu'\lambda'$$

gives

$$\frac{c-u}{\nu} = \lambda' = \frac{c}{\nu'}$$

Hence

$$\nu' = \frac{\nu c}{c-u}$$

This observed change of frequency is called the Doppler Effect.

Suppose that the source S is now stationary but that an observer O′ moves with a velocity v away from S. If we superimpose a velocity $-v$ on observer, source and waves, we bring the observer to rest;

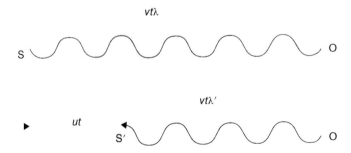

Figure 7.7 *If waves from a stationary source S are received by a stationary observer 0 at frequency ν and wavelength λ the frequency is observed as ν' and the wavelength as λ' at 0 if the source S′ moves during transmission. This is the Doppler effect.*

the source now has a velocity $-v$ and waves a velocity of $c - v$. Using these values in the expression for ν' gives a new observed frequency

$$\nu'' = \frac{\nu(c - v)}{c}$$

Worked Example

Light from a star of wavelength 6×10^{-7}m is found to be shifted 10^{-11}m towards the red when compared with the same wavelength from a laboratory source. If the velocity of light is 3×10^8 m s^{-1} show that the earth and the star are separating at a velocity of 5 km s^{-1}.

Solution

$$\nu' = \frac{\nu c}{c - u} = \frac{c}{\lambda'} = \frac{c^2}{\lambda(c - u)} \quad \therefore \lambda' = \frac{c - u}{c}\lambda. \text{ Red shift } \therefore \lambda' > \lambda$$

$$\Delta\lambda = \lambda' - \lambda = -\frac{u\lambda}{c} \quad \therefore u = -\frac{c\Delta\lambda}{\lambda} = \frac{3 \times 10^8 \times 10^{-11}}{6 \times 10^{-7}} = 5\,\text{Km}\cdot\text{s}^{-1}$$

Problem 7.1. Show that the displacement amplitude of an air molecule at a painful sound level of 10 W m^{-2} at 500 Hz $\approx 6.9 \times 10^{-5}$m.

Problem 7.2. Barely audible sound in air has an intensity of $10^{-10}I_0$. Show that the displacement amplitude of an air molecule for sound at this level at 500 Hz is $\approx 10^{-10}$m; that is, about the size of the molecular diameter.

Problem 7.3. Hi-fi equipment is played very loudly at an intensity of 100 I_0 in a small room of cross section 3 m $\times$ 3 m. Show that this audio output is about 10 W.

Problem 7.4. Two sound waves, one in water and one in air, have the same intensity. Show that the ratio of their pressure amplitudes (p_{water}/p_{air}) is about 60. When the pressure amplitudes are equal show that the intensity ratio is $\approx 3 \times 10^{-4}$.

Problem 7.5. A spring of mass m, stiffness s and length L is stretched to a length $L+l$. When longitudinal waves propagate along the spring the equation of motion of a length dx may be written

$$\rho\,\text{d}x\frac{\partial^2\eta}{\partial t^2} = \frac{\partial F}{\partial x}\text{d}x$$

where ρ is the mass per unit length of the spring, η is the longitudinal displacement and F is the restoring force. Derive the wave equation to show that the wave velocity v is given by

$$v^2 = s(L + l)/\rho$$

Problem 7.6. A solid has a Poisson's ratio $\sigma = 0.25$. Show that the ratio of the longitudinal wave velocity to the transverse wave velocity is $\sqrt{3}$. Use the values of these velocities given in the text to derive an appropriate value of σ for the earth.

Problem 7.7. Show that when sound waves are normally incident on a plane steel water interface 86% of the energy is reflected.

$$(\rho c \text{ values in kg m}^{-2}\text{s}^{-1})$$

$$\text{water} = 1.43 \times 10^6$$

$$\text{steel} = 3.9 \times 10^7$$

Problem 7.8. Use the boundary conditions for standing acoustic waves in a tube to confirm the following:

	Particle displacement		Pressure	
	closed end	open end	closed end	open end
Phase change on reflection	180°	0	0	180°
	node	antinode	antinode	node

Problem 7.9. Standing acoustic waves are formed in a tube of length l with (a) both ends open and (b) one end open and the other closed. If the particle displacement

$$\eta = (A \cos kx + B \sin kx) \sin \omega t$$

and the boundary conditions are as shown in the diagrams below, show that for

$$\text{(a)} \quad \eta = A \cos kx \sin \omega t \quad \text{with} \quad \lambda = 2l/n$$

and for

$$\text{(b)} \quad \eta = A \cos kx \sin \omega t \quad \text{with} \quad \lambda = 4l/(2n+1)$$

Sketch the first three harmonics for each case.

Problem 7.10. Some longitudinal waves in a plasma exhibit a combination of electrical and acoustical phenomena. They obey a dispersion relation at temperature T of $\omega^2 = \omega_e^2 + 3aTk^2$, where ω_e is the constant electron plasma frequency (See Problem 6.5) and the Boltzmann constant is written as a to avoid confusion with the wave number k. Show that the product of the phase and group velocities is related to the average thermal energy of an electron (found from $pV = RT$).

Problem 7.11. Waves near the surface of a non-viscous incompressible liquid of density ρ have a phase velocity given by

$$v^2(k) = \left[\frac{g}{k} + \frac{Tk}{\rho}\right] \tanh kh \tag{7.4}$$

where g is the acceleration due to gravity, T is the surface tension, k is the wave number and h is the liquid depth. When $h \ll \lambda$ the liquid is shallow; when $h \gg \lambda$ the liquid is deep.

(a) The condition $\lambda \gg \lambda_c$ defines a gravity wave, and surface tension is negligible. Show that gravity waves in a shallow liquid are non-dispersive with a velocity $v = \sqrt{gh}$.

(b) Show that gravity waves in a deep liquid have a phase velocity $v = \sqrt{g/k}$ and a group velocity of half this value.

(c) The condition $\lambda < \lambda_c$ defines a ripple (dominated by surface tension). Show that short ripples in a deep liquid have a phase velocity $v = \sqrt{Tk/\rho}$ and a group velocity of $\frac{3}{2}v$. (Note the anomalous dispersion.)

Problem 7.12. Show that, in the Doppler effect, the change of frequency noted by a stationary observer O as a moving source S′ passes him is given by

$$\Delta\nu = \frac{2\nu c u}{(c^2 - u^2)}$$

where $c = \nu\lambda$, the signal velocity and u is the velocity of S′.

Problem 7.13. Suppose, in the Doppler effect, that a source S′ and an observer O′ move in the same direction with velocities u and v, respectively. Bring the observer to rest by superimposing a velocity $-v$ on the system to show that O′ now registers a frequency

$$\nu''' = \frac{\nu(c - v)}{(c - u)}$$

Problem 7.14. An aircraft flying on a level course transmits a signal of 3×10^9 Hz which is reflected from a distant point ahead on the flight path and received by the aircraft with a frequency difference of 15 kHz. What is the aircraft speed?

Problem 7.15. Light from hot sodium atoms is centred about a wavelength of 6×10^{-7}m but spreads 2×10^{-12}m on either side of this wavelength due to the Doppler effect as radiating atoms move towards and away from the observer. Calculate the thermal velocity of the atoms to show that the gas temperature is ~ 900 K.

Problem 7.16. Show that in the Doppler effect when the source and observer are not moving in the same direction the frequencies

$$\nu' = \frac{\nu c}{c - u'}, \quad \nu'' = \frac{\nu(c - v)}{c}$$

and

$$\nu''' = \nu\left(\frac{c - v}{c - u}\right)$$

are valid if u and v are not the actual velocities but the components of these velocities along the direction in which the waves reach the observer.

8

Waves on Transmission Lines

Introduction

In the wave motion discussed so far four major points have emerged. They are

(1) Individual particles in the medium oscillate about their equilibrium positions with simple harmonic motion but do not propagate through the medium.
(2) Crests and troughs and all planes of equal phase are transmitted through the medium to give the wave motion.
(3) The wave or phase velocity is governed by the product of the inertia of the medium and its capacity to store potential energy; that is, its elasticity.
(4) The impedance of the medium to this wave motion is governed by the ratio of the inertia to the elasticity (see Appendix 8).

In this chapter we wish to investigate the wave propagation of voltages and currents and we shall see that the same physical features are predominant. Voltage and current waves are usually sent along a geometrical configuration of wires and cables known as transmission lines. The physical scale or order of magnitude of these lines can vary from that of an oscilloscope cable on a laboratory bench to the electric power distribution lines supported on pylons over hundreds of miles or the submarine telecommunication cables lying on an ocean bed.

Any transmission line can be simply represented by a pair of parallel wires into one end of which power is fed by an a.c. generator. Figure 8.1a shows such a line at the instant when the generator terminal A is positive with respect to terminal B, with current flowing out of the terminal A and into terminal B as the generator is doing work. A half cycle later the position is reversed and B is the positive terminal, the net result being that along each of the two wires there will be a distribution of charge as shown, reversing in sign at each half cycle due to the oscillatory simple harmonic motion of the charge carriers (Figure 8.1b). These carriers move a distance equal to a fraction of a wavelength on either side of their equilibrium positions. As the charge moves current flows, having a maximum value where the product of charge density and velocity is greatest.

Introduction to Vibrations and Waves, First Edition. H. J. Pain and Patricia Rankin.
© 2015 John Wiley & Sons, Ltd. Published 2015 by John Wiley & Sons, Ltd.
Companion website: http://booksupport.wiley.com

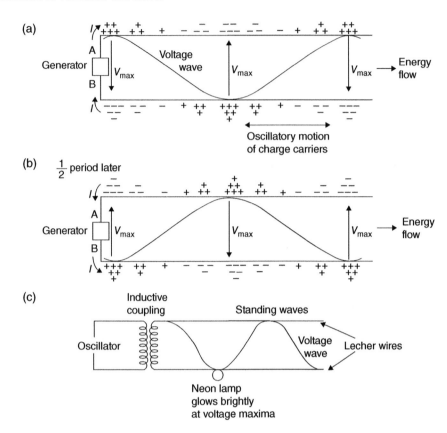

Figure 8.1 *Power fed continuously by a generator into an infinitely long transmission line. Charge distribution and voltage waves for (a) generator terminal positive at A and (b) a half period later, generator terminal positive at B. Laboratory demonstration (c) of voltage maxima along a Lecher wire system. The neon lamp glows when held near a position of V_{max} of a standing wave.*

The existence along the cable of maximum and minimum current values varying simple harmonically in space and time describes a current wave along the cable. Associated with these currents there are voltage waves (Figure 8.1a), and if the voltage and current at the generator are always in phase then power is continuously fed into the transmission line and the waves will always be carrying energy away from the generator. In a laboratory the voltage and current waves may be shown on a Lecher wire sysem (Figure 8.1c).

In deriving the wave equation for both voltage and current to obtain the velocity of wave propagation we shall concentrate our attention on a short element of the line having a length very much less than that of the waves. Over this element we may consider the variables to change linearly to the first order and we can use differentials.

The currents which flow will generate magnetic flux lines which thread the region between the cables, giving rise to a self inductance L_0 per unit length measured in henries per metre. Between the lines, which form a capacitor, there is an electrical capacitance C_0 per unit length measured in farads per metre. In the absence of any resistance in the line these two parameters completely describe the line, which is known as ideal or lossless.

8.1 Ideal or Lossless Transmission Line

Figure 8.2 represents a short element of zero resistance of an ideal transmission line length $dx \ll \lambda$ (the voltage or current wavelength). The self inductance of the element is $L_0\, dx$ and its capacitance is $C_0\, dx$ F.

If the rate of change of voltage per unit length at constant time is $\partial V/\partial x$, then the voltage difference between the ends of the element dx is $\partial V/\partial x\, dx$, which equals the voltage drop from the self inductance $-(L_0\, dx)\partial I/\partial t$.

Thus

$$\frac{\partial V}{\partial x}dx = -(L_0\, dx)\frac{\partial I}{\partial t}$$

or

$$\frac{\partial V}{\partial x} = -L_0\frac{\partial I}{\partial t} \tag{8.1}$$

If the rate of change of current per unit length at constant time is $\partial I/\partial x$ there is a loss of current along the length dx of $-\partial I/\partial x\, dx$ because some current has charged the capacitance $C_0\, dx$ of the line to a voltage V.

If the amount of charge is $q = (C_0\, dx)V$,

$$dI = \frac{dq}{dt} = \frac{\partial}{\partial t}(C_0\, dx)V$$

so that

$$\frac{-\partial I}{\partial x}dx = \frac{\partial}{\partial t}(C_0\, dx)V$$

or

$$\frac{-\partial I}{\partial x} = C_0\frac{\partial V}{\partial t} \tag{8.2}$$

Since $\partial^2/\partial x\partial t = \partial^2/\partial t\, \partial x$ it follows, by taking $\partial/\partial x$ of equation (8.1) and $\partial/\partial t$ of equation (8.2), that

$$\frac{\partial^2 V}{\partial x^2} = L_0 C_0\frac{\partial^2 V}{\partial t^2} \tag{8.3}$$

a pure wave equation for the voltage with a velocity of propagation given by $v^2 = 1/L_0 C_0$.

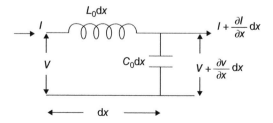

Figure 8.2 *Representation of element of an ideal transmission line of inductance L_0 H per unit length and capacitance C_0 F per unit length. The element length $\ll \lambda$, the voltage and current wavelength.*

Similarly $\partial/\partial t$ of (8.1) and $\partial/\partial x$ of (8.2) gives

$$\frac{\partial^2 I}{\partial x^2} = L_0 C_0 \frac{\partial^2 I}{\partial t^2} \tag{8.4}$$

showing that the current waves propagate with the same velocity $v^2 = 1/L_0 C_0$. We must remember here, in checking dimensions, that L_0 and C_0 are defined per unit length.

So far then, the oscillatory motion of the charge carriers (our particles in a medium) has led to the propagation of voltage and current waves with a velocity governed by the product of the magnetic inertia or inductance of the medium and its capacity to store potential energy.

8.2   Coaxial Cables

Very high frequency current and voltage waves are transmitted along coaxial cables which consist of a conducting wire of radius r_1 acting as the central axis of a hollow cylindrical conductor of radius r_2 made from plaited fine copper wire. The inductance and capacitance per unit length of this configuration are respectively

$$L_0 = \frac{\mu_0}{2\pi} \log \frac{r_2}{r_1} \text{ H}$$

and

$$C_0 = \frac{2\pi\epsilon_0}{\log r_2/r_1} \text{ F}$$

where μ_0 is the magnetic permeability and ϵ_0 is the permittivity of free space. The velocity of the waves along this air cored coaxial cable is

$$c = \frac{1}{(L_0 C_0)^{1/2}} = \frac{1}{(\mu_0\epsilon_0)^{1/2}} = 3 \times 10^8 \text{ m} \cdot \text{s}^{-1}$$

Even for larger ratios of r_2/r_1, $\log_e r_2/r_1$ remains a small factor. The frequency of the transmitted waves must be enough to reduce the wavelength to suitable values. At a frequency of 30 MHz the wavelength is still 10 metres. Such high frequencies require very low values of L_0 and C_0 which are typically $\approx \mu$H and pF respectively. As we shall see in the next section the ratio of the voltage to the current in the waves on the cable is

$$\frac{V}{I} = Z_0 = \left(\frac{L_0}{C_0}\right)^{1/2} \text{ ohms}$$

where Z_0 defines the impedance seen by the waves moving down an infinitely long cable. It is called the *Characteristic Impedance*.

A continuous air-cored cable is impractical and the space between the conductors is filled with a dielectric such as polythene with a permittivity $\epsilon = 2\epsilon_0$ and a permeability $\mu = \mu_0$. We write $\epsilon = \epsilon_r\epsilon_0$

where ϵ_r is the relative permittivity (dielectric constant) of a material and $\mu = \mu_r\mu_0$ where μ_r is the relative permittivity. Hence for a polythene filled cable where $\mu_r \approx 1$

$$Z_0 = \sqrt{\frac{L_0}{C_0}} = \frac{1}{2\pi}\sqrt{\frac{\mu}{\epsilon}}\log_e\frac{r_2}{r_1} = \frac{1}{2\pi}\sqrt{\frac{1}{\epsilon_r}}\log_e\frac{r_2}{r_1}\sqrt{\frac{\mu_0}{\epsilon_0}}$$

$$\text{where } Z_0 \text{ for free space} = \sqrt{\frac{\mu_0}{\epsilon_0}} = 376.6\,\Omega$$

Dielectric filled cables can be made to very high degrees of precision with Z_0 values of 50–100 Ω and signal velocities $= \frac{2}{3}c$ where c is the velocity of light. This precision allows the time for an electrical signal to travel a given length of the cable to be accurately calculated and such a cable is used as a delay line to separate the arrival of signals at a given point by very small intervals of time. Such a line is not short. A pulse of $\lambda/2$ at a frequency of 30 MHz is 5 metres long and the delay line must exceed this length. Delays are therefore measured in nanoseconds.

8.3 Characteristic Impedance of a Transmission Line

The solutions to equations (8.3) and (8.4) are,

$$V_+ = V_{0+}\sin\frac{2\pi}{\lambda}(vt - x)$$

and

$$I_+ = I_{0+}\sin\frac{2\pi}{\lambda}(vt - x)$$

where V_0 and I_0 are the maximum values and where the subscript $+$ refers to a wave moving in the positive x direction. Equation (8.1), $\partial V/\partial x = -L_0\partial I/\partial t$, therefore gives $-V'_+ = -vL_0I'_+$, where the superscript refers to differentiation with respect to the bracket $(vt - x)$.
   Integration of this equation gives

$$V_+ = vL_0I_+$$

where the constant of integration has no significance because we are considering only oscillatory values of voltage and current whilst the constant will change merely the d.c. level.
   The ratio

$$\frac{V_+}{I_+} = vL_0 = \sqrt{\frac{L_0}{C_0}}\,\Omega$$

and the value of $\sqrt{L_0/C_0}$, written as Z_0, is a constant for a transmission line of given properties and is called the characteristic impedance. Note that it is a pure resistance (no dimensions of length are involved) and it is the impedance seen by the wave system propagating along an infinitely long line, just as an acoustic wave experiences a specific acoustic impedance ρc. The physical correspondence between ρc and $L_0v = \sqrt{L_0/C_0} = Z_0$ is immediately evident.

The value of Z_0 for the coaxial cable considered earlier can be shown to be

$$Z_0 = \frac{1}{2\pi}\sqrt{\frac{\mu}{\varepsilon}}\log_e\frac{r_2}{r_1}$$

Electromagnetic waves in free space experience an impedance $Z_0 = \sqrt{\mu_0/\varepsilon_0} = 376.6\,\Omega$.

So far we have considered waves travelling only in the x direction. Waves which travel in the negative x direction will be represented (from solving the wave equation) by

$$V_- = V_{0-}\sin\frac{2\pi}{\lambda}(vt+x)$$

and

$$I_- = I_{0-}\sin\frac{2\pi}{\lambda}(vt+x)$$

where the negative subscript denotes the negative x direction of propagation.

Equation (8.1) then yields the results that

$$\frac{V_-}{I_-} = -vL_0 = -Z_0$$

so that, in common with the specific acoustic impedance, a negative sign is introduced into the ratio when the waves are travelling in the negative x direction.

When waves are travelling in both directions along the transmission line the total voltage and current at any point will be given by

$$V = V_+ + V_-$$

and

$$I = I_+ + I_-$$

When a transmission line has waves only in the positive direction the voltage and current waves are always in phase, energy is propagated and power is being fed into the line by the generator at all times. This situation is destroyed when waves travel in both directions; waves in the negative x direction are produced by reflection at a boundary when a line is terminated or mismatched; we shall now consider such reflections.

Worked Example

A signal of peak amplitude of 30 volts travels down a cable of $Z_0 = 100\,\Omega$. What is the average power travelling down the cable?

Solution

$$\frac{V_0^2}{2}\frac{1}{Z_0} = \frac{900}{200} = 4.5\,\text{watts}$$

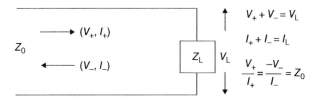

Figure 8.3 *Transmission line terminated by impedance Z_L to produce reflected waves unless $Z_L = Z_0$, the characteristic impedance.*

8.4 Reflections from the End of a Transmission Line

Suppose that a transmission line of characteristic impedance Z_0 has a finite length and that the end opposite that of the generator is terminated by a load of impedance Z_L as shown in Figure 8.3.

A wave travelling to the right (V_+, I_+) may be reflected to produce a wave (V_-, I_-).

The boundary conditions at Z_L must be $V_+ + V_- = V_L$, where V_L is the voltage across the load and $I_+ + I_- = I_L$. In addition $V_+/I_+ = Z_0$, $V_-/I_- = -Z_0$ and $V_L/I_L = Z_L$. These equations yield

$$\frac{V_-}{V_+} = \frac{Z_L - Z_0}{Z_L + Z_0}$$

(the voltage amplitude reflection coefficient),

$$\frac{I_-}{I_+} = \frac{Z_0 - Z_L}{Z_L + Z_0}$$

(the current amplitude reflection coefficient),

$$\frac{V_L}{V_+} = \frac{2Z_L}{Z_L + Z_0}$$

and

$$\frac{I_L}{I_+} = \frac{2Z_0}{Z_L + Z_0}$$

in complete correspondence with the reflection and transmission coefficients we have met so far. (See Summary in Appendix 8.)

We see that if the line is terminated by a load $Z_L = Z_0$, its characteristic impedance, the line is matched, all the energy propagating down the line is absorbed and there is no reflected wave. When $Z_L = Z_0$, therefore, the wave in the positive direction continues to behave as though the transmission line were infinitely long.

8.5 Short Circuited Transmission Line ($Z_L = 0$)

If the ends of the transmission line are short circuited (Figure 8.4), $Z_L = 0$, and we have

$$V_L = V_+ + V_- = 0$$

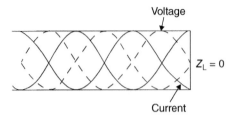

Voltage

$Z_L = 0$

Current

Figure 8.4   *Short circuited transmission line of length* $(2n + 1)\lambda/4$ *produces a standing wave with a current maximum and zero voltage at end of line.* **Note that V and I amplitudes double in standing waves.**

so that $V_+ = -V_-$, and there is total reflection with a phase change of π. But this is the condition, as we saw in an earlier chapter, for the existence of standing waves; we shall see that such waves exist on the transmission line.

At any position x on the line we may express the two voltage waves by

$$V_+ = Z_0 I_+ = V_{0+}\, e^{i(\omega t - kx)}$$

and

$$V_- = -Z_0 I_- = V_{0-}\, e^{i(\omega t + kx)}$$

where, with total reflection and π phase change, $V_{0+} = -V_{0-}$. The total voltage at x is

$$V_x = (V_+ + V_-) = V_{0+}(e^{-ikx} - e^{ikx})e^{i\omega t} = (-i)2V_{0+}\sin kx\, e^{i\omega t}$$

and the total current at x is

$$I_x = (I_+ + I_-) = \frac{V_{0+}}{Z_0}(e^{-ikx} + e^{ikx})\, e^{i\omega t} = \frac{2V_{0+}}{Z_0}\cos kx\, e^{i\omega t}$$

We see then that at any point x along the line the voltage V_x varies as $\sin kx$ and the current I_x varies as $\cos kx$, so that voltage and current are $90°$ out of phase in space. In addition the $-i$ factor in the voltage expression shows that the voltage lags the current $90°$ in time, so that if we take the voltage to vary with $\cos \omega t$ from the $e^{i\omega t}$ term, then the current will vary with $-\sin \omega t$. If we take the time variation of voltage to be as $\sin \omega t$ the current will change with $\cos \omega t$.

Voltage and current at all points are $90°$ out of phase in space and time, and the power factor $\cos \phi = \cos 90° = 0$, so that no power is consumed. A standing wave system exists with equal energy propagated in each direction and the total energy propagation equal to zero. Nodes of voltage and current are spaced along the transmission line as shown in Figure 8.4, with I always a maximum where $V = 0$ and vice versa.

If the current I varies with $\cos \omega t$ it will be at a maximum when $V = 0$; when V is a maximum the current is zero. The energy of the system is therefore completely exchanged each quarter cycle between the magnetic inertial energy $\frac{1}{2}L_0 I^2$ and the electric potential energy $\frac{1}{2}C_0 V^2$.

Cable Resonator

Figure 1.1h showed an electrical LC circuit capable of resonating at different frequencies by varying the value of C. A highly accurate resonator at a fixed high frequency can be made by short circuiting the end of a transmission line to create a standing wave. Figure 8.4 shows that $Z_L = 0$ produced a current antinode and an open end $Z_L = \infty$ produces a voltage antinode. The length of the cable is chosen to be exactly $\lambda/4$ of the desired resonance which produces a first harmonic $\nu_1 = c/4L$ where L is the length of the cable. Figure 8.1 shows how the voltage antinodes are detected.

Worked Example

How long is a cable with a velocity $1/(L_0 C_0)^{1/2} = c/2$ where $c = 3 \times 10^8$ metres which resonates at $\nu_1 = 10$ MHz.

Solution

$$\text{Length } L = \frac{1.5 \times 10^8}{4 \times 10^7} = 3.75 \text{ metres}$$

8.6 The Transmission Line as a Filter

The transmission line is a continuous network of impedances in series and parallel combination. The unit section is shown in Figure 8.5(a) and the continuous network in Figure 8.5(b).

If we add an infinite series of such sections a wave travelling down the line will meet its characteristic impedance Z_0. Figure 8.6 shows that adding an extra section to the beginning of the line does not change Z_0. The impedance in Figure 8.6 is

$$Z = Z_1 + \left(\frac{1}{Z_2} + \frac{1}{Z_0} \right)^{-1}$$

or

$$Z = Z_1 + \frac{Z_2 Z_0}{Z_2 + Z_0} = Z_0$$

so the characteristic impedance is

$$Z_0 = \frac{Z_1}{2} + \sqrt{\frac{Z_1^2}{4} + Z_1 Z_2}$$

Note that $Z_1/2$ is half the value of the first impedance in the line so if we measure the impedance from a point half way along this impedance we have

$$Z_0 = \left(\frac{Z_1^2}{4} + Z_1 Z_2 \right)^{1/2}$$

We shall, however, use the larger value of Z_0 in what follows.

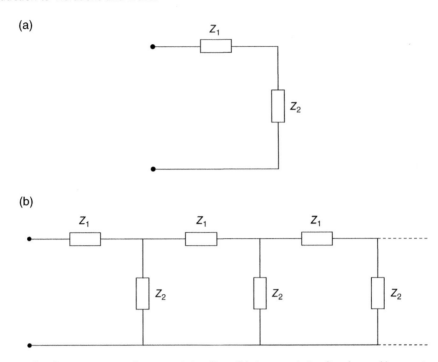

Figure 8.5   *(a) The elementary unit of a transmission line. (b) A transmission line formed by a series of such units.*

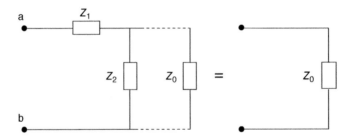

Figure 8.6   *A infinite series of elementary units presents a characteristic impedance Z_0 to a wave travelling down the transmission line. Adding an extra unit at the input terminal leaves Z_0 unchanged.*

In Figure 8.7 we now consider the currents and voltages at the far end of the transmission line. Any V_n since it is across Z_0 is given by $V_n = I_n Z_0$

Moreover

$$V_n - V_{n+1} = I_n Z_1 = V_n \frac{Z_1}{Z_0}$$

So

$$\frac{V_{n+1}}{V_n} = 1 - \frac{Z_1}{Z_0} = \frac{Z_0 - Z_1}{Z_0}$$

a result which is the same for all sections of the line.

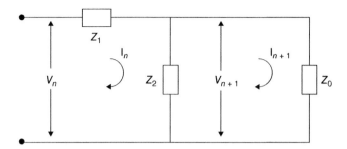

Figure 8.7   *The propagation constant $\alpha = V_{n+1}/V_n = Z_0 - 1/Z_0$ for all sections of the transmission line.*

We define a propagation factor

$$\alpha = \frac{V_{n+1}}{V_n} = \frac{Z_0 - Z_1}{Z_0}$$

which, with

$$Z_0 = \frac{Z_1}{2} + \left(\frac{Z_1^2}{4} + Z_1 Z_2\right)^{1/2}$$

gives

$$\alpha = \frac{\left(\sqrt{Z_0} - \frac{Z_1}{2}\right)}{\left(\sqrt{Z_0} + \frac{Z_1}{2}\right)}$$

$$= 1 + \frac{Z_1}{2Z_2} - \left[\left(1 + \frac{Z_1}{2Z_2}\right)^2 - 1\right]^{1/2}$$

In all practical cases Z_1/Z_2 is real since

1. there is either negligible resistance so that Z_1 and Z_2 are imaginary or
2. the impedances are purely resistive.

So, given (1) or (2) we see that if

(a) $\left(1 + \frac{Z_1}{2Z_2}\right)^2 = \left[1 + \frac{Z_1}{Z_2}\left(1 + \frac{Z_1}{4Z_2}\right)\right] \geq 1$ then α is real, and

(b) $\left(1 + \frac{Z_1}{2Z_2}\right)^2 < 1$ then α is complex.

For α real we have $Z_1/4Z_2 \geq 0$ or ≤ -1.

   If $Z_1/4Z_2 \geq 0$, then $0 < \alpha < 1$, the currents in successive sections decrease progressively and since α is real and positive there is no phase change from one section to another.

   If $Z_1/4Z_2 \leq -1$, then $\alpha \leq 0$, and there is again a progressive decrease in current amplitudes along the network but here α is negative and there is a π phase change for each successive section.

When α is complex we have

$$-1 < \frac{Z_1}{4Z_2} < 0$$

and

$$\alpha = 1 + \frac{Z_1}{2Z_2} - i \left[1 - \left(1 + \frac{Z_1}{2Z_2}\right)^2\right]^{1/2}$$

Note that $|\alpha| = 1$ so we can write

$$\alpha = \cos\beta - i\sin\beta = e^{-i\beta}$$

where

$$\cos\beta = 1 + \frac{Z_1}{2Z_2}$$

The current amplitude remains constant along the transmission line but the phase is retarded by β with each section. If β is constant then $\beta = k = 2\pi/\lambda$. If Z_1 and Z_2 are purely resistive α is fixed and the attenuation is constant for all voltage inputs.

If Z_1 is an inductance with Z_2 a capacitance (or vice versa) the division between α real and α complex occurs at certain frequencies governed by their relative magnitudes.

If $Z_1 = i\omega L$ and $Z_2 = 1/i\omega C$ for an input voltage $V = V_0 e^{i\omega t}$ then $|\alpha| = 1$ when $0 \le \omega^2 LC \le 4$.

So the line behaves as a low pass filter with a cut-off frequency $\omega_c = 2/\sqrt{LC}$. Above this frequency there is a progressive decrease in amplitude with a phase change of π in each section (Figure 8.8a).

If the positions of Z_1 and Z_2 are now interchanged so that $Z_1 = 1/i\omega C$ is now a capacitance and Z_2 is now an inductance with $Z_2 = i\omega L$ the transmisson line becomes a high pass filter with zero attenuation for $0 \le 1/\omega^2 LC \le 4$, that is for all frequencies above $\omega_C = (1/2\sqrt{LC})$ (Figure 8.8b).

8.7 Effect of Resistance in a Transmission Line

The discussion so far has concentrated on a transmission line having only inductance and capacitance, i.e. wattless components which consume no power. In practice, of course, no such line exists: there is always some resistance in the wires which will be responsible for energy losses. We shall take this resistance into account by supposing that the transmission line has a series resistance $R_0\Omega$ per unit length and a short circuiting or shunting resistance between the wires, which we express as a shunt conductance (inverse of resistance) written as G_0, where G_0 has the dimensions of siemens per metre. Our model of the short element of length dx of the transmission line now appears in Figure 8.9, with a resistance $R_0\,dx$ in series with $L_0\,dx$ and the conductance $G_0\,dx$ shunting the capacitance $C_0\,dx$. Current will now leak across the transmission line because the dielectric is not perfect. We have seen that the time-dependence of the voltage and current variations along a transmission line may be written

$$V = V_0\,e^{i\omega t} \quad \text{and} \quad I = I_0\,e^{i\omega t}$$

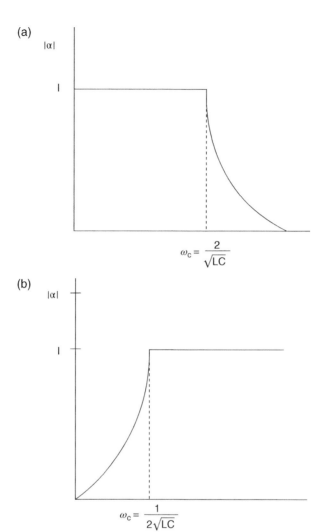

Figure 8.8   (a) When $Z_1 = i\omega L$ and $Z_2 = (i\omega C)^{-1}$ the transmission line acts as a low-pass filter. (b) Reversing the positions of Z_1 and Z_2 changes the transmission line into a high-pass filter.

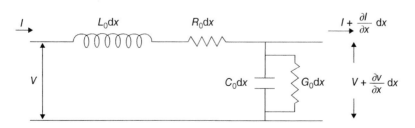

Figure 8.9   Real transmission line element includes a series resistance $R_0\,\Omega$ per unit length and a shunt conductance G_0 S per unit length.

so that

$$L_0 \frac{\partial I}{\partial t} = i\omega L_0 I \quad \text{and} \quad C_0 \frac{\partial V}{\partial t} = i\omega C_0 V$$

The voltage and current changes across the line element length dx are now given by

$$\frac{\partial V}{\partial x} = -L_0 \frac{\partial I}{\partial t} - R_0 I = -(R_0 + i\omega L_0)I \tag{8.1a}$$

$$\frac{\partial I}{\partial x} = -C_0 \frac{\partial V}{\partial t} - G_0 V = -(G_0 + i\omega C_0)V \tag{8.2a}$$

since $(G_0 \, dx)V$ is the current shunted across the condenser. Inserting $\partial/\partial x$ of equation (8.1a) into equation (8.2a) gives

$$\frac{\partial^2 V}{\partial x^2} = -(R_0 + i\omega L_0)\frac{\partial I}{\partial x} = (R_0 + i\omega L_0)(G_0 + i\omega C_0)V = \gamma^2 V \tag{8.3a}$$

where $\gamma^2 = (R_0 + i\omega L_0)(G_0 + i\omega C_0)$, so that γ is a complex quantity which may be written

$$\gamma = \alpha + ik$$

Inserting $\partial/\partial x$ of equation (8.2a) into equation (8.1a) gives

$$\frac{\partial^2 I}{\partial x^2} = -(G_0 + i\omega C_0)\frac{\partial V}{\partial x} = (R_0 + i\omega L_0)(G_0 + i\omega C_0)I = \gamma^2 I \tag{8.4a}$$

an equation similar to that for V.

The equation

$$\frac{\partial^2 V}{\partial x^2} - \gamma^2 V = 0 \tag{8.5}$$

has solutions for the x-dependence of V of the form

$$V = Ae^{-\gamma x} \quad \text{or} \quad V = Be^{+\gamma x}$$

where A and B are constants.

We know already that the time-dependence of V is of the form $e^{i\omega t}$, so that the complete solution for V may be written

$$V = (Ae^{-\gamma x} + Be^{\gamma x})e^{i\omega t}$$

or, since $\gamma = \alpha + ik$,

$$V = (Ae^{-\alpha x}e^{-ikx} + Be^{\alpha x}e^{+ikx})e^{i\omega t}$$
$$= Ae^{-\alpha x}e^{i(\omega t - kx)} + Be^{\alpha x}e^{i(\omega t + kx)}$$

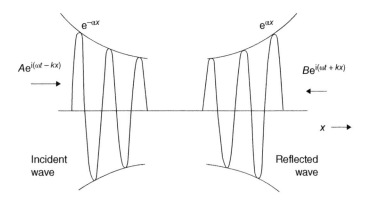

Figure 8.10 *Voltage and current waves in both directions along a transmission line with resistance. The effect of the dissipation term is shown by the exponentially decaying wave in each direction.*

The behaviour of V is shown in Figure 8.10 – a wave travelling to the right with an amplitude decaying exponentially with distance because of the term $\mathrm{e}^{-\alpha x}$ and a wave travelling to the left with an amplitude decaying exponentially with distance because of the term $\mathrm{e}^{\alpha x}$.

In the expression $\gamma = \alpha + \mathrm{i}k$, γ is called the propagation constant, α is called the attenuation or absorption coefficient and k is the wave number.

The behaviour of the current wave I is exactly similar and since power is the product VI, the power loss with distance varies as $(\mathrm{e}^{-\alpha x})^2$; that is, as $\mathrm{e}^{-2\alpha x}$.

We would expect this behaviour from our discussion of damped simple harmonic oscillations. When the transmission line properties are purely inductive (inertial) and capacitative (elastic), a pure wave equation with a sine or cosine solution will follow. The introduction of a resistive or loss element produces an exponential decay with distance along the transmission line in exactly the same way as an oscillator is damped with time.

From equations 8.3a and 8.4a we can calculate the values of α and k in $\gamma = \alpha + \mathrm{i}k$. For light damping we assume $R_0 \ll \omega L_0$ and $G_0 \ll \omega C_0$. We write

$$\gamma = [(R_0 + \mathrm{i}\omega L_0)(G_0 + \mathrm{i}\omega C_0)]^{1/2} = \mathrm{i}\omega(L_0 C_0)^{1/2}\left[1 - \mathrm{i}\left(\frac{R_0}{\omega L_0} + \frac{G_0}{\omega C_0}\right)\right]^{\frac{1}{2}}$$

and expand this as a binomial expression with $n = \frac{1}{2}$ to give

$$\gamma = \frac{(L_0 C_0)^{1/2}}{2}\left(\frac{R_0}{L_0} + \frac{G_0}{C_0}\right) + \mathrm{i}\omega(L_0 C_0)^{1/2}$$

where the wave velocity c is given by

$$\frac{1}{c} = (L_0 C_0)^{1/2}.$$

The first term in this last expression is real, equal to α and the second term is $i\omega/c = ik$. R_0 and G_0 should be as small as possible with $G_0 \ll R$, so

$$\alpha = \frac{(L_0 C_0)^{1/2}}{2} \frac{R_0}{L_0} = \frac{1}{c} \frac{R_0}{2L_0} \quad \text{with} \quad \frac{G_0}{C_0} \ll 1$$

Recalling from section 6.6 (a few lines above the subsection Energy Loss on a Transmission Line) that $1/\alpha$ is the distance over which Φ_m, the maximum of the wave amplitude, attenuates to $\Phi_m e^{-1}$ we have

$$\frac{1}{\alpha} = 2c \frac{L_0}{R_0}$$

where $\frac{L_0}{R_0}$ is $\frac{1}{\text{time}}$ which modern manufacturers can produce to be $\approx 10^{-5}$ sec for an air-cored cable. This gives

$$1/\alpha = 6 \times 10^8 \times 10^{-5} = 6000\,\text{km}$$

For a dielectric cored cable with $c = 1.5 \times 10^8\,\text{m} \cdot \text{s}^{-1}$

$$1/\alpha \approx 3 \times 10^8 \times 10^{-5} = 3000\,\text{km}.$$

Note that to this approximation $1/\alpha$ is independent of frequency.

8.8   Characteristic Impedance of a Transmission Line with Resistance

In a lossless line we saw that the ratio $V_+/I_+ = Z_0 = \sqrt{L_0/C_0} = Z_0\,\Omega$, a purely resistive term. In what way does the introduction of the resistance into the line affect the characteristic impedance?

The solution to the equation $\partial^2 I/\partial x^2 = \gamma^2 I$ may be written (for the x-dependence of I) as

$$I = (A' e^{-\gamma x} + B' e^{\gamma x}) \tag{8.6}$$

so that equation (8.2a)

$$\frac{\partial I}{\partial x} = -(G_0 + i\omega C_0)V$$

gives

$$-\gamma(A' e^{-\gamma x} - B' e^{\gamma x}) = -(G_0 + i\omega C_0)V$$

or

$$\frac{\sqrt{(R_0 + i\omega L_0)(G_0 + i\omega C_0)}}{G_0 + i\omega C_0}(A' e^{-\gamma x} - B' e^{\gamma x}) = V = V_+ + V_-$$

But, except for the $e^{i\omega t}$ term,

$$A' e^{-\gamma x} = I_+$$

the current wave in the positive x direction, so that

$$\sqrt{\frac{R_0 + i\omega L_0}{G_0 + i\omega C_0}} I_+ = V_+$$

or

$$\frac{V_+}{I_+} = \sqrt{\frac{R_0 + i\omega L_0}{G_0 + i\omega C_0}} = Z_0' \tag{8.6a}$$

for a transmission line with resistance. Similarly $B' e^{\gamma x} = I_-$ and

$$\frac{V_-}{I_-} = -\sqrt{\frac{R_0 + i\omega L_0}{G_0 + i\omega C_0}} = -Z_0' \tag{8.6b}$$

The presence of the resistance term in the complex characteristic impedance means that power will be lost through joule dissipation and that energy will be absorbed from the wave system.

Expanding the expression $Z_0' = [(R + i\omega/L_0)/(G_0 + i\omega C_0)]^{1/2}$ as a binomial expression with $n = \frac{1}{2}$ with $G_0 \ll R_0$ and $G_0 \ll \omega C_0$ we have

$$Z_0' = \left(\frac{R + i\omega L_0}{i\omega C_0}\right)^{\frac{1}{2}} = \sqrt{\frac{L_0}{C_0}}\left(1 - \frac{iR_0}{\omega L_0}\right)^{\frac{1}{2}} = \sqrt{Z_0}\left(1 - \frac{iR_0}{2\omega L_0}\right) \approx \sqrt{Z_0}$$

a pure resistance. Thus the characteristic impedance of the cable is approximately a pure resistance independent of the line resistance which affects only the small reactive part of the impedance. The attenuation

$$\alpha = \frac{R_0}{2}\sqrt{\frac{C_0}{L_0}} = \frac{R_0}{2Z_0} \quad \text{i.e.} \ \propto R_0 \ \text{and} \ \propto 1/Z_0.$$

This holds at frequencies where R_0 is not a function of ω but this is lost in small cables where L_0 is not $\gg R_0$.

At high frequencies the 'skin effect' (Chapter 9) becomes important. A very high conductivity has a large damping effect which restricts the current to the outer surface of the conductor.

Worked Example

Show that the impedance of a real transmission line with resistance seen from a position x in the line is given by

$$Z_x = Z_0' \cdot \frac{Ae^{-\gamma x} - Be^{+\gamma x}}{Ae^{-\gamma x} + Be^{+\gamma x}}$$

Solution

Equation 8.6 gives the x dependence of $I = (Ae^{-\gamma x} + Be^{+\gamma x})$ where A is the value of I_+ at $x = 0$ for the right-going wave and B is the value of I_- at $x = 0$ for the left-going wave. Equation 8.6a gives $V_+ = Z'_0 I_+$ and equation 8.6b gives $V_- = -Z'_0 I_-$. Hence, $V_+ + V_- = Z'_0(I_+ - I_-)$ so

$$Z_x = Z'_0 \left(\frac{V_+ + V_-}{I_+ + I_-} \right) = Z'_0 \left(\frac{Ae^{-\gamma x} - Be^{+\gamma x}}{Ae^{-\gamma x} + Be^{+\gamma x}} \right)$$

8.9   Matching Impedances

Proof that two cables with impedances Z_0 and Z_L are matched by the insertion between them of a cable with impedance Z_m where $Z_m^2 = Z_0 Z_L$. The length of Z_m is $\lambda/4$ measured in Z_m. This result is true for the impedances of all media capable of propagating waves. Note Z_m is loss free.

The boundary condition at $Z_0 Z_m$ junction gives:

$$V_{0+} + V_{0-} = V_{m0+} + V_{m0-}$$

$$I_{0+} + I_{0-} = I_{m0+} + I_{m0-}$$

where V_{0+}, V_{0-} are the voltages of forward and backward waves on Z_0 side of $Z_0 Z_m$ junction; I_{0+}, I_{0-} are the currents of forward and backward waves on Z_0 side of $Z_0 Z_m$ junction; V_{m0+}, V_{m0-} are the voltages of forward and backward waves on Z_m side of $Z_0 Z_m$ junction; I_{m0+}, I_{m0-} are the currents of forward and backward waves on Z_m side of $Z_0 Z_m$ junction.

The boundary condition at $Z_m Z_L$ junction gives:

$$V_{mL+} + V_{mL-} = V_L \quad \text{and} \quad I_{mL+} + I_{mL-} = I_L$$

where V_{mL+}, V_{mL-} are the voltages of forward and backward waves on Z_m side of $Z_m Z_L$ junction; I_{mL+}, I_{mL-} are the currents of forward and backward waves on Z_m side of $Z_m Z_L$ junction; V_L, I_L are the voltage and current across the load.

If the length of the matching line is l, we have:

$$V_{m0+} = V_{mL+}e^{ikl} \quad \text{and} \quad I_{m0+} = I_{mL+}e^{ikl}$$

$$V_{m0-} = V_{mL-}e^{-ikl} \quad \text{and} \quad I_{m0-} = I_{mL-}e^{-ikl}$$

In addition, we have the relations:

$$\frac{V_L}{I_L} = Z_L \quad \text{and} \quad \frac{V_0}{I_0} = Z_0$$

$$\frac{V_{m0+}}{I_{m0+}} = -\frac{V_{m0-}}{I_{m0-}} = \frac{V_{mL+}}{I_{mL+}} = -\frac{V_{mL-}}{I_{mL-}} = Z_m$$

The above conditions yield:

$$V_{mL+} = V_{m0+}e^{-ikl} \quad \text{and} \quad I_{mL+} = I_{m0+}e^{-ikl}$$

$$V_{mL-} = \frac{Z_L - Z_m}{Z_L + Z_m}V_{mL+}$$

$$V_{mL-} = \frac{Z_m - Z_L}{Z_m + Z_L}I_{mL+}$$

$$V_{m0-} = V_{mL-}e^{-ikl} = \frac{Z_L - Z_m}{Z_L + Z_m}V_{mL+}e^{-ikl} = V_{m0+}\frac{Z_L - Z_m}{Z_L + Z_m}e^{-i2kl}$$

$$I_{m0-} = I_{mL-}e^{-ikl} = \frac{Z_m - Z_L}{Z_L + Z_m}I_{mL+}e^{-ikl} = I_{m0+}\frac{Z_m - Z_L}{Z_L + Z_m}e^{-i2kl}$$

Impedance matching requires $V_{0-} = 0$ and $I_{0-} = 0$, i.e.

$$V_{0+} = V_{m0+} + V_{m0-} \quad \text{and} \quad I_{0+} = I_{m0+} + I_{m0-}$$

i.e.

$$V_{0+} = V_{m0+}\left(1 + \frac{Z_L - Z_m}{Z_L + Z_m}e^{-i2kl}\right)$$

$$I_{0+} = I_{m0+}\left(1 + \frac{Z_m - Z_L}{Z_L + Z_m}e^{-i2kl}\right)$$

By dividing the above equations we have:

$$Z_0 = Z_m\frac{(Z_L + Z_m)e^{ikl} + (Z_L - Z_m)e^{-ikl}}{(Z_L + Z_m)e^{ikl} + (Z_m - Z_L)e^{-ikl}} = Z_m\frac{Z_L \cos kl + iZ_m \sin kl}{Z_m \cos kl + iZ_L \sin kl}$$

which for $kl = \pi/2$, or $l = \lambda/4$ yields:

$$Z_m^2 = Z_0 Z_L$$

Problem 8.1.   In a short-circuited lossless transmission line integrate the magnetic (inductive) energy $\frac{1}{2}L_0I^2$ and the electric (potential) energy $\frac{1}{2}C_0V^2$ over the last quarter wavelength (0 to $-\lambda/4$) to show that they are equal.

Problem 8.2.   Show, in Problem 8.1, that the sum of the instantaneous values of the two energies over the last quarter wavelength is equal to the maximum value of either.

Problem 8.3.   The impedance of a real transmission line seen from a position x on the line is given by

$$Z_x = Z_0 \frac{Ae^{-\gamma x} - Be^{+\gamma x}}{Ae^{-\gamma x} + Be^{+\gamma x}}$$

where γ is the propagation constant and A and B are the current amplitudes at $x = 0$ of the waves travelling in the positive and negative x directions respectively. If the line has a length l and is terminated by a load Z_L, show that

$$Z_L = Z_0 \frac{Ae^{-\gamma l} - Be^{\gamma l}}{Ae^{-\gamma l} + Be^{\gamma l}}$$

Problem 8.4.   Show that the input impedance of the line of Problem 8.3; that is, the impedance of the line at $x = 0$, is given by

$$Z_i = Z_0 \left(\frac{Z_0 \sinh \gamma l + Z_L \cosh \gamma l}{Z_0 \cosh \gamma l + Z_L \sinh \gamma l} \right)$$

$$(Note : 2 \cosh \gamma l = e^{\gamma l} + e^{-\gamma l}$$
$$2 \sinh \gamma l = e^{\gamma l} - e^{-\gamma l})$$

Problem 8.5.   If the transmission line of Problem 8.4 is short-circuited, show that its input impedance is given by

$$Z_{sc} = Z_0 \tanh \gamma l$$

and when it is open-circuited the input impedance is

$$Z_{0c} = Z_0 \coth \gamma l$$

By taking the product of these quantities, suggest a method for measuring the characteristic impedance of the line.

Problem 8.6.   Show that the input impedance of a short-circuited loss-free line of length l is given by

$$Z_i = i\sqrt{\frac{L_0}{C_0}} \tan \frac{2\pi l}{\lambda}$$

and by sketching the variation of the ratio $Z_i/\sqrt{L_0/C_0}$ with l, show that for l just greater than $(2n+1)\lambda/4$, Z_i is capacitative, and for l just greater than $n\lambda/2$ it is inductive. (This provides a positive or negative reactance to match another line.)

Problem 8.7.   Show that a short-circuited quarter wavelength loss-free line has an infinite impedance and that if it is bridged across another transmission line it will not affect the fundamental wavelength but will short-circuit any undesirable second harmonic.

Problem 8.8.   Show that a loss-free line of characteristic impedance Z_0 and length $n\lambda/2$ may be used to couple two high frequency circuits without affecting other impedances.

Problem 8.9.   A transmission line has $Z_1 = i\omega L$ and $Z_2 = (i\omega C)^{-1}$. If, for a range of frequencies ω, the phase shift per section β is very small show that $\beta = k$ the wave number and that the phase velocity is independent of the frequency.

Problem 8.10.   In a transmission line with losses where $R_0/\omega L_0$ and $G_0/\omega C_0$ are both small quantities the expression for the propagation constant is

$$\gamma = [(R_0 + i\omega L_0)(G_0 + i\omega C_0)]^{1/2}.$$

If $\gamma = \alpha + ik$ where the attenuation constant

$$\alpha = \frac{R_0}{2}\sqrt{\frac{C_0}{L_0}} + \frac{G_0}{2}\sqrt{\frac{L_0}{C_0}}$$

and the wave number

$$k = \omega\sqrt{L_0 C_0} = \frac{\omega}{v}$$

Show that for $G_0 = 0$ the Q value of such a line is given by $k/2\alpha$.

Problem 8.11.   Expand the expression for the characteristic impedance of the transmission line of Problem 8.10 in terms of the characteristic impedance of a lossless line to show that if

$$\frac{R_0}{L_0} = \frac{G_0}{C_0}$$

the impedance remains real because the phase effects introduced by the series and shunt losses are equal but opposite.

Problem 8.12.   The wave description of an electron of total energy E in a potential well of depth V over the region $0 < x < l$ is given by Schrödinger's time-independent wave equation

$$\frac{\partial^2 \psi}{\partial x^2} + \frac{8\pi^2 m}{h^2}(E - V)\psi = 0$$

where m is the electron mass and h is Planck's constant. (Note that $V = 0$ within the well.)

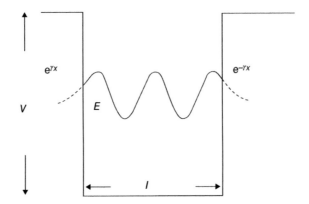

Show that for $E > V$ (inside the potential well) the solution for ψ is a standing wave solution but for $E < V$ (outside the region $0 < x < l$) the x dependence of ψ is $e^{\pm \gamma x}$, where

$$\gamma = \frac{2\pi}{h}\sqrt{2m(V - E)}$$

9

Electromagnetic Waves

Introduction

Earlier chapters have shown that the velocity of waves through a medium is determined by the inertia and the elasticity of the medium. These two properties are capable of storing wave energy in the medium, and in the absence of energy dissipation they also determine the impedance presented by the medium to the waves. In addition, when there is no loss mechanism a pure wave equation with a sine or cosine solution will always be obtained, but this equation will be modified by any resistive or loss term to give an oscillatory solution which decays with time or distance.

These physical processes describe exactly the propagation of electromagnetic waves through a medium. The magnetic inertia of the medium, as in the case of the transmission line, is provided by the inductive property of the medium, i.e. the permeability μ, which has the units of henries per metre. The elasticity or capacitive property of the medium is provided by the permittivity ε, with units of farads per metre. The storage of magnetic energy arises through the permeability μ; the potential or electric field energy is stored through the permittivity ε.

If the material is defined as a dielectric, only μ and ε are effective and a pure wave equation for both the magnetic field vector H and the electric field vector E will result. If the medium is a conductor, having conductivity σ (the inverse of resistivity) with dimensions of siemens per metre or $(\text{ohm m})^{-1}$, in addition to μ and ε, then some of the wave energy will be dissipated and absorption will take place.

In this chapter we will consider first the propagation of electromagnetic waves in a medium characterized by μ and ε only, and then treat the general case of a medium having μ, ε and σ properties.

9.1 Maxwell's Equations

Electromagnetic waves arise whenever an electric charge changes its velocity. Electrons moving from a higher to a lower energy level in an atom will radiate a wave of a particular frequency and wavelength. A very hot ionized gas consisting of charged particles will radiate waves over a continuous spectrum as the paths of individual particles are curved in mutual collisions. This radiation is called 'Bremsstrahlung'.

Introduction to Vibrations and Waves, First Edition. H. J. Pain and Patricia Rankin.
© 2015 John Wiley & Sons, Ltd. Published 2015 by John Wiley & Sons, Ltd.
Companion website: http://booksupport.wiley.com

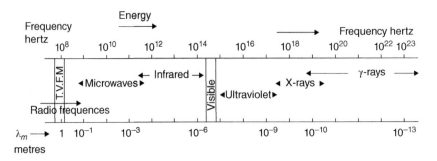

Figure 9.1   *Wavelengths and frequencies in the electromagnetic spectrum.*

The radiation of electromagnetic waves from an aerial is due to the oscillatory motion of charges in an alternating current flowing in the aerial.

Figure 9.1 shows the frequency spectrum of electromagnetic waves. All of these waves exhibit the same physical characteristics.

It is quite remarkable that the whole of electromagnetic theory can be described by the four vector relations in Maxwell's equations. In examining these relations in detail we shall see that two are steady state; that is, independent of time, and that two are time-varying.

The two time-varying equations are mathematically sufficient to produce separate wave equations for the electric and magnetic field vectors, E and H, but the steady state equations help to identify the wave nature as transverse.

The first time-varying equation relates the time variation of the magnetic induction, $\mu H = B$, with the space variation of E; that is

$$\frac{\partial}{\partial t}(\mu H) \text{ is connected with } \frac{\partial E}{\partial z}(\text{say})$$

This is nothing but a form of Lenz's or Faraday's Law, as we shall see.

The second time-varying equation states that the time variation of εE defines the space variation of H, that is

$$\frac{\partial}{\partial t}(\varepsilon E) \text{ is connected with } \frac{\partial H}{\partial z}(\text{say})$$

Again we shall see that this is really a statement of Ampere's Law.

These equations show that the variations of E in time and space affect those of H and vice versa. E and H cannot be considered as isolated quantities but are interdependent.

The product εE has dimensions

$$\frac{\text{farads}}{\text{metre}} \times \frac{\text{volts}}{\text{metre}} = \frac{\text{charge}}{\text{area}}$$

This charge per unit area is called the displacement charge $D = \varepsilon E$.

Physically it appears in a dielectric when an applied electric field polarizes the constituent atoms or molecules and charge moves across any plane in the dielectric which is normal to the applied field direction. If the applied field is varying or alternating with time we see that the dimensions of

$$\frac{\partial \mathbf{D}}{\partial t} = \frac{\partial}{\partial t}(\varepsilon \mathbf{E}) = \frac{\text{charge}}{\text{time} \times \text{area}}$$

current per unit area. This current is called the displacement current. It is comparatively simple to visualize this current in a dielectric where physical charges may move – it is not easy to associate a displacement current with free space in the absence of a material but it may always be expressed as $I_d = \varepsilon(\partial \phi_E / \partial t)$, where ϕ_E is the electric field flux through a surface.

Consider what happens in the electric circuit of Figure 9.2 when the switch is closed and the battery begins to charge the capacitor C to a potential V. A current I obeying Ohm's Law ($V = IR$) will flow through the connecting leads as long as the condenser is charging and a compass needle or other magnetic field detector placed near the leads will show the presence of the magnetic field associated with that current. But suppose a magnetic field detector (shielded from all outside effects) is placed in the region between the condenser plates where no ohmic or conduction current is flowing. Would it detect a magnetic field? The answer is yes; all the magnetic field effects from a current exist in this region as long as the condenser is charging, that is, as long as the potential difference and the electric field between the capacitor plates are changing.

It was Maxwell's major contribution to electromagnetic theory to assert that the existence of a time-changing electric field in free space gave rise to a displacement current. The same result follows from considering the conservation of charge. The flow of charge into any small volume in space must equal that flowing out. If the volume includes the top plate of the capacitor the ohmic current through the leads produces the flow into the volume, while the displacement current represents the flow out.

In future, therefore, two different kinds of current will have to be considered:

1. The familar conduction current obeying Ohm's Law ($V = IR$) and
2. The displacement current of density $\partial \mathbf{D}/\partial t$.

In a medium of permeability μ and permittivity ε, but where the conductivity $\sigma = 0$, the displacement current will be the only current flowing. In this case a pure wave equation for $\mathbf{E}$ and $\mathbf{H}$ will follow and there will be no energy loss or attenuation.

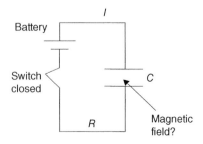

Figure 9.2   *In this circuit, when the switch is closed the conduction current charges the capacitor. Throughout charging the quantity $\varepsilon \mathbf{E}$ in the volume of the condenser is changing and the displacement current per unit area $\partial / \partial t(\varepsilon \mathbf{E})$ is associated with the magnetic field present between the capacitor plates.*

When $\sigma \neq 0$ a resistive element allows the conduction current to flow, energy loss will follow, a diffusion term is added to the wave equation and the wave amplitude will attenuate exponentially with distance. We shall see that the relative magnitude of these two currents is frequency-dependent and that their ratio governs whether the medium behaves as a conductor or as a dielectric.

9.2   Electromagnetic Waves in a Medium having Finite Permeability μ and Permittivity ϵ but with Conductivity $\sigma = 0$

We shall consider a system of plane waves and choose the plane xy as that region over which the wave properties are constant. These properties will not vary with respect to x and y and all derivatives $\partial/\partial x$ and $\partial/\partial y$ will be zero.

The first time-varying equation of Maxwell is written in vector notation as

$$\operatorname{curl} \boldsymbol{E} = \nabla \times \boldsymbol{E} = -\frac{\partial}{\partial t}\boldsymbol{B} = -\mu\frac{\partial}{\partial t}\boldsymbol{H} \qquad *$$

This represents three component equations:

$$\left.\begin{array}{ll}
-\mu\frac{\partial}{\partial t}H_x &= \frac{\partial}{\partial y}E_z \quad -\frac{\partial}{\partial z}E_y \\[2mm]
-\mu\frac{\partial}{\partial t}H_y &= \frac{\partial}{\partial z}E_x \quad -\frac{\partial}{\partial x}E_z \\[2mm]
-\mu\frac{\partial}{\partial t}H_z &= \frac{\partial}{\partial x}E_y \quad -\frac{\partial}{\partial y}E_x
\end{array}\right\} \qquad (9.1)$$

where the subscripts represent the component directions. E_x, E_y and E_z are, respectively, the magnitudes of vectors E_x, E_y and E_z. Similarly, H_x, H_y and H_z are the magnitudes of vectors H_x, H_y and H_z. The dimensions of these equations may be written

$$-\frac{\mu H}{\text{time}} = \frac{E}{\text{length}}$$

and multiplying each side by (length)^2 gives

$$-\frac{\mu H}{\text{time}} \times \text{area} = E \times \text{length}$$

i.e.

$$\frac{\text{total magnetic flux}}{\text{time}} = \text{volts}$$

This is dimensionally of the form of Lenz's or Faraday's Law.

The second time-varying equation of Maxwell is written in vector notation as

$$\operatorname{curl} \boldsymbol{H} = \nabla \times \boldsymbol{H} = \frac{\partial \boldsymbol{D}}{\partial t} = \varepsilon\frac{\partial \boldsymbol{E}}{\partial t}$$

---

*The electromagnetic wave equations are derived using vector methods in Appendix 5.

This represents three component equations:

$$
\left.
\begin{aligned}
\varepsilon \frac{\partial}{\partial t} E_x &= \frac{\partial}{\partial y} H_z - \frac{\partial}{\partial z} H_y \\[2mm]
\varepsilon \frac{\partial}{\partial t} E_y &= \frac{\partial}{\partial z} H_x - \frac{\partial}{\partial x} H_z \\[2mm]
\varepsilon \frac{\partial}{\partial t} E_z &= \frac{\partial}{\partial x} H_y - \frac{\partial}{\partial y} H_x
\end{aligned}
\right\}
\tag{9.2}
$$

The dimensions of these equations may be written

$$
\frac{\text{current } I}{\text{area}} = \frac{H}{\text{length}}
$$

and multiplying both sides by length gives

$$
\frac{\text{current}}{\text{length}} = \frac{I}{\text{length}} = H
$$

which is dimensionally of the form of Ampere's Law (i.e. the circular magnetic field at radius r due to the current I flowing in a straight wire is given by $H = I/2\pi r$). Maxwell's first steady state equation may be written

$$
\operatorname{div} \boldsymbol{D} = \nabla \cdot \boldsymbol{D} = \varepsilon \left(\frac{\partial E_x}{\partial x} + \frac{\partial E_y}{\partial y} + \frac{\partial E_z}{\partial z} \right) = \rho
\tag{9.3}
$$

where ε is constant and ρ is the charge density. This states that over a small volume element $dx\,dy\,dz$ of charge density ρ the change of displacement depends upon the value of ρ.

When $\rho = 0$ the equation becomes

$$
\varepsilon \left(\frac{\partial E_x}{\partial x} + \frac{\partial E_y}{\partial y} + \frac{\partial E_z}{\partial z} \right) = 0
\tag{9.3a}
$$

so that if the displacement $\boldsymbol{D} = \varepsilon \boldsymbol{E}$ is graphically represented by flux lines which must begin and end on electric charges, the number of flux lines entering the volume element $dx\,dy\,dz$ must equal the number leaving it.

The second steady state equation is written

$$
\operatorname{div} \boldsymbol{B} = \nabla \cdot \boldsymbol{B} = \mu \left(\frac{\partial H_x}{\partial x} + \frac{\partial H_y}{\partial y} + \frac{\partial H_z}{\partial z} \right) = 0
\tag{9.4}
$$

Again this states that an equal number of magnetic induction lines enter and leave the volume $dx\,dy\,dz$. This is a physical consequence of the non-existence of isolated magnetic poles, i.e. a single north pole or south pole.

Whereas the charge density ρ in equation (9.3) can be positive, i.e. a source of flux lines (or displacement), or negative, i.e. a sink of flux lines (or displacement), no separate source or sink of magnetic induction can exist in isolation, every source being matched by a sink of equal strength.

9.3   The Wave Equation for Electromagnetic Waves

Since, with these plane waves, all derivatives with respect to x and y are zero, equations (9.1) and (9.4) give

$$\mu \frac{\partial H_z}{\partial t} = 0 \quad \text{and} \quad \frac{\partial H_z}{\partial z} = 0$$

therefore, H_z is constant in space and time and because we are considering only the oscillatory nature of H a constant H_z can have no effect on the wave motion. We can therefore put $H_z = 0$. A similar consideration of equations (9.2) and (9.3a) leads to the result that $E_z = 0$.

The absence of variation in H_z and E_z means that the oscillations or variations in H and E occur in directions perpendicular to the z direction. We shall see that this leads to the conclusion that electromagnetic waves are transverse waves.

In addition to having plane waves we shall simplify our picture by considering only plane-polarized waves (see Figure 9.3).

We can choose the electric field vibration to be in either the x or y direction. Let us consider E_x only, with $E_y = 0$. In this case equations (9.1) give

$$-\mu \frac{\partial H_y}{\partial t} = \frac{\partial E_x}{\partial z} \tag{9.1a}$$

and equations (9.2) give

$$\varepsilon \frac{\partial E_x}{\partial t} = -\frac{\partial H_y}{\partial z} \tag{9.2a}$$

Using the fact that

$$\frac{\partial^2}{\partial z \partial t} = \frac{\partial^2}{\partial t \partial z}$$

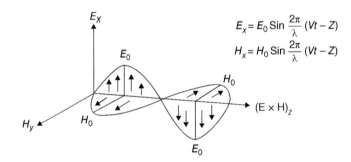

$$E_x = E_0 \operatorname{Sin} \frac{2\pi}{\lambda} (Vt - Z)$$

$$H_x = H_0 \operatorname{Sin} \frac{2\pi}{\lambda} (Vt - Z)$$

Figure 9.3   *In a plane-polarized electromagnetic wave the electric field vector E_x and magnetic field vector H_y are perpendicular to each other and vary sinusoidally. In a non-conducting medium they are in phase. The vector product, **E** × **H**, gives the direction of energy flow and the instantaneous value of energy flow. It is called the Poynting vector.*

it follows by taking $\partial/\partial t$ of equation (9.1a) and $\partial/\partial z$ of equation (9.2a) that

$$\frac{\partial^2}{\partial z^2}H_y = \mu\varepsilon\frac{\partial^2}{\partial t^2}H_y \quad \text{(the wave equation for } H_y)$$

Similarly, by taking $\partial/\partial t$ of (9.2a) and $\partial/\partial z$ of (9.1a), we obtain

$$\frac{\partial^2}{\partial z^2}E_x = \mu\varepsilon\frac{\partial^2}{\partial t^2}E_x \quad \text{(the wave equation for } E_x)$$

Thus, the vectors E_x and H_y both obey the same wave equation, propagating in the z direction with the same velocity $v^2 = 1/\mu\varepsilon$. In free space the velocity is that of light, that is, $c^2 = 1/\mu_0\varepsilon_0$, where μ_0 is the permeability of free space and ε_0 is the permittivity of free space.

The solutions to these wave equations may be written, for plane waves, as

$$E_x = E_0 \sin\frac{2\pi}{\lambda}(vt - z)$$

$$H_y = H_0 \sin\frac{2\pi}{\lambda}(vt - z)$$

where E_0 and H_0 are the maximum amplitude values of E and H. Note that the sine (or cosine) solutions means that no attenuation occurs: only displacement currents are involved and there are no conductive or ohmic currents.

We can represent the electromagnetic wave (E_x, H_y) travelling in the z direction in Figure 9.3, and recall that because E_z and H_z are constant (or zero) the electromagnetic wave is a transverse wave.

The direction of propagation of the waves will always be in the $\boldsymbol{E} \times \boldsymbol{H}$ direction; in this case, $\boldsymbol{E} \times \boldsymbol{H}$ has magnitude, $E_x H_y$ and is in the z direction. E_x and H_y are plane polarized.

This product has the dimensions

$$\frac{\text{voltage} \times \text{current}}{\text{length} \times \text{length}} = \frac{\text{electrical power}}{\text{area}}$$

measured in units of watts per square metre. $\boldsymbol{E} \times \boldsymbol{H}$ is called the Poynting vector. It gives the direction and instantaneous value of energy flow.

The time averaged energy flow per second across unit area is given:

$$S_{\text{av}} = \frac{1}{2}\boldsymbol{E} \times \boldsymbol{H}^*$$

where $\boldsymbol{H}^*$ is the complex conjugate of $\boldsymbol{H}$.

9.4   Illustration of Poynting Vector

We can illustrate the flow of electromagnetic energy in terms of the Poynting vector by considering the simple circuit of Figure 9.4, where the parallel plate capacitor of area A and separation d, containing a dielectric of permittivity ε, is being charged to a voltage V.

Throughout the charging process current flows, and the electric and magnetic field vectors show that the Poynting vector is always directed into the volume Ad occupied by the dielectric.

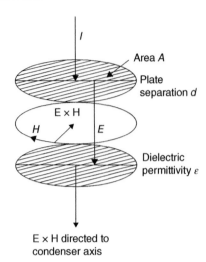

I

Area A

Plate separation d

E x H

H   E

Dielectric permittivity ε

E x H directed to condenser axis

Figure 9.4   *During charging the vector $\mathbf{E} \times \mathbf{H}$ is directed into the capacitor volume. At the end of the charging the energy is totally electrostatic and equals the product of the capacitor volume, Ad, and the electrostatic energy per unit volume, $\frac{1}{2}\varepsilon E^2$.*

The capacitance C of the capacitor is $\varepsilon A/d$ and the total energy of the capacitor at potential V is $\frac{1}{2}CV^2$ joules, which is stored as electrostatic energy. But $V = Ed$, where E is the final value of the electric field, so that the total energy

$$\frac{1}{2}CV^2 = \frac{1}{2}\left(\frac{\varepsilon A}{d}\right)E^2 d^2 = \frac{1}{2}(\varepsilon E^2)Ad$$

where Ad is the volume of the capacitor.

The electrostatic energy per unit volume stored in the capacitor is therefore $\frac{1}{2}\varepsilon E^2$ and results from the flow of electromagnetic energy during charging.

Worked Example

Show that when a current is increasing in a long uniformly wound solenoid of radius r the total energy flow rate over a length l (the Poynting vector times the surface area $2\pi rl$) gives the time rate of change of the magnetic energy stored in that length of the solenoid.

Solution

The electric field driving the current around the solenoid wire is azimuthal, that is E_θ. The solenoid consists of many turns and when the current is increasing the e.m.f. around a single turn is

$$E_0 2\pi r = -\mu\frac{\partial H_z}{\partial t}\pi r^2$$

where πr^2 is the cross-sectional area of the solenoid. The value of

$$E_0 = -\frac{\mu r}{2}\frac{\partial H_z}{\partial t}.$$

The Poynting vector $\mathbf{S} = \mathbf{E} \times \mathbf{H} = -(E_0 H_z)_r$ in the -ve radial direction from the solenoid surface to its central axis. So

$$S = \frac{\mu r}{2} H_z \frac{\partial H_z}{\partial t}.$$

Over the length of the solenoid the Poynting vector is

$$S \times 2\pi r l = \mu \pi r^2 l H_z \frac{\partial H_z}{\partial t}$$

where the change in the magnetic energy from the current increase is

$$\frac{\partial}{\partial t} \left(\frac{1}{2} \mu H^2 \times \pi r^2 l \right) = \mu \pi r^2 l H_z \frac{\partial H_z}{\partial t}$$

9.5   Impedance of a Dielectric to Electromagnetic Waves

If we put the solutions

$$E_x = E_0 \sin \frac{2\pi}{\lambda} (vt - z)$$

and

$$H_y = H_0 \sin \frac{2\pi}{\lambda} (vt - z)$$

in equation (9.1a) where

$$-\mu \frac{\partial H_y}{\partial t} = \frac{\partial E_x}{\partial z}$$

then

$$-\mu v H_y = -E_x, \quad \text{and since} \quad v^2 = \frac{1}{\mu\varepsilon}$$
$$\sqrt{\mu} H_y = \sqrt{\varepsilon} E_x$$

that is

$$\frac{E_x}{H_y} = \sqrt{\frac{\mu}{\varepsilon}} = \frac{E_0}{H_0}$$

which has the dimensions of ohms.

The value $\sqrt{\mu/\varepsilon}$ therefore represents the characteristic impedance of the medium to electromagnetic waves (compare this with the equivalent result $V/I = \sqrt{L_0/C_0} = Z_0$ for the transmission line of the previous chapter).

Worked Example

A plane-polarized electromagnetic wave propagates along a transmission line consisting of two 1 metre wide parallel strips of a perfect conductor containing a medium of permeability μ and permittivity ϵ. The planes are separated by 1 metre. The wave propagation is in the z direction, the magnetic field H_y lies in the plane of the transmission line and the electric vector E_x is normal to (and joins) the planes of the conductors. The electric field E_x generates equal but opposite surface charges on the conductors of magnitude $\epsilon E_x C \, \mathrm{m}^{-2}$, the motion of which in the direction of wave propagation gives rise to a surface current. Show that the magnitude of this current is H_y and the characteristic impedance of the transmission line is

$$\frac{E_x}{H_y} = \sqrt{\frac{\mu}{\epsilon}}$$

Solution

The induced displaced charge

$$q = \epsilon E_x C \, \mathrm{m}^2$$

so the current

$$I = qv = \frac{q}{(\mu\epsilon)^{1/2}} = \frac{\epsilon E_x}{(\mu\epsilon)^{1/2}}$$

Since

$$\sqrt{\mu} H_y = \sqrt{\epsilon} E_x$$

then

$$I = \sqrt{\frac{\epsilon}{\mu}} E_x = \sqrt{\frac{\epsilon}{\mu}} \sqrt{\frac{\mu}{\epsilon}} H_y = H_y.$$
$$V = E_x(\text{plane separation}) = E_x$$
$$\therefore Z = \frac{V}{I} = \frac{E_x}{H_y} = \sqrt{\frac{\mu}{\epsilon}}.$$

___

In free space

$$\frac{E_x}{H_y} = \sqrt{\frac{\mu_0}{\varepsilon_0}} = 376.7\Omega$$

so that free space presents an impedance of 376.7Ω to electromagnetic waves travelling through it.
   It follows from

$$\frac{E_x}{H_y} = \sqrt{\frac{\mu}{\varepsilon}} \quad \text{that} \quad \frac{E_x^2}{H_y^2} = \frac{\mu}{\varepsilon}$$

and therefore

$$\varepsilon E_x^2 = \mu H_y^2$$

Both of these quantities have the dimensions of energy per unit volume, for instance εE_x^2 has dimensions

$$\frac{\text{farads}}{\text{metre}} \times \frac{\text{volts}^2}{\text{metres}^2} = \frac{\text{joules}}{\text{metres}^3}$$

as we saw in the illustration of the Poynting vector. Thus, for a dielectric the electrostatic energy $\frac{1}{2}\varepsilon E_x^2$ per unit volume in an electromagnetic wave equals the magnetic energy per unit volume $\frac{1}{2}\mu H_y^2$ and the total energy is the sum $\frac{1}{2}\varepsilon E_x^2 + \frac{1}{2}\mu H_y^2$.

This gives the instantaneous value of the energy per unit volume and we know that, in the wave,

$$E_x = E_0 \sin(2\pi/\lambda)\,(vt - z)$$

and

$$H_y = H_0 \sin(2\pi/\lambda)\,(vt - z)$$

so that the time average value of the energy per unit volume is

$$\frac{1}{2}\varepsilon \bar{E}_x^2 + \frac{1}{2}\mu \bar{H}_y^2 = \frac{1}{4}\varepsilon E_0^2 + \frac{1}{4}\mu H_0^2$$
$$= \frac{1}{2}\varepsilon E_0^2 \ \text{J}\,\text{m}^{-3}$$

Now the amount of energy in an electromagnetic wave which crosses unit area in unit time is called the intensity, I, of the wave and is evidently $\left(\frac{1}{2}\varepsilon E_0^2\right) v$ where v is the velocity of the wave.

This gives the time averaged value of the Poynting vector and, for an electromagnetic wave in free space we have

$$I = \frac{1}{2}c\varepsilon_0 E_0^2 = \frac{1}{2}c\mu_0 H_0^2 \ \text{W}\,\text{m}^{-2}$$

9.6 Electromagnetic Waves in a Medium of Properties μ, ϵ and σ (where $\sigma \neq 0$)

From a physical point of view the electric vector in electromagnetic waves plays a much more significant role than the magnetic vector, e.g. most optical effects are associated with the electric vector. We shall therefore concentrate our discussion on the electric field behaviour.

In a medium of conductivity $\sigma = 0$ we have obtained the wave equation

$$\frac{\partial^2 E_x}{\partial z^2} = \mu\varepsilon \frac{\partial^2 E_x}{\partial t^2}$$

where the right-hand term, rewritten

$$\mu \frac{\partial}{\partial t} \left[\frac{\partial}{\partial t} (\varepsilon E_x) \right]$$

shows that we are considering a term

$$\mu \frac{\partial}{\partial t} \left[\frac{\text{displacement current}}{\text{area}} \right]$$

When $\sigma \neq 0$ we must also consider the conduction currents which flow. These currents are given by Ohm's Law as $I = V/R$, and we define the current density; that is, the current per unit area, as

$$\boldsymbol{J} = \frac{I}{\text{Area}} = \frac{1}{R \times \text{Length}} \times \frac{V}{\text{Length}} = \sigma \boldsymbol{E}$$

where σ is the conductivity $1/(R \times \text{Length})$ and $\boldsymbol{E}$ is the electric field. $\boldsymbol{J} = \sigma \boldsymbol{E}$ is another form of Ohm's Law.

With both displacement and conduction currents flowing, Maxwell's second time-varying equation reads, in vector form,

$$\nabla \times \boldsymbol{H} = \frac{\partial}{\partial t} \boldsymbol{D} + \boldsymbol{J} \tag{9.5}$$

each term on the right-hand side having dimensions of current per unit area. The presence of the conduction current modifies the wave equation by adding a second term of the same form to its right-hand side, namely

$$\mu \frac{\partial}{\partial t} \left(\frac{\text{current}}{\text{area}} \right) \text{ which is } \mu \frac{\partial}{\partial t} (\boldsymbol{J}) = \mu \frac{\partial}{\partial t} (\sigma \boldsymbol{E})$$

The final equation is therefore given by

$$\boxed{\frac{\partial^2}{\partial z^2} E_x = \mu \varepsilon \frac{\partial^2}{\partial t^2} E_x + \mu \sigma \frac{\partial}{\partial t} E_x} \tag{9.6}$$

and this equation may be derived formally by writing the component equation of (9.5) as

$$\varepsilon \frac{\partial E_x}{\partial t} + \sigma E_x = -\frac{\partial H_y}{\partial z} \tag{9.5a}$$

together with

$$-\mu \frac{\partial H_y}{\partial t} = \frac{\partial E_x}{\partial z} \tag{9.1a}$$

and taking $\partial/\partial t$ of (9.5a) and $\partial/\partial z$ of (9.1a). **We see immediately that the presence of the resistive or dissipation term, which allows conduction currents to flow, will add a diffusion term of the type discussed in Chapter 6 to the pure wave equation.** The product $(\mu\sigma)^{-1}$ is called the magnetic diffusivity, and has the dimensions $L^2 T^{-1}$, as we expect of all diffusion coefficients (see section 6.6).

Worked Example

Show that equation 9.6 has dimensions of

$$V = L\frac{\partial I_1}{\partial t} + L\frac{\partial I_2}{\partial t}$$

where I_1 is the displacement current and I_2 is the ohmic current.

Solution

$$\frac{\partial^2 E_x}{\partial z^2} = \mu\epsilon\frac{\partial^2 E_x}{\partial t^2} + \mu\sigma\frac{\partial E_x}{\partial t}$$

has dimensions

$$\frac{V/l}{l^2} = \frac{\text{inductance}}{l}\frac{\text{displacement current}}{t \times l^2} + \frac{\text{inductance}}{l}\frac{\text{ohmic current}}{t \times l^2}$$

giving

$$V = L\frac{\partial I_1}{\partial t} + L\frac{\partial I_2}{\partial t}$$

We are now going to look for the behaviour of E_x in this new equation (9.6), with the assumption that its time variation is simple harmonic, so that $E_x = E_0 e^{i\omega t}$. Using this value in equation (9.6) gives

$$\frac{\partial^2 E_x}{\partial z^2} - (i\omega\mu\sigma - \omega^2\mu\varepsilon)E_x = 0$$

which is in the form of equation (8.5), written

$$\frac{\partial^2 E_x}{\partial z^2} - \gamma^2 E_x = 0$$

where $\gamma^2 = i\omega\mu\sigma - \omega^2\mu\varepsilon$.

We saw in Chapter 8 that this produced a solution with the term $e^{-\gamma z}$ or $e^{+\gamma z}$, but we concentrate on the E_x oscillation in the positive z direction by writing

$$E_x = E_0 e^{i\omega t} e^{-\gamma z}$$

In order to assign a suitable value to γ we must go back to equation (9.6) and consider the relative magnitudes of the two right-hand side terms. If the medium is a dielectric, only displacement currents will flow. When the medium is a conductor, the ohmic currents of the second term on the right-hand side will be dominant. The ratio of the magnitudes of the conduction current density to the displacement current density is the ratio of the two right-hand side terms. This ratio is

$$\frac{J}{\partial D/\partial t} = \frac{\sigma E_x}{\partial/\partial t(\varepsilon E_x)} = \frac{\sigma E_x}{\partial/\partial t(\varepsilon E_0 \, e^{i\omega t})} = \frac{\sigma E_x}{i\omega\varepsilon E_x} = \frac{\sigma}{i\omega\varepsilon}$$

We see immediately from the presence of i that the phase of the displacement current is 90° ahead of that of the ohmic or conduction current. It is also 90° ahead of the electric field E_x so the displacement current dissipates no power.

For a conductor, where $\boldsymbol{J} \gg \partial\boldsymbol{D}/\partial t$, we have $\sigma \gg \omega\varepsilon$, and $\gamma^2 = i\sigma(\omega\mu) - \omega\varepsilon(\omega\mu)$ becomes

$$\gamma^2 \approx i\sigma\omega\mu$$

to a high order of accuracy.

Now

$$\sqrt{i} = \frac{1+i}{\sqrt{2}}$$

so that

$$\gamma = (1+i)\left(\frac{\omega\mu\sigma}{2}\right)^{1/2}$$

and

$$E_x = E_0\,e^{i\omega t}\,e^{-\gamma z}$$
$$= E_0\,e^{-(\omega\mu\sigma/2)^{1/2}z}\,e^{i[\omega t-(\omega\mu\sigma/2)^{1/2}z]}$$

a progressive wave in the positive z direction with an amplitude decaying with the factor $e^{-(\omega\mu\sigma/2)^{1/2}z}$.

Note that the product $\omega\mu\sigma$ has dimensions L^{-2}, where L is the short distance associated with very strong damping by magnetic diffusivity. The electric field is effectively short circuited.

9.7   Skin Depth

After travelling a distance

$$\delta = \left(\frac{2}{\omega\mu\sigma}\right)^{1/2}$$

in the conductor the electric field vector has decayed to a value $E_x = E_0\,e^{-1}$; this distance is called the skin depth (Figure 9.5).

For copper, with $\mu \approx \mu_0$ and $\sigma = 5.8 \times 10^7$ S m^{-1} at a frequency of 60 Hz, $\delta \approx 9$ mm; at 1 MHz, $\delta \approx 6.6 \times 10^{-5}$m and at 30 000 MHz (radar wavelength of 1 cm), $\delta \approx 3.8 \times 10^{-7}$m.

Thus, high frequency electromagnetic waves propagate only a very small distance in a conductor. The electric field is confined to a very small region at the surface; significant currents will flow only at the surface and the resistance of the conductor therefore increases with frequency. We see also why a conductor can act to 'shield' a region from electromagnetic waves.

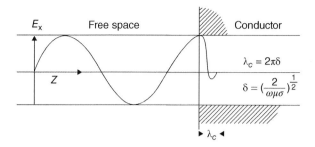

Figure 9.5 *Electromagnetic waves in a dielectric strike the plane surface of a conductor, and the electric field vector E_0 is damped to a value $E_0 e^{-1}$ in a distance of $(2/\omega\mu\sigma)^{1/2}$, the 'skin depth'. This explains the electrical shielding properties of a conductor. λ_c is the wavelength in the conductor.*

9.8 Electromagnetic Wave Velocity in a Conductor and Anomalous Dispersion

The phase velocity of the wave v is given by

$$v = \frac{\omega}{k} = \frac{\omega}{(\omega\mu\sigma/2)^{1/2}} = \omega\delta = \left(\frac{2\omega}{\mu\sigma}\right)^{1/2} = \nu\lambda_c$$

When δ is small, v is small, and the refractive index c/v of a conductor can be very large. We shall see later that this can explain the high optical reflectivities of good conductors. The velocity $v = \omega\delta = 2\pi\nu\delta$, so that λ_c in the conductor is $2\pi\delta$ and can be very small. Since v is a function of the frequency an electrical conductor is a dispersive medium to electromagnetic waves. Moreover, as the table below shows us, $\partial v/\partial\lambda$ is negative, so that the conductor is anomalously dispersive and the group velocity is greater than the wave velocity. Since $c^2/v^2 = \mu\varepsilon/\mu_0\varepsilon_0 = \mu_r\varepsilon_r$, where the subscript r defines non-dimensional relative values; that is, $\mu/\mu_0 = \mu_r, \varepsilon/\varepsilon_0 = \varepsilon_r$, then for $\mu_r \approx 1$

$$\varepsilon_r v^2 = c^2$$

and

$$\frac{\partial}{\partial\lambda}\varepsilon_r = -\frac{2}{v}\varepsilon_r\frac{\partial v}{\partial\lambda}$$

which confirms our statement in the chapter on group velocity that for $\partial\varepsilon_r/\partial\lambda$ positive a medium is anomalously dispersive. We see too that $c^2/v^2 = \varepsilon_r = n^2$, where n is the refractive index, so that the curve in Figure 3.11 showing the reactive behaviour of the oscillator impedance at displacement resonance is also showing the behaviour of n. This relative value of the permittivity is familiarly known as the dielectric constant when the frequency is low. This identity is lost at higher frequencies because the permittivity is frequency-dependent.

Note that $\lambda_c = 2\pi\delta$ is very small, and that when an electromagnetic wave strikes a conducting surface the electric field vector will drop to about 1% of its surface value in a distance equal to $\frac{3}{4}\lambda_c = 4.6\delta$. Effectively, therefore, the electromagnetic wave travels less than one wavelength into the conductor.

Frequency	$\lambda_{\text{free space}}$	δ (m)	$v_{\text{conductor}} = \omega\delta$ (m/s)	Refractive index $(c/v_{\text{conductor}})$
60	5000 km	9×10^{-3}	3.2	9.5×10^{7}
10^{6}	300 km	6.6×10^{-5}	4.1×10^{2}	7.3×10^{5}
3×10^{10}	10^{-2} m	3.9×10^{-7}	7.1×10^{4}	4.2×10^{3}

9.9   When is a Medium a Conductor or a Dielectric?

We have already seen that in any medium having $\mu\varepsilon$ and σ properties the magnitude of the ratio of the conduction current density to the displacement current density

$$\frac{J}{\partial D/\partial t} = \frac{\sigma}{\omega\varepsilon}$$

a non-dimensional quantity.

We may therefore represent the medium by the simple circuit in Figure 9.6 where the total current is divided between the two branches, a capacitive branch of reactance $1/\omega\varepsilon$ (ohm-metres) and a resistive branch of conductance σ (siemens/metre). If σ is large the resistivity is small, and most of the current flows through the σ branch and is conductive. If the capacitive reactance $1/\omega\varepsilon$ is so small that it takes most of the current, this current is the displacement current and the medium behaves as a dielectric.

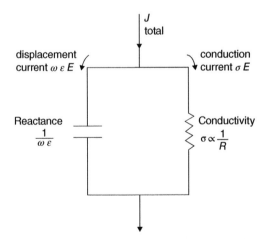

Figure 9.6   *A simple circuit showing the response of a conducting medium to an electromagnetic wave. The total current density J is divided by the parallel circuit in the ratio $\sigma/\omega\varepsilon$(the ratio of the conduction current density to the displacement current density). A large conductance σ (small resistance) gives a large conduction current while a small capacitive reactance $1/\omega\varepsilon$ allows a large displacement current to flow. For a conductor $\sigma/\omega\varepsilon \geq 100$; for a dielectric $\omega\varepsilon/\sigma \geq 100$. Note the frequency dependence of this ratio. At $\omega \approx 10^{20}$rad/s copper is a dielectric to X-rays.*

Quite arbitrarily we say that if

$$\frac{J}{\partial D/\partial t} = \frac{\sigma}{\omega\varepsilon} > 100$$

then conduction currents dominate and the medium is a conductor. If

$$\frac{\partial D/\partial t}{J} = \frac{\omega\varepsilon}{\sigma} > 100$$

then displacement currents dominate and the material behaves as a dielectric. Between these values exist a range of quasi-conductors; some of the semi-conductors fall into this category.

The ratio $\sigma/\omega\varepsilon$ is, however, frequency dependent, and a conductor at one frequency may be a dielectric at another.

For copper, which has $\sigma = 5.8 \times 10^7$ S m^{-1} and $\varepsilon \approx \varepsilon_0 = 9 \times 10^{-12}$ F m^{-1},

$$\frac{\sigma}{\omega\varepsilon} \approx \frac{10^{18}}{\text{frequency}}$$

so up to a frequency of 10^{16} Hz (the frequency of ultraviolet light) $\sigma/\omega\varepsilon > 100$, and copper is a conductor. At a frequency of 10^{20} Hz, however (the frequency of X-rays), $\omega\varepsilon/\sigma > 100$, and copper behaves as a dielectric. This explains why X-rays travel distances equivalent to many wavelengths in copper.

Typically, an insulator has $\sigma \approx 10^{-15}$ S m^{-1} and $\varepsilon \approx 10^{-11}$ F m^{-1}, which gives

$$\frac{\omega\varepsilon}{\sigma} \approx 10^4\omega$$

so the conduction current is negligible at all frequencies.

9.10   Why will an Electromagnetic Wave not Propagate into a Conductor?

To answer this question we need only consider the simple circuit where a capacitor C discharges through a resistance R. The voltage equation gives

$$\frac{q}{C} + IR = 0$$

and since $I = dq/dt$, we have

$$\frac{dq}{dt} = -\frac{q}{RC} \quad \text{or} \quad q = q_0 e^{-t/RC}$$

where q_0 is the initial charge.

We see that an electric field will exist between the plates of the capacitor only for a time $t \sim RC$ and will disappear when the charge has had time to distribute itself uniformly throughout the circuit. An electric field can only exist in the presence of a non-uniform charge distribution.

If we take a slab of any medium and place a charge of density q at a point within the slab, the medium will behave as an RC circuit and the equation

$$q = q_0 e^{-t/RC}$$

becomes

$$\boxed{q = q_0\, e^{-\sigma/\omega\varepsilon} \rightarrow q_0\, e^{-\sigma t/\varepsilon} \quad \left(\begin{array}{c} \varepsilon \equiv C \\ \sigma \equiv 1/R \end{array} \right)}$$

The charge will distribute itself uniformly in a time $t \sim \varepsilon/\sigma$, and the electric field will be maintained for that time only. The time ε/σ is called the relaxation time of the medium (RC time of the electrical circuit) and it is a measure of the maximum time for which an electric field can be maintained before the charge distribution becomes uniform.

Any electric field of a frequency ν, where $1/\nu = t > \varepsilon/\sigma$, will not be maintained; only a high frequency field where $1/\nu = t < \varepsilon/\sigma$ will establish itself.

9.11 Impedance of a Conducting Medium to Electromagnetic Waves

The impedance of a lossless medium is a real quantity. For the transmission line of Chapter 8 the characteristic impedance

$$Z_0 = \frac{V_+}{I_+} = \sqrt{\frac{L_0}{C_0}}\,\Omega;$$

for an electromagnetic wave in a dielectric

$$Z = \frac{E_x}{H_y} = \sqrt{\frac{\mu}{\varepsilon}}\,\Omega$$

with E_x and H_y in phase.

We saw in the case of the transmission line that when the loss mechanisms of a series resistance R_0 and a shunt conductance G_0 were introduced the impedance became the complex quantity

$$\mathbf{Z} = \sqrt{\frac{R_0 + i\omega L_0}{G_0 + i\omega C_0}}$$

We now ask what will be the impedance of a conducting medium of properties μ, ε and σ to electromagnetic waves? If the ratio of E_x to H_y is a complex quantity, it implies that a phase difference exists between the two field vectors.

We have already seen that in a conductor

$$E_x = E_0 e^{i\omega t} e^{-\gamma z}$$

where $\gamma = (1 + \mathrm{i})(\omega\mu\sigma/2)^{1/2}$, and we shall now write $H_y = H_0 \mathrm{e}^{\mathrm{i}(\omega t - \phi)}\mathrm{e}^{-\gamma z}$, suggesting that H_y lags E_x by a phase angle ϕ. This gives the impedance of the conductor as

$$\mathbf{Z}_c = \frac{E_x}{H_y} = \frac{E_0}{H_0}\mathrm{e}^{\mathrm{i}\phi}\ \mathrm{ohms}$$

Equation (9.1a) gives

$$\frac{\partial E_x}{\partial z} = -\mu\frac{\partial H_y}{\partial t}$$

so that

$$-\gamma E_x = -\mathrm{i}\omega\mu H_y$$

and

$$
\begin{aligned}
\mathbf{Z}_c &= \frac{E_x}{H_y} = \frac{\mathrm{i}\omega\mu}{\gamma} = \frac{\mathrm{i}(\omega\mu)}{(1 + \mathrm{i})(\omega\mu\sigma/2)^{1/2}} = \frac{\mathrm{i}(1 - \mathrm{i})}{(1 + \mathrm{i})(1 - \mathrm{i})}\left(\frac{2\omega\mu}{\sigma}\right)^{1/2}\\
&= \frac{(1 + \mathrm{i})}{2}\left(\frac{2\omega\mu}{\sigma}\right)^{1/2} = \frac{1 + \mathrm{i}}{\sqrt{2}}\left(\frac{\omega\mu}{\sigma}\right)^{1/2}\\
&= \left(\frac{\omega\mu}{\sigma}\right)^{1/2}\left(\frac{1}{\sqrt{2}} + \mathrm{i}\frac{1}{\sqrt{2}}\right) = \left(\frac{\omega\mu}{\sigma}\right)^{1/2}\mathrm{e}^{\mathrm{i}\phi}\ \mathrm{ohms}
\end{aligned}
$$

a vector of magnitude $(\omega\mu/\sigma)^{1/2}$ and phase angle $\phi = 45°$. Thus the magnitude

$$Z_c = \frac{E_0}{H_0} = \left(\frac{\omega\mu}{\sigma}\right)^{1/2}\ \mathrm{ohms} \tag{9.7}$$

and H_y lags E_x by $45°$ giving

$$S_{\mathrm{av}} = \frac{1}{2}E_0 H_0 \cos 45°$$

We can also express Z_c by

$$\mathbf{Z}_c = R + \mathrm{i}X = \left(\frac{\omega\mu}{2\sigma}\right)^{1/2} + \mathrm{i}\left(\frac{\omega\mu}{2\sigma}\right)^{1/2} \tag{9.8}$$

and also write it

$$
\begin{aligned}
\mathbf{Z}_c &= \frac{1 + \mathrm{i}}{\sqrt{2}}\left(\frac{\omega\mu}{\sigma}\right)^{1/2}\\
&= \sqrt{\frac{\mu_0}{\varepsilon_0}\frac{\varepsilon_0}{\varepsilon}\frac{\mu}{\mu_0}\frac{\omega\varepsilon}{\sigma}}\mathrm{e}^{\mathrm{i}\phi}
\end{aligned}
\tag{9.9}
$$

of magnitude

$$|Z_c| = 376.6\,\Omega \sqrt{\frac{\mu_r}{\varepsilon_r}} \sqrt{\frac{\omega\varepsilon}{\sigma}}$$

At a wavelength $\lambda = 10^{-1}$m, i.e. at a frequency $\nu = 3000$ MHz, the value of $\omega\varepsilon/\sigma$ for copper is 2.9×10^{-9} and $\mu_r \approx \varepsilon_r \approx 1$. This gives a magnitude $Z_c = 0.02\,\Omega$ at this frequency; for $\sigma = \infty$, $Z_c = 0$, and the electric field vector E_x vanishes, so we can say that when Z_c is small or zero the conductor behaves as a short circuit to the electric field. This sets up large conduction currents and the magnetic energy is increased.

Worked Example

Using equations (9.7) and (9.9) and information from the paragraph immediately above this worked example show that for a plane 1000 MHz wave travelling in air with $E_0 = 1V \cdot m^{-1}$ incident normally on a large copper sheet, the real part of the copper impedance is $R(Z_c) = 8.2 \times 10^{-3}\,\Omega$.

Solution

Equation (9.7) gives

$$S_{\mathrm{av}} = \frac{1}{2}E_0 H_0 \cos 45° = \frac{1}{2}H_0^2 R(Z_c) \quad \text{(from 9.7)}$$
$$\therefore R(Z_c) = (E_0/H_0)\cos 45°$$

or

$$E_0 = \frac{H_0}{\cos 45°}R(Z_c).$$

Using equation (9.9) and a value of $\omega\varepsilon/\sigma = 1/3$rd of that in the paragraph so that $\omega\varepsilon/\sigma = 9.7 \times 10^{-10}$ we have

$$R(Z_c) = \frac{1}{\sqrt{2}}377.6\sqrt{\frac{\mu_r}{\varepsilon_r}} = \sqrt{9.7 \times 10^{-10}} \approx 8.2 \times 10^{-3}\,\text{ohms} \quad \text{where } \mu_r \approx \varepsilon_r \approx 1.$$

In a dielectric, the impedance

$$Z = \frac{E_x}{H_y} = \sqrt{\frac{\mu}{\varepsilon}}\ \text{ohms}$$

led to the equivalence of the electric and magnetic field energy densities; that is, $\frac{1}{2}\mu H_y^2 = \frac{1}{2}\varepsilon E_x^2$. In a conductor, the magnitude of the impedance

$$Z_c = \left|\frac{E_x}{H_y}\right| = \left(\frac{\omega\mu}{\sigma}\right)^{1/2}$$

so that the ratio of the magnetic to the electric field energy density in the wave is

$$\frac{\frac{1}{2}\mu H_y^2}{\frac{1}{2}\varepsilon E_x^2} = \frac{\mu}{\varepsilon}\frac{\sigma}{\omega\mu} = \frac{\sigma}{\omega\varepsilon}$$

We already know that this ratio is very large for a conductor for it is the ratio of conduction to displacement currents, so that in a conductor the magnetic field energy dominates the electric field energy and increases as the electric field energy decreases.

9.12 Reflection and Transmission of Electromagnetic Waves at a Boundary

9.12.1 Normal Incidence

An infinite plane boundary separates two media of impedances Z_1 and Z_2 (real or complex) in Figure 9.7.

The electromagnetic wave normal to the boundary has the components shown where subscripts i, r and t denote incident, reflected and transmitted, respectively. Note that the vector direction $(\boldsymbol{E_r} \times \boldsymbol{H_r})$ must be opposite to that of $(\boldsymbol{E_i} \times \boldsymbol{H_i})$ to satisfy the energy flow condition of the Poynting vector.

The boundary conditions, from electromagnetic theory, are that the components of the field vectors $\boldsymbol{E}$ and $\boldsymbol{H}$ tangential or parallel to the boundary are continuous across the boundary.

Thus

$$E_i + E_r = E_t$$

and

$$H_i + H_r = H_t$$

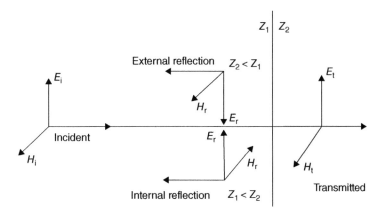

Figure 9.7 *Reflection and transmission of an electromagnetic wave incident normally on a plane between media of impedances Z_1 and Z_2. The Poynting vector of the reflected wave $(\mathbf{E} \times \mathbf{H})_r$ shows that either $\mathbf{E}$ or $\mathbf{H}$ may be reversed in phase, depending on the relative magnitudes of Z_1 and Z_2.*

where

$$\frac{E_i}{H_i} = Z_1, \quad \frac{E_r}{H_r} = -Z_1 \quad \text{and} \quad \frac{E_t}{H_t} = Z_2$$

From these relations it is easy to show that the amplitude reflection coefficient

$$R = \frac{E_r}{E_i} = \frac{Z_2 - Z_1}{Z_2 + Z_1} \tag{9.10}$$

and the amplitude transmission coefficient

$$T = \frac{E_t}{E_i} = \frac{2Z_2}{Z_2 + Z_1} \tag{9.11}$$

in agreement with the reflection and transmission coefficients we have found for the acoustic pressure p (Chapter 7) and voltage V (Chapter 8). If the wave is travelling in air and strikes a perfect conductor of $Z_2 = 0$ at normal incidence then

$$\frac{E_r}{E_i} = \frac{Z_2 - Z_1}{Z_2 + Z_1} = -1$$

giving complete reflection and

$$\frac{E_t}{E_i} = \frac{2Z_2}{Z_2 + Z_1} = 0$$

Thus, good conductors are very good reflectors of electromagnetic waves, e.g. lightwaves are well reflected from metal surfaces.

Worked Example

Show that when a plane electromagnetic wave travelling in air is reflected normally from a plane conducting surface the transmitted magnetic field value $H_t \approx 2H_i$ and that a magnetic standing wave exists in air with a very large standing wave ratio. This is analogous to a short-circuited transmission line. Show that the doubled value $2H_i$ in the wave of the worked example at the end of section 9.11 delivers a power of $S_{av} = 1.16 \times 10^{-7}$ watts·m^{-2} to the copper sheet.

Solution

From equation (9.11) we can calculate $T_H = H_t/H_i$. Writing Z_c as the conductor impedance we have $E_t = Z_c H_t$ and $E_i = Z_{air} H_i$. With $T = E_t/E_i = 2Z_c/Z_c + Z_{air}$ we have

$$T_H = \frac{H_t}{H_i} = \frac{E_t Z_{air}}{E_i Z_c} = \frac{Z_{air}}{Z_c} \frac{2Z_c}{Z_c + Z_{air}} = \frac{2Z_{air}}{Z_c + Z_{air}}$$

For large σ, $Z_c = 0$ and $T_H = 2Z_{air}/Z_{air} = 2$ $\therefore H_t = 2H_i$.
   Since $H = 2H_i$ we may write

$$S_{av} = \frac{1}{2}H_0^2 R(Z_c) = \frac{1}{2}H_{copper}^2 R(Z_c)$$

For the wave in the worked example at the end of section 9.11, $E_c = 1\,\mathrm{V \cdot m^{-1}}$ $\therefore$ $E_0/H_0 = 377.6$ ohm, and $2H^2_{\mathrm{copper}} = 2E_0^2/(377.6)^2 = 2/(377.6)^2$ so

$$S_{\mathrm{av}} = \frac{2}{(377.6)^2} R(Z_{\mathrm{copper}}) = \frac{2}{(377.6)^2} \times (8.2 \times 10^{-3}) = 1.16 \cdot 10^{-7}\,\mathrm{watts \cdot m^{-2}}.$$

9.13   Reflection from a Conductor (Normal Incidence)

For Z_2 a conductor and Z_1 free space, the refractive index

$$n = \frac{Z_1}{Z_2} = \frac{\beta}{\alpha + i\alpha}$$

is complex, where

$$\beta = \sqrt{\frac{\mu_0}{\varepsilon_0}}\ \mathrm{ohms}$$

and

$$\alpha = \left(\frac{\omega\mu}{2\sigma}\right)^{1/2}\ \mathrm{ohms}$$

A complex refractive index must always be interpreted in terms of absorption because a complex impedance is determined by a complex propagation constant, e.g. here $Z_2 = i\omega\mu/\gamma$, so that

$$n = \frac{Z_1}{Z_2} = \sqrt{\frac{\mu_0}{\varepsilon_0}\frac{1}{i\omega\mu}}(1+i)\left(\frac{\omega\mu\sigma}{2}\right)^{1/2} = (1-i)\left(\frac{\sigma}{2\omega\varepsilon_0}\right)^{1/2}$$

where

$$\frac{(\mu\mu_0)^{1/2}}{\mu} \approx 1$$

The ratio E_r/E_i is therefore complex (there is a phase difference between the incident and reflected vectors) with a value

$$\frac{E_r}{E_i} = \frac{Z_2 - Z_1}{Z_2 + Z_1} = \frac{\alpha + i\alpha - \beta}{\alpha + i\alpha + \beta} = \frac{1 - \beta/\alpha + i}{1 + \beta/\alpha + i}$$

where $\beta/\alpha \gg 1$.

Since E_r/E_i is complex, the value of the reflected intensity $I_r = (E_r/E_i)^2$ is found by taking the ratio the squares of the moduli of the numerator and the denominator, so that

$$I_r = \frac{|E_r|^2}{|E_i|^2} = \frac{|Z_2 - Z_1|^2}{|Z_2 + Z_1|^2} = \frac{(1 - \beta/\alpha)^2 + 1}{(1 + \beta/\alpha)^2 + 1}$$

$$= 1 - \frac{4\beta/\alpha}{2 + 2\beta/\alpha + (\beta/\alpha)^2} \rightarrow 1 - \frac{4\alpha}{\beta} \quad (\text{for } \beta/\alpha \gg 1)$$

so that

$$I_r = 1 - 4\left(\frac{\omega\mu}{2\sigma}\right)^{1/2}\left(\frac{\varepsilon_0}{\mu_0}\right)^{1/2} \approx 1 - 2\sqrt{\frac{2\omega\varepsilon_0}{\sigma}} \approx 1 - \sqrt{\frac{8\omega\varepsilon_0}{\sigma}}$$

For copper $\sigma = 6 \times 10^7$ (s·m^{-1}) and $(2\omega\varepsilon_0/\sigma)^{1/2} \approx 0.01$ at infra-red frequencies. The emission from an electric heater at 10^3 K has a peak at $\lambda \approx 2.5 \times 10^{-6}$ m. A metal reflector behind the heater filament reflects $\approx 97\%$ of these infra-red rays with 3% entering the metal to be lost as joule heating between the metal surface and the skin depth. Thus the factor $(8\omega\varepsilon_0/\sigma)^{1/2}$, the ratio of displacement to ohmic current, gives direct information on the reflectivity of a metal as well as indicating the amount of energy absorbed.

Problem 9.1.   The solutions to the e.m. wave equations are given in Figure 9.3 as

$$E_x = E_0 \sin \frac{2\pi}{\lambda}(vt - z)$$

and

$$H_y = H_0 \sin \frac{2\pi}{\lambda}(vt - z)$$

Use equations (9.1a) and (9.2a) to prove that they have the same wavelength and phase as shown in Figure 9.3.

Problem 9.2.   Show that the concept of $B^2/2\mu$ (magnetic energy per unit volume) as a magnetic pressure accounts for the fact that two parallel wires carrying currents in the same direction are forced together and that reversing one current will force them apart. (Consider a point midway between the two wires.) Show that it also explains the motion of a conductor carrying a current which is situated in a steady externally applied magnetic field.

Problem 9.3.   At a distance r from a charge e on a particle of mass m the electric field value is $E = e/4\pi\varepsilon_0 r^2$. Show by integrating the electrostatic energy density over the spherical volume of radius a to infinity and equating it to the value mc^2 that the 'classical' radius of the electron is given by

$$a = 1.41 \times 10^{-15}\text{m}$$

Problem 9.4.   The rate of generation of heat in a long cylindrical wire carrying a current I is I^2R, where R is the resistance of the wire. Show that this joule heating can be described in terms of the flow of energy into the wire from surrounding space and is equal to the product of the Poynting vector and the surface area of the wire.

Problem 9.5.   The plane-polarized electromagnetic wave (E_x, H_y) of this chapter travels in free space. Show that its Poynting vector (energy flow in watts per square metre) is given by

$$S = E_x H_y = c \left(\frac{1}{2} \varepsilon_0 E_x^2 + \frac{1}{2} \mu_0 H_y^2 \right) = c \varepsilon_0 E_x^2$$

where c is the velocity of light. The intensity in such a wave is given by

$$I = \bar{S}_{av} = c \varepsilon_0 \overline{E^2} = \frac{1}{2} c \varepsilon_0 E_{max}^2$$

Show that

$$\bar{S}_{av} = 1.327 \times 10^{-3} E_{max}^2$$
$$E_{max} = 27.45 \bar{S}^{1/2} \, \text{V m}^{-1}$$
$$H_{max} = 7.3 \times 10^{-2} \bar{S}_{av}^{1/2} \, \text{A m}^{-1}$$

Problem 9.6.   A light pulse from a ruby laser consists of a linearly polarized wavetrain of constant amplitude lasting for 10^{-4} s and carrying energy of 0.3 J. The diameter of the circular cross section of the beam is 5×10^{-3} m. Use the results of Problem 9.5 to calculate the energy density in the beam to show that the root mean square value of the electric field in the wave is

$$2.4 \times 10^5 \text{V m}^{-1}$$

Problem 9.7.   One square metre of the earth's surface is illuminated by the sun at normal incidence by an energy flux of 1.35 kW. Show that the amplitude of the electric field at the earth's surface is 1010 V m^{-1} and that the associated magnetic field in the wave has an amplitude of 2.7 A m^{-1} (See Problem 9.5). The electric field energy density $\frac{1}{2} \varepsilon E^2$ has the dimensions of a pressure. Calculate the radiation pressure of sunlight upon the earth.

Problem 9.8.   If the total power lost by the sun is equal to the power received per unit area of the earth's surface multiplied by the surface area of a sphere of radius equal to the Earth–Sun distance $(15 \times 10^7 \text{ km})$, show that the mass per second converted to radiant energy and lost by the sun is 4.2×10^9 kg. (See Problem 9.5.)

Problem 9.9.   A radio station radiates an average power of 10^5 W uniformly over a hemisphere concentric with the station. Find the magnitude of the Poynting vector and the amplitude of the electric and magnetic fields of the plane electromagnetic wave at a point 10 km from the station. (See Problem 9.5.)

Problem 9.10.   Show that when a group of electromagnetic waves of nearly equal frequencies propagates in a conducting medium the group velocity is twice the wave velocity. Use wave number $k = \left(\frac{\omega \mu \sigma}{2}\right)^{1/2}$

Problem 9.11.   A medium has a conductivity $\sigma = 10^{-1}$ S m^{-1} and a relative permittivity $\varepsilon_r = 50$, which is constant with frequency. If the relative permeability $\mu_r = 1$, is the medium a conductor or a dielectric at a frequency of (a) 50 kHz, and (b) 10^4 MHz?

$$[\varepsilon_0 = (36\pi \times 10^9)^{-1} \, \text{F m}^{-1}; \quad \mu_0 = 4\pi \times 10^{-7} \, \text{H m}^{-1}]$$

$$\text{Answer}: \quad \text{(a)} \quad \sigma/\omega\varepsilon = 720 \,(\text{conductor})$$
$$\text{(b)} \quad \sigma/\omega\varepsilon = 3.6 \times 10^{-3} \,(\text{dielectric}).$$

Problem 9.12.   The electrical properties of the Atlantic Ocean are given by

$$\varepsilon_r = 81, \quad \mu_r = 1, \quad \sigma = 4.3 \,\text{S m}^{-1}$$

Show that it is a conductor up to a frequency of about 10 MHz. What is the longest electromagnetic wavelength you would expect to propagate under water?

Problem 9.13.   For a good conductor $\varepsilon_r = \mu_r = 1$. Show that when an electromagnetic wave is reflected normally from such a conducting surface its fractional loss of energy $(1 - \text{reflection coefficient } I_r)$ is $\approx \sqrt{8\omega\varepsilon}/\sigma$. Note that the ratio of the displacement current density to the conduction current density is therefore a direct measure of the reflectivity of the surface.

Problem 9.14.   Show that when light travelling in free space is normally incident on the surface of a dielectric of refractive index n the reflected intensity

$$I_r = \left(\frac{E_r}{E_i}\right)^2 = \left(\frac{1-n}{1+n}\right)^2$$

and the transmitted intensity

$$I_t = \frac{Z_i E_t^2}{Z_t E_i^2} = \frac{4n}{(1+n)^2}$$

(Note $I_r + I_t = 1$.)

Problem 9.15.   Show that if the medium of Problem 9.14 is glass $(n = 1.5)$ then $I_r = 4\%$ and $I_t = 96\%$. If an electromagnetic wave of 100 MHz is normally incident on water $(\varepsilon_r = 81)$ show that $I_r = 65\%$ and $I_t = 35\%$.

Problem 9.16.   Light passes normally through a glass plate suffering only one air-to-glass and one glass-to-air reflection. What is the loss of intensity?

Problem 9.17.   A radiating antenna in simplified form is just a length x_0 of wire in which an oscillating current is maintained. The expression for the radiating power of an oscillating electron is

$$P = \frac{dE}{dt} = \frac{q^2\omega^4 x_0^2}{12\pi\varepsilon_0 c^3}$$

where $c = 3 \times 10^8$ m·s^{-1}, q is the electron charge and ω is the oscillation frequency. The current I in the antenna may be written $I_0 = \omega q$. If $P = \frac{1}{2}RI_0^2$ show that the radiation resistance of the antenna is given by

$$R = \frac{2\pi}{3}\sqrt{\frac{\mu_0}{\varepsilon_0}}\left(\frac{x_0}{\lambda}\right)^2 = 787\left(\frac{x_0}{\lambda}\right)^2 \Omega$$

where λ is the radiated wavelength (an expression valid for $\lambda \gg x_0$).

If the antenna is 30 m long and transmits at a frequency of 5×10^5 Hz with a root mean square current of 20 A, show that its radiation resistance is 1.97 Ω and that the power radiated is 400 W. (Verify that $\lambda \gg x_0$.)

10

Waves in More Than One Dimension

Introduction

This chapter extends our treatment of one-dimensional waves into two- and three-dimensional waves and discusses how waves behave in wave guides. You will see how *in wave guides the solution consists of a travelling wave in one dimension together with standing wave patterns in the other dimensions.* A calculation of the number of normal modes in three-dimensional space and the energy of each mode leads to a breakdown of classical physics in the 'ultra violet' catastrophe and its solution via Planck's radiation law. Planck's constant h, together with Heisenberg's Uncertainty Principle, places a limit on the smallest space which a particle of given energy may occupy.

10.1 Plane Wave Representation in Two and Three Dimensions

Figure 10.1 shows that in two dimensions waves of velocity c may be represented by lines of constant phase propagating in a direction $\boldsymbol{k}$ which is normal to each line, where the magnitude of $\boldsymbol{k}$ is the wave number $k = 2\pi/\lambda$.

The direction cosines of $\boldsymbol{k}$ are given by

$$l = \frac{k_1}{k}, \quad m = \frac{k_2}{k} \quad \text{where} \quad k^2 = k_1^2 + k_2^2$$

and any point $r(x, y)$ on the line of constant phase satisfies the equation

$$lx + my = p = ct$$

Introduction to Vibrations and Waves, First Edition. H. J. Pain and Patricia Rankin.
© 2015 John Wiley & Sons, Ltd. Published 2015 by John Wiley & Sons, Ltd.
Companion website: http://booksupport.wiley.com

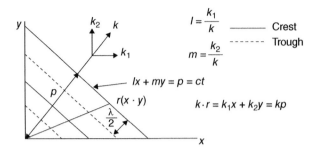

Figure 10.1   *Crests and troughs of a two-dimensional plane wave propagating in a general direction **k** (direction cosines l and m). The wave is specified by lx + my = p = ct, where p is its perpendicular distance from the origin, travelled in a time t at a velocity c.*

where p is the perpendicular distance from the line to the origin. The displacements at all points $r(x, y)$ on a given line are in phase and the phase difference ϕ between the origin and a given line is

$$\phi = \frac{2\pi}{\lambda}(\text{path difference}) = \frac{2\pi}{\lambda}p = \boldsymbol{k} \cdot \boldsymbol{r} = k_1 x + k_2 y$$
$$= kp$$

Hence, the bracket $(\omega t - \phi) = (\omega t - kx)$ used in a one-dimensional wave is replaced by $(\omega t - \boldsymbol{k} \cdot \boldsymbol{r})$ in waves of more than one dimension, e.g. we shall use the exponential expression

$$e^{\mathrm{i}(\omega t - \boldsymbol{k} \cdot \boldsymbol{r})}$$

In three dimensions all points $r(x, y, z)$ in a given wavefront will lie on planes of constant phase satisfying the equation

$$lx + my + nz = p = ct$$

where the vector $\boldsymbol{k}$ which is normal to the plane and in the direction of propagation has direction cosines

$$l = \frac{k_1}{k}, \quad m = \frac{k_2}{k}, \quad n = \frac{k_3}{k}$$

(so that $k^2 = k_1^2 + k_2^2 + k_3^2$) and the perpendicular distance p is given by

$$kp = \boldsymbol{k} \cdot \boldsymbol{r} = k_1 x + k_2 y + k_3 z$$

10.2   Wave Equation in Two Dimensions

We shall consider waves propagating on a stretched plane membrane of negligible thickness having a mass ρ per unit area and stretched under a uniform tension S. This means that if a line of unit length is drawn in the surface of the membrane, then the material on one side of this line exerts a force S (per unit length) on the material on the other side in a direction perpendicular to that of the line.

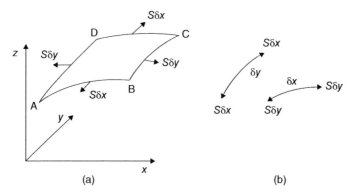

Figure 10.2   *Rectangular element of a uniform membrane vibrating in the z direction subject to one restoring force, Sδx, along its sides of length δy and another, Sδy, along its sides of length δx.*

If the equilibrium position of the membrane is the xy plane the vibration displacements perpendicular to this plane will be given by z where z depends on the position x, y. In Figure 10.2a where the small rectangular element ABCD of sides δx and δy is vibrating, forces $S\delta x$ and $S\delta y$ are shown acting on the sides in directions which tend to restore the element to its equilibrium position.

In deriving the equation for waves on a string we saw that the tension T along a curved element of string of length $\mathrm{d}x$ produced a force perpendicular to x of

$$S\frac{\partial^2 y}{\partial x^2}\mathrm{d}x$$

where y was the perpendicular displacement. Here in Figure 10.2b by exactly similar arguments we see that a force $S\delta y$ acting on a membrane element of length δx produces a force

$$S\delta y\frac{\partial^2 z}{\partial x^2}\delta x,$$

where z is the perpendicular displacement, whilst another force $S\delta x$ acting on a membrane element of length δy produces a force

$$S\delta x\frac{\partial^2 z}{\partial y^2}\delta y$$

The sum of these restoring forces which act in the z direction is equal to the mass of the element $\rho\delta x\delta y$ times its perpendicular acceleration in the z direction, so that

$$S\frac{\partial^2 z}{\partial x^2}\delta x\delta y + S\frac{\partial^2 z}{\partial y^2}\delta x\delta y = \rho\,\delta x\delta y\frac{\partial^2 y}{\partial t^2}$$

giving the wave equation in two dimensions as

$$\frac{\partial^2 z}{\partial x^2} + \frac{\partial^2 z}{\partial y^2} = \frac{\rho}{S}\frac{\partial^2 z}{\partial t^2} = \frac{1}{c^2}\frac{\partial^2 z}{\partial t^2}$$

where

$$c^2 = \frac{S}{\rho}$$

The displacement of waves propagating on this membrane will be given by

$$z = A\,\mathrm{e}^{\mathrm{i}(\omega t - \boldsymbol{k}\cdot\boldsymbol{r})} = A\,\mathrm{e}^{\mathrm{i}[\omega t - (k_1 x + k_2 y)]}$$

where

$$k^2 = k_1^2 + k_2^2$$

The reader should verify that this expression for z is indeed a solution to the two-dimensional wave equation when $\omega = ck$.

10.3 Wave Guides

10.3.1 Reflection of a 2D Wave at Rigid Boundaries

Let us first consider a 2D wave propagating in a vector direction $\boldsymbol{k}(k_1, k_2)$ in the xy plane along a membrane of width b stretched under a tension S between two long rigid rods which present an infinite impedance to the wave.

We see from Figure 10.3 that upon reflection from the line $y = b$ the component k_1 remains unaffected whilst k_2 is reversed to $-k_2$. Reflection at $y = 0$ leaves k_1 unaffected whilst $-k_2$ is reversed to its original value k_2. The wave system on the membrane will therefore be given by the superposition of the incident and reflected waves; that is, by

$$z = A_1\,\mathrm{e}^{\mathrm{i}[\omega t - (k_1 x + k_2 y)]} + A_2\,\mathrm{e}^{\mathrm{i}[\omega t - (k_1 x - k_2 y)]} \tag{10.1}$$

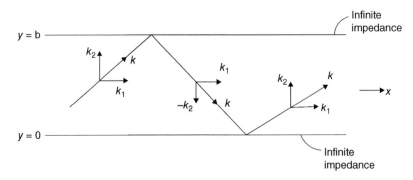

Figure 10.3 *Propagation of a two-dimensional wave along a stretched membrane with infinite impedances at $y = 0$ and $y = b$ giving reversal of k_2 at each reflection.*

subject to the boundary conditions that

$$z = 0 \quad \text{at} \quad y = 0 \quad \text{and} \quad y = b$$

the positions of the frame of infinite impedance.

The condition $z = 0$ at $y = 0$ requires

$$A_2 = -A_1$$

and $z = 0$ at $y = b$ gives

$$\sin k_2 b = 0$$

or

$$k_2 = \frac{n\pi}{b}$$

With these values of A_2 and k_2 the displacement of the wave system is given by the real part of z, i.e.

$$z = +2A_1 \sin k_2 y \sin(\omega t - k_1 x)$$

which represents a wave travelling along the x direction with a phase velocity

$$v_{\mathrm{p}} = \frac{\omega}{k_1} = \left(\frac{k}{k_1} \right) v$$

where v, the velocity on an infinitely wide membrane, is given by

$$v = \frac{\omega}{k} \quad \text{which is} \quad < v_{\mathrm{p}}$$

because

$$k^2 = k_1^2 + k_2^2$$

Now

$$k^2 = k_1^2 + \frac{n^2 \pi^2}{b^2}$$

so

$$k_1 = \left(k^2 - \frac{n^2 \pi^2}{b^2} \right)^{1/2} = \left(\frac{\omega^2}{v^2} - \frac{n^2 \pi^2}{b^2} \right)^{1/2}$$

and the group velocity for the wave in the x direction

$$v_{\mathrm{g}} = \frac{\partial \omega}{\partial k_1} = \frac{k_1}{\omega} v^2 = \left(\frac{k_1}{k} \right) v$$

giving the product

$$v_{\mathrm{p}} v_{\mathrm{g}} = v^2$$

Since k_1 must be real for the wave to propagate we have, from

$$k_1^2 = k^2 - \frac{n^2 \pi^2}{b^2}$$

the condition that

$$k^2 = \frac{\omega^2}{v^2} \geq \frac{n^2 \pi^2}{b^2}$$

that is

$$\omega \geq \frac{n \pi v}{b}$$

or

$$\nu \geq \frac{nv}{2b},$$

where n defines the mode number in the y direction. Thus, only waves of certain frequencies ν are allowed to propagate along the membrane which acts as a wave guide.

There is a cut-off frequency $n\pi v/b$ for each mode of number n and the wave guide acts as a frequency filter (recall the discussion on similar behaviour in wave propagation on the loaded string in Chapter 4). If $\nu = nv/2b$, the presence of the $\sin k_2 y$ term in the expression for the displacement z shows that the amplitude varies across the transverse y direction as shown in Figure 10.4 for the mode values $n = 1, 2, 3$. Thus, along any direction in which the waves meet rigid boundaries a standing wave system will be set up analogous to that on a string of fixed length and we shall discuss the implication of this in the section on normal modes and the method of separation of variables.

Wave guides are used for all wave systems, particularly in those with acoustical and electromagnetic applications. Fibre optics is based on wave guide principles, but the major use of wave guides has been with electromagnetic waves in telecommunications. Here the reflecting surfaces are the sides of a copper

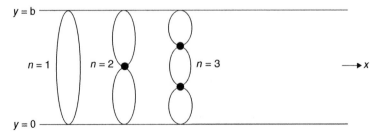

Figure 10.4 *Variation of amplitude with y direction for two-dimensional wave propagating along the membrane of Figure 9.3. Normal modes (n = 1, 2 and 3 shown) are set up along any axis bounded by infinite impedances.*

tube of circular or rectangular cross section. Note that in this case the free space velocity becomes the velocity of light

$$c = \frac{\omega}{k} < v_p$$

the phase velocity, but the relation $v_p v_g = c^2$ ensures that energy in the wave always travels with a group velocity $v_g < c$.

If $v < nv/2b$, k_1^2 becomes negative to give $\pm i k_1$ and the travelling wave in equation 10.1 has the terms

$$(A_1 + A_2)e^{i(\omega t - k_1 x)} = (A_1 + A_2)e^{-k_1 x}e^{i\omega t}$$

which rapidly extinguishes the wave in the x direction.

Worked Example

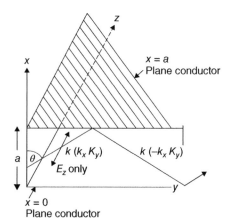

An electromagnetic wave loses negligible energy when reflected from a highly conducting surface. With repeated reflections it may travel along a transmission line or wave guide consisting of two parallel, infinitely conducting planes (separation a). If the wave in the Figure above is plane polarized, so that only E_z exists, then the propagating direction k lies wholly in the xy plane. The boundary conditions require that the total tangential electric field E_z is zero at the conducting surfaces $x = 0$ and $x = a$. Show that the first boundary condition allows E_z to be written $E_z = E_0(e^{ik_x x} - e^{-ik_x x})e^{i(k_y y - \omega t)}$, where $k_x = k \cos \theta$ and $k_y = k \sin \theta$ and the second boundary condition requires $k_x = n\pi/a$.

Solution

This problem is the electrical equivalent of the two-dimensional wave guide in Problem 10.3 (Figure 10.3). The electrical field E_z between the two plane boundaries is the superposition of the incident and reflected waves written as

$$E_z = E_1 e^{i(\omega t - (k_1 x + k_2 y))} + E_2 e^{i(\omega t - (-k_1 x + k_2 y))}$$

where $k_1 x = k_x = k \cos \theta$ and $k_2 y = k_y = k \sin \theta$. The boundary condition $E_z = 0$ at $x = 0$ gives

$$(E_1 + E_2)e^{i(\omega t - k_y y)} = 0$$

which is true for any t and y if $E_1 = -E_2 = E_0$. Thus

$$E_z = E_0(e^{-ik_x a} - e^{-ik_x a})e^{i(\omega t - k_y y)} = 0$$

i.e.

$$E_z = E_0 \sin k_x a\, e^{i(\omega t - k_y y)} = 0$$

which is true for any t and y if $\sin k_x a = 0$ i.e. $k_x = n\pi/a$. We then have a travelling wave

$$E_z = E_0\, e^{i(\omega t - k_y y)}$$

in the y direction and a standing wave $n = 1$ between $x = 0$ and $x = a$ in the x direction.

10.4   Normal Modes and the Method of Separation of Variables

We have just seen that when waves propagate in more than one dimension a standing wave system will be set up along any axis which is bounded by infinite impedances.

In Chapter 5 we found that standing waves could exist on a string of fixed length l where the displacement was of the form

$$y = A \genfrac{}{}{0pt}{}{\sin}{\cos}\} kx \genfrac{}{}{0pt}{}{\sin}{\cos}\} \omega_n t,$$

where A is constant and where $\genfrac{}{}{0pt}{}{\sin}{\cos}\}$ means that either solution may be used to fit the boundary conditions in space and time. When the string is fixed at both ends, the condition $y = 0$ at $x = 0$ removes the $\cos kx$ solution, and $y = 0$ at $x = l$ requires $k_n l = n\pi$ or $k_n = n\pi/l = 2\pi/\lambda_n$, giving $l = n\lambda_n/2$. Since the wave velocity $c = \nu_n \lambda_n$, this permits frequencies $\omega_n = 2\pi\nu_n = \pi nc/l$, defined as normal modes of vibration or eigenfrequencies.

We can obtain this solution in a way which allows us to extend the method to waves in more than one dimension. We have seen that the wave equation

$$\frac{\partial^2 \phi}{\partial x^2} = \frac{1}{c^2}\frac{\partial^2 \phi}{\partial t^2}$$

has a solution which is the product of two terms, one a function of x only and the other a function of t only.

Let us write $\phi = X(x)T(t)$ and apply the method known as separation of variables.

The wave equation then becomes

$$\frac{\partial^2 X}{\partial x^2} \cdot T = \frac{1}{c^2}X\frac{\partial^2 T}{\partial t^2}$$

or

$$X_{xx}T = \frac{1}{c^2}XT_{tt}$$

where the double subscript refers to double differentiation with respect to the variables. Dividing by $\phi = X(x)T(t)$ we have

$$\frac{X_{xx}}{X} = \frac{1}{c^2}\frac{T_{tt}}{T}$$

where the left-hand side depends on x only and the right-hand side depends on t only. However, both x and t are independent variables and the equality between both sides can only be true when both sides are independent of x and t and are equal to a constant, which we shall take, for convenience, as $-k^2$. Thus

$$\frac{X_{xx}}{X} = -k^2, \quad \text{giving} \quad X_{xx} + k^2 X = 0$$

and

$$\frac{1}{c^2}\frac{T_{tt}}{T} = -k^2, \quad \text{giving} \quad T_{tt} + c^2 k^2 T = 0$$

$X(x)$ is therefore of the form $e^{\pm ikx}$ and $T(t)$ is of the form $e^{\pm ickt}$, so that $\phi = A\,e^{\pm ikx}e^{\pm ickt}$, where A is constant, and we choose a particular solution in a form already familiar to us by writing

$$\phi = A\,e^{i(ckt-kx)}$$
$$= A\,e^{i(\omega t-kx)},$$

where $\omega = ck$, or we can write

$$\phi = A{\textstyle{\sin\atop\cos}\}}\, kx{\textstyle{\sin\atop\cos}\}}\, ckt$$

as above, where for fixed ends only a sine solution is allowed for x.

10.5 Two-Dimensional Case

In extending this method to waves in two dimensions we consider the wave equation in the form

$$\frac{\partial^2 \phi}{\partial x^2} + \frac{\partial^2 \phi}{\partial y^2} = \frac{1}{c^2}\frac{\partial^2 \phi}{\partial t^2}$$

and we write $\phi = X(x)Y(y)T(t)$, where $Y(y)$ is a function of y only.
   Differentiating twice and dividing by $\phi = XYT$ gives

$$\frac{X_{xx}}{X} + \frac{Y_{yy}}{Y} = \frac{1}{c^2}\frac{T_{tt}}{T}$$

where the left-hand side depends on x and y only and the right-hand side depends on t only. Since x, y and t are independent variables each side must be equal to a constant, $-k^2$ say. This means that the left-hand side terms in x and y differ by only a constant for all x and y, so that each term is itself equal to a constant. Thus we can write

$$\frac{X_{xx}}{X} = -k_1^2, \quad \frac{Y_{yy}}{Y} = -k_2^2$$

and

$$\frac{1}{c^2}\frac{T_{tt}}{T} = -(k_1^2 + k_2^2) = -k^2$$

giving

$$X_{xx} + k_1^2 X = 0$$
$$Y_{yy} + k_2^2 Y = 0$$
$$T_{tt} + c^2 k^2 T = 0$$

or

$$\phi = A e^{\pm ik_1 x} e^{\pm ik_2 y} e^{\pm ickt}$$

where $k^2 = k_1^2 + k_2^2$. Typically we may write

$$\phi = A {\sin \atop \cos} \} k_1 x {\sin \atop \cos} \} k_2 y {\sin \atop \cos} \} ckt.$$

for example $\phi = A \sin k_2 y \sin(\omega t - k_1 x)$, a standing wave in the y direction and a travelling wave in the x direction. The standing wave is a normal mode.

10.6   Three-Dimensional Case

The three-dimensional treatment is merely a further extension. The wave equation is

$$\frac{\partial^2 \phi}{\partial x^2} + \frac{\partial^2 \phi}{\partial y^2} + \frac{\partial^2 \phi}{\partial z^2} = \frac{1}{c^2}\frac{\partial^2 \phi}{\partial t^2}$$

with a solution

$$\phi = X(x)Y(y)Z(z)T(t)$$

yielding

$$\phi = A {\sin \atop \cos} \} k_1 x {\sin \atop \cos} \} k_2 y {\sin \atop \cos} \} k_3 z {\sin \atop \cos} \} ckt,$$

where $k_1^2 + k_2^2 + k_3^2 = k^2$; for example $\phi = A \sin k_2 y \sin k_3 z \cos(\omega t - k_1 x)$, standing waves in the y and z directions and a travelling wave in the x direction. The standing waves are normal modes.

Using vector notation we may write

$$\phi = A\, e^{i(\omega t - \boldsymbol{k} \cdot \boldsymbol{r})}, \quad \text{where} \quad \boldsymbol{k} \cdot \boldsymbol{r} = k_1 x + k_2 y + k_3 z$$

10.7 Normal Modes in Two Dimensions on a Rectangular Membrane

Suppose waves proceed in a general direction k on the rectangular membrane of sides a and b shown in Figure 10.5. Each dotted wave line is separated by a distance $\lambda/2$ and a standing wave system will exist whenever $a = n_1 \text{AA}'$ and $b = n_2 \text{BB}'$, where n_1 and n_2 are integers.

But

$$\text{AA}' = \frac{\lambda}{2\cos\alpha} = \frac{\lambda}{2}\frac{k}{k_1} = \frac{\lambda}{2}\frac{2\pi}{\lambda}\frac{1}{k_1} = \frac{\pi}{k_1}$$

so that

$$a = \frac{n_1 \pi}{k_1} \quad \text{and} \quad k_1 = \frac{n_1 \pi}{a}.$$

Similarly

$$k_2 = \frac{n_2 \pi}{b}$$

Hence

$$k^2 = k_1^2 + k_2^2 = \frac{4\pi^2}{\lambda^2} = \pi^2\left(\frac{n_1^2}{a^2} + \frac{n_2^2}{b^2}\right)$$

or

$$\frac{2}{\lambda} = \sqrt{\frac{n_1^2}{a^2} + \frac{n_2^2}{b^2}}$$

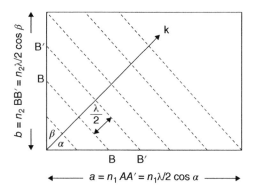

Figure 10.5 *Normal modes on a rectangular membrane in a direction $\boldsymbol{k}$ satisfying boundary conditions of zero displacement at the edges of length $a = n_1\lambda/2\cos\alpha$ and $b = n_2\lambda/2\cos\beta$.*

defining the frequency of the n_1th mode on the x axis and the n_2th mode on the y axis, that is, the $(n_1 n_2)$ normal mode, as

$$v = \frac{c}{2}\sqrt{\frac{n_1^2}{a^2} + \frac{n_2^2}{b^2}}, \quad \text{where} \quad c^2 = \frac{S}{\rho}$$

If k is not normal to the direction of either a or b we can write the general solution for the waves as

$$z = A \left\{ \begin{smallmatrix} \sin \\ \cos \end{smallmatrix} \right\} k_1 x \left\{ \begin{smallmatrix} \sin \\ \cos \end{smallmatrix} \right\} k_2 y \left\{ \begin{smallmatrix} \sin \\ \cos \end{smallmatrix} \right\} ckt.$$

with the boundary conditions $z = 0$ at $x = 0$ and a; $z = 0$ at $y = 0$ and b.

The condition $z = 0$ at $x = y = 0$ requires a $\sin k_1 x \sin k_2 y$ term, and the condition $z = 0$ at $x = a$ defines $k_1 = n_1 \pi / a$. The condition $z = 0$ at $y = b$ gives $k_2 = n_2 \pi / b$, so that

$$z = A \sin \frac{n_1 \pi x}{a} \sin \frac{n_2 \pi y}{b} \sin ckt$$

The fundamental vibration is given by $n_1 = 1$, $n_2 = 1$, so that

$$v = \sqrt{\left(\frac{1}{a^2} + \frac{1}{b^2} \right) \frac{S}{4\rho}}$$

In the general mode $(n_1 n_2)$ zero displacement or nodal lines occur at

$$x = 0, \quad \frac{a}{n_1}, \quad \frac{2a}{n_1}, \ldots a$$

and

$$y = 0, \quad \frac{b}{n_2}, \quad \frac{2b}{n_2}, \ldots b$$

Some of these normal modes are shown in Figure 10.6, where the shaded and plain areas have opposite displacements as shown.

The complete solution for a general displacement would be the sum of individual normal modes, as with the simpler case of waves on a string (see Chapter 11 on Fourier Methods) where boundary conditions of space and time would have to be met. Several modes of different values $(n_1 n_2)$ may have the same frequency, e.g. in a square membrane the modes (4,7) (7,4) (1,8) and (8,1) all have equal frequencies. If the membrane is rectangular and $a = 3b$, modes (3,3) and (9,1) have equal frequencies.

These modes are then said to be degenerate, a term used in describing equal energy levels for electrons in an atom which are described by different quantum numbers.

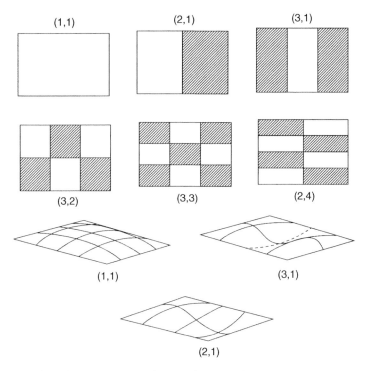

Figure 10.6   *Some normal modes on a rectangular membrane with shaded and clear sections having opposite sinusoidal displacements as indicated.*

10.8   Normal Modes in Three Dimensions

In three dimensions a normal mode is described by the numbers n_1, n_2, n_3, with a frequency

$$\nu = \frac{c}{2}\sqrt{\frac{n_1^2}{l_1^2} + \frac{n_2^2}{l_2^2} + \frac{n_3^2}{l_3^2}}, \tag{10.2}$$

where l_1, l_2 and l_3 are the lengths of the sides of the rectangular enclosure. If we now form a rectangular lattice with the x, y and z axes marked off in units of

$$\frac{c}{2l_1}, \frac{c}{2l_2} \quad \text{and} \quad \frac{c}{2l_3}$$

respectively (Figure 10.7), we can consider a vector of components n_1 units in the x direction, n_2 units in the y direction and n_3 units in the z direction to have a length

$$\nu = \frac{c}{2}\sqrt{\frac{n_1^2}{l_1^2} + \frac{n_2^2}{l_2^2} + \frac{n_3^2}{l_3^2}}$$

Each frequency may thus be represented by a line joining the origin to a point $cn_1/2l_1$, $cn_2/2l_2$, $cn_3/2l_3$ in the rectangular lattice.

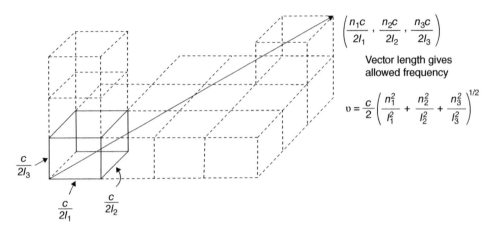

Figure 10.7 *Lattice of rectangular cells in frequency space. The length of the vector joining the origin to any cell corner is the value of the frequency of an allowed normal mode. The vector direction gives the propagation direction of that particular mode.*

The length of the line gives the magnitude of the frequency, and the vector direction gives the direction of the standing waves.

Each point will be at the corner of a rectangular unit cell of sides $c/2l_1$, $c/2l_2$ and $c/2l_3$ with a volume $c^3/8l_1l_2l_3$. There are as many cells as points (i.e. as frequencies) since each cell has eight points at its corners and each point serves as a corner to eight cells.

A very important question now arises: how many normal modes (stationary states in quantum mechanics) can exist in the frequency range ν to $\nu + d\nu$?

The answer to this question is the total number of all those positive integers n_1, n_2, n_3 for which, from equation (10.2),

$$\nu^2 < \frac{c^2}{4}\left(\frac{n_1^2}{l_1^2} + \frac{n_2^2}{l_2^2} + \frac{n_3^2}{l_3^2}\right) < (\nu + d\nu)^2$$

This total is the number of possible points (n_1, n_2, n_3) lying in the positive octant between two concentric spheres of radii ν and $\nu + d\nu$. The other octants will merely repeat the positive octant values because the n's appear as squared quantities.

Hence the total number of possible points or cells will be

$$\frac{1}{8}\frac{(\text{volume of spherical shell})}{\text{volume of cell}}$$

$$= \frac{4\pi\nu^2 d\nu}{8} \cdot \frac{8l_1l_2l_3}{c^3}$$

$$= 4\pi l_1l_2l_3 \cdot \frac{\nu^2 d\nu}{c^3}$$

so that the number of possible normal modes in the frequency range ν to $\nu + d\nu$ *per unit volume* of the enclosure

$$= \frac{4\pi\nu^2 d\nu}{c^3}$$

Note that this result, *per unit volume of the enclosure*, is independent of any particular system; we shall consider two very important applications after the next section.

10.9   3D Normal Frequency Modes and the de Broglie Wavelength

The number of normal modes in the frequency range ν to $\nu + d\nu$ *per unit volume* is $n = 4\pi\nu^2 d\nu/c^3$ where c is the velocity. In a sense, Figure 10.7 combines frequency ν and volume (xyz) spaces because the number of frequencies equals the number of unit cells in the volume V. Such a combination of ν and V spaces is called 'phase space'.

Each dimension of 'phase space' represents a coordinate capable of taking up energy. Another form of 'phase space' is pV space where p is the momentum. A quantum particle of momentum $p = h\nu/c$ where h is Planck's constant and c is the velocity of light allows a straightfoward conversion from νV space to pV space where $n = 4\pi p^2 dp/h^3$ *per unit volume* and where the momentum range is p to $p + dp$. These expressions give the number of unit phase space cells *per unit volume* available to be occupied in statistical distributions.

A statistical distribution answers the question: 'How many gas particles in a gas at temperature T are there in the three-dimensional velocity $v(v_x, v_y, v_z)$ to $dv(v_x, v_y, v_z)$ range and the volume range $dxdydz$?' There are two parts to every statistical distribution. The first part is the number of unit phase cells *per unit volume* available: the second part is the average occupation of each unit cell by a particle. The number $4\pi p^2 dp/h^3$ gives the first part of all statistical distributions.

There are three statistical distributions, the classical distribution, Maxwell–Boltzmann and the quantum distributions Fermi–Dirac and Bose–Einstein. The Maxwell–Boltzmann is classical because there are always many phase space cells available to a particle in the given momentum range. Fermi–Dirac is highly restricted in phase space. Bose–Einstein is not restrictive in phase space but has a quantum, not a classical, occupation number. Planck's Radiation Law is a example of a Bose–Einstein distribution.

The first part of each distribution is n_i where n_i the number of particles equals the number of space cells per unit volume in the momentum range p to $p + dp$, that is $n_i = 4\pi p^2 dp/h^3$. Note that $4\pi p^2 dp$ is the volume of the shell in momentum space between spheres of radii p and $p + dp$.

Over the space volume V there are $4\pi p^2 dpV/h^3$ phase space cells in the momentum range p to $p + dp$. But Heisenberg's Uncertainty Principle (section 6.3) tells us that $(\Delta x \Delta p) \approx h$ so $(\Delta x \Delta p_x)(\Delta y \Delta p_y)(\Delta z \Delta p_z) \approx h^3$ that is, the 'volume' of a cell in pV phase space.

This volume is the smallest acceptable volume which a particle with momentum p in this range may occupy for it defines the volume associated with a particle as $(h/\Delta p_x)^3 \approx (\Delta x)^3 \approx \lambda_{dB}^3$ where λ_{dB} is the de Broglie wavelength. λ_{dB} must not exceed $h/\Delta p_x$ because particles in statistical distributions must be free, that is, have only kinetic and no potential energy arising from interactions with other particles which would occur from the overlap of de Broglie matter waves. The average occupation factor for each of the three distributions is different and a cell may or may not be occupied.

10.10   Frequency Distribution of Energy Radiated from a Hot Body. Planck's Law

The electromagnetic energy radiated from a hot body at temperature T in the small frequency interval ν to $\nu + d\nu$ may be written $E_\nu d\nu$. If this quantity is measured experimentally over a wide range of ν a curve T_1 in Figure 10.8 will result. The general shape of the curve is independent of the temperature, but as T is increased the maximum of the curve increases and shifts towards a higher frequency.

The early attempts to describe the shape of this curve were based on two results we have already used.

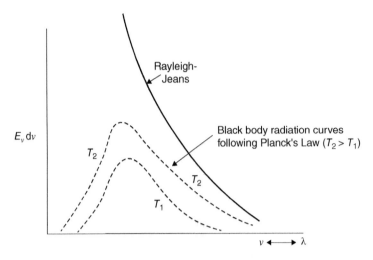

Figure 10.8   *Planck's black body radiation curve plotted for two different temperatures $T_2 > T_1$, together with the curve of the classical Rayleigh–Jeans explanation leading to the 'ultra-violet catastrophe'.*

In the chapter on coupled oscillations we associated normal modes with 'degrees of freedom', the number of ways in which a system could take up energy. In kinetic theory, assigning an energy $\frac{1}{2}kT$ to each degree of freedom of a monatomic gas at temperature T leads to the gas law $pV = RT = NkT$ where N is Avogadro's number, k is Boltzmann's constant and R is the gas constant.

If we assume that each frequency radiated from a hot body is associated with the normal mode of an oscillator with two degrees of freedom and two transverse planes of polarization, the energy radiated per frequency interval $\mathrm{d}\nu$ may be considered as the product of the number of normal modes or oscillators in the interval $\mathrm{d}\nu$ and an energy contribution of kT from each oscillator for each plane of polarization. This gives

$$E_\nu \mathrm{d}\nu = \frac{4\pi\nu^2\,\mathrm{d}\nu\,2kT}{c^3} = \frac{8\pi\nu^2 kT\,\mathrm{d}\nu}{c^3}$$

a result known as the Rayleigh–Jeans Law.

This, however, gives the energy density proportional to ν^2 which, as the solid curve in Figure 10.8 shows, becomes infinite at very high frequencies, a physically absurd result known as the *ultraviolet catastrophe*.

The correct solution to the problem was a major advance in physics. Planck had introduced the quantum theory, which predicted that the average energy value kT should be replaced by the factor $h\nu/(\mathrm{e}^{h\nu/kT} - 1)$, where h is Planck's constant (the unit of action). The experimental curve is thus accurately described by Planck's Radiation Law

$$\boxed{E_\nu \mathrm{d}\nu = \frac{8\pi\nu^2}{c^3}\frac{h\nu}{\mathrm{e}^{h\nu/kT} - 1}\mathrm{d}\nu} \qquad \text{(see Appendix 6)} \qquad (10.3)$$

10.11 Debye Theory of Specific Heats

The success of the modern theory of the specific heats of solids owes much to the work of Debye, who considered the thermal vibrations of atoms in a solid lattice in terms of a vast complex of standing waves over a great range of frequencies. This picture corresponds in three dimensions to the problem of atoms spaced along a one-dimensional line (Chapter 6). In the specific heat theory each atom was allowed two transverse vibrations (perpendicular planes of polarization) and one longitudinal vibration.

The number of possible modes or oscillations per unit volume in the frequency interval ν to $\nu + d\nu$ is then given by

$$dn = 4\pi\nu^2\, d\nu \left(\frac{2}{c_T^3} + \frac{1}{c_L^3} \right) \tag{10.4}$$

where c_T and c_L are respectively the transverse and longitudinal wave velocities.

Each mode has an average energy (from Planck's Law) of $\bar{\varepsilon} = h\nu/(e^{h\nu/kT} - 1)$ and the total energy in the frequency range ν to $\nu + d\nu$ for a gram atom of the solid of volume V_A is then

$$V_A \bar{\varepsilon}\, dn = 4\pi V_A \left(\frac{2}{c_T^3} + \frac{1}{c_L^3} \right) \frac{h\nu^3}{e^{h\nu/kT} - 1}\, d\nu$$

The total energy per gram atom over all permitted frequencies is then

$$E_A = \int V_A \bar{\varepsilon}\, dn = 4\pi V_A \left(\frac{2}{c_T^3} + \frac{1}{c_L^3} \right) \int_0^{\nu_m} \frac{h\nu^3}{e^{h\nu/kT} - 1}\, d\nu$$

where ν_m is the maximum frequency of the oscillations.

There are N atoms per gram atom of the solid (N is Avogadro's number) and each atom has three allowed oscillation modes, so an approximation to ν_m is found by writing the integral of equation (9.2) for a gram atom as

$$\int dn = 3N = 4\pi V_A \left(\frac{2}{c_T^3} + \frac{1}{c_L^3} \right) \int_0^{\nu_m} \nu^2 d\nu = \frac{4\pi V_A}{3} \left(\frac{2}{c_T^3} + \frac{1}{c_L^3} \right) \nu_m^3$$

The values of c_T and c_L can be calculated from the elastic constants of the solid (see Chapter 7 on longitudinal waves) and ν_m can then be found.

The values of E_A thus becomes

$$E_A = \frac{9N}{\nu_m^3} \int_0^{\nu_m} \frac{h\nu}{e^{h\nu/kT} - 1} \nu^2\, d\nu$$

and the variation of E_A with the temperature T is the molar specific heat of the substance at constant volume. The specific heat of aluminium calculated by this method is compared with experimental results in Figure 10.9.

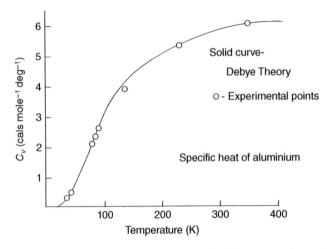

Figure 10.9   *Debye theory of specific heat of solids. Experimental values versus theoretical curve for aluminium.*

Problem 10.1.   A square membrane with sides of length 1 metre each side held under a tension of $2\ \mathrm{N \cdot m^{-1}}$ weighs 100 grams. What is the velocity of waves on the membrane?

Problem 10.2.   Show that

$$z = A\mathrm{e}^{\mathrm{i}\{\omega t-(k_1 x+k_2 y)\}}$$

where $k^2 = \omega^2/c^2 = k_1^2 + k_2^2$ is a solution of the two-dimensional wave equation

$$\frac{\partial^2 z}{\partial x^2} + \frac{\partial^2 z}{\partial y^2} = \frac{1}{c^2}\frac{\partial^2 z}{\partial t^2}$$

Problem 10.3.   Show that if the displacement of the waves on the membrane of width b of Figure 9.3 is given by the superposition

$$z = A_1\, \mathrm{e}^{\mathrm{i}[\omega t-(k_1 x+k_2 y)]} + A_2\, \mathrm{e}^{\mathrm{i}[\omega t-(k_1 x-k_2 y)]}$$

with the boundary conditions

$$z = 0 \quad \text{at} \quad y = 0 \quad \text{and} \quad y = b$$

then the real part of z is

$$z = +2A_1 \sin k_2 y \sin(\omega t - k_1 x)$$

where

$$k_2 = \frac{n\pi}{b}$$

Problem 10.4.   Consider now the extension of Problem 10.3 where the waves are reflected at the rigid edges of the rectangular membrane of sides length a and b as shown in the diagram. The final displacement is the result of the superposition

$$z = A_1 e^{i[\omega t - (k_1 x + k_2 y)]}$$
$$+ A_2 e^{i[\omega t - (k_1 x - k_2 y)]}$$
$$+ A_3 e^{i[\omega t - (-k_1 x - k_2 y)]}$$
$$+ A_4 e^{i[\omega t - (-k_1 x + k_2 y)]}$$

with the boundary conditions

$$z = 0 \quad \text{at} \quad x = 0 \quad \text{and} \quad x = a$$

and

$$z = 0 \quad \text{at} \quad y = 0 \quad \text{and} \quad y = b$$

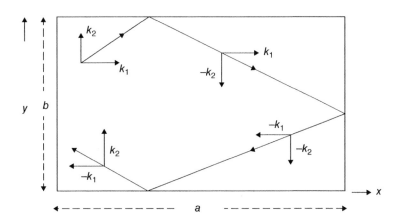

Show that this leads to a displacement

$$z = -4 A_1 \sin k_1 x \sin k_2 y \cos \omega t$$

(the real part of z), where

$$k_1 = \frac{n_1 \pi}{a} \quad \text{and} \quad k_2 = \frac{n_2 \pi}{b}$$

Problem 10.5.   Referring to the figure at the start of the worked example in section 10.3.1, if $\lambda_0 = 2\pi c/\omega$, $\lambda_c = 2\pi/k_x$ and $\lambda_g = 2\pi/k_y$ are the wavelengths propagating in the x and y directions respectively show that

$$\frac{1}{\lambda_c^2} + \frac{1}{\lambda_g^2} = \frac{1}{\lambda_0^2}$$

We see that for $n = 1$, $k_x = \pi/a$ and $\lambda_c = 2a$, and that as ω decreases and λ_0 increases, $k_y = k\sin\theta$ becomes imaginary and the wave is damped. Thus, $n = 2(k_x = 2\pi/a)$ gives $\lambda_c = a$, the 'critical wavelength', i.e. the longest wavelength propagated by a wave guide of separation a. Such cut-off wavelengths and frequencies are a feature of wave propagation in periodic structures, transmission lines and wave guides.

Problem 10.6.   An electromagnetic wave guide has a rectangular cross section of width a in the y direction and height b in the z direction. The wave propagation is in the x direction and there are standing waves normal to it. If E_x has an amplitude A but is 0 at $y = 0$ and a and at $z = 0$ and b, what is the complex expression for E_x?

Problem 10.7.   In problem 10.6 what is the lowest possible of ω (cut-off frequency) for k_x to be real?

Problem 10.8.   The dispersion relation for problem 10.6 and 10.7 is given by $k^2 = \omega^2/c^2 - k_x^2$. Show that the product of the phase velocity ω/k_x and the group velocity $\partial\omega/\partial k_x$ of the wave is c^2 where c is the velocity of light.

Problem 10.9.   A wave guide consists of a pair of parallel conducting plates of width b and separation a. A dielectric of permeability μ and permittivity ϵ fills its volume. If an electromagnetic wave of amplitude E_0 travels down it use either the Poynting vector or the energy per unit volume to show that the power transmitted is $\frac{1}{2}abE_0^2\sqrt{\frac{\epsilon}{\mu}}$.

Problem 10.10.   An electron (mass 9.1×10^{-31} kg) is accelerated through 1 volt to an energy $1\text{eV} = 1.6 \times 10^{-19}$ joules. Its energy $E = p^2/2$ m. Show that its de Broglie wavelength $\lambda_{dB} \approx 1$ nm. Planck's constant $h = 6.63 \times 10^{-34}$ J$\cdot$s^{-1}.

Problem 10.11.   By expanding the term $e^{h\nu/kT} - 1$ in the denominator of Planck's Radiation Law by a Binomial series for $h\nu \ll kT$, show that for long wavelengths Planck's Law becomes the Rayleigh–Jeans expression.

Problem 10.12.   Planck's Radiation Law expressed in terms of E_λ the energy per unit range of wavelength has a maximum λ_m, given by $ch/\lambda_m = 5kT$. Show that if the sun's temperature is about 6000 K, then $\lambda_m \approx 4.7 \times 10^{-7}$ m, the green region of the visible spectrum where the human eye is most sensitive (evolution ?). $c = 3 \times 10^8$ m$\cdot$s^{-1}, h is Planck's constant and k is Boltzmann's constant.

Problem 10.13.   The tungsten filament of an electric light bulb has a temperature of ≈ 2000 K. Show that in this case $\lambda_m \approx 14 \times 10^{-7}$ m, well into the infrared. Such a lamp is therefore a good heat source but an inefficient light source.

11

Fourier Methods

11.1 Fourier Series

In this chapter we are going to look in more detail at the implications of the principles of superposition which we met at the beginning of the book when we added the two separate solutions of the simple harmonic motion equation. Our discussion of monochromatic waves has led to the idea of repetitive behaviour in a simple form. Now we consider more complicated forms of repetition which arise from superposition.

Any function which repeats itself regularly over a given interval of space or time is called a periodic function. This may be expressed by writing it as $f(x) = f(x \pm \alpha)$ where α is the interval or period.

The simplest examples of a periodic function are sines and cosines of fixed frequency and wavelength, where α represents the period τ, the wavelength λ or the phase angle 2π rad, according to the form of x. Most periodic functions, for example the square wave system of Figure 11.1, although quite simple to visualize are more complicated to represent mathematically. Fortunately this can be done for almost all periodic functions of interest in physics using the method of Fourier Series, which states that any periodic function may be represented by the series

$$
\begin{aligned}
f(x) = \frac{1}{2}a_0 + a_1 \cos x + a_2 \cos 2x \ldots + a_n \cos nx \\
+ b_1 \sin x + b_2 \sin 2x \ldots + b_n \sin nx,
\end{aligned}
\tag{11.1}
$$

that is, a constant $\frac{1}{2}a_0$ plus sine and cosine terms of different amplitudes, having frequencies which increase in discrete steps. Such a series must, of course, satisfy certain conditions, chiefly those of convergence. These convergence criteria are met for a function with discontinuities which are not too severe and with first and second differential coefficients which are well behaved. At such discontinuities, for instance in the square wave where $f(x) = \pm h$ at $x = 0, \pm 2\pi$, etc., the series represents the mean of the values of the function just to the left and just to the right of the discontinuity.

Introduction to Vibrations and Waves, First Edition. H. J. Pain and Patricia Rankin.
© 2015 John Wiley & Sons, Ltd. Published 2015 by John Wiley & Sons, Ltd.
Companion website: http://booksupport.wiley.com

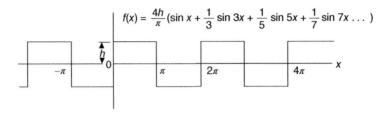

Figure 11.1   *Square wave of height h and its Fourier sine series representation (odd function).*

We may write the series in several equivalent forms:

$$f(x) = \frac{1}{2}a_0 + \sum_{n=1}^{\infty}(a_n \cos nx + b_n \sin nx)$$

$$= \frac{1}{2}a_0 + \sum_{n=1}^{\infty} c_n \cos(nx - \theta_n)$$

where

$$c_n^2 = a_n^2 + b_n^2$$

and

$$\tan \theta_n = b_n/a_n$$

or

$$f(x) = \sum_{n=-\infty}^{\infty} d_n\, e^{inx}$$

where

$$2d_n = a_n - ib_n(n \geq 0)$$

and

$$2d_n = a_{-n} + ib_{-n}(n < 0)$$

To find the values of the coefficients a_n and b_n let us multiply both sides of equation (11.1) by $\cos nx$ and integrate with respect to x over the period 0 to 2π (say).

Every term

$$\int_0^{2\pi} \cos mx \cos nx\, dx = \left\{ \begin{array}{l} 0 \text{ if } m \neq n \\ \pi \text{ if } m = n \end{array} \right.$$

whilst every term

$$\int_0^{2\pi} \sin mx \cos nx \, dx = 0 \text{ for all } m \text{ and } n.$$

Thus for $m = n$,

$$a_n \int_0^{2\pi} \cos^2 nx \, dx = \pi a_n$$

so that

$$a_n = \frac{1}{\pi} \int_0^{2\pi} f(x) \cos nx \, dx$$

Similarly, by multiplying both sides of equation (11.1) by $\sin nx$ and integrating from 0 to 2π we have, since

$$\int_0^{2\pi} \sin mx \sin nx \, dx = \begin{cases} 0 \text{ if } m \neq n \\ \pi \text{ if } m = n \end{cases}$$

that

$$b_n = \frac{1}{\pi} \int_0^{2\pi} f(x) \sin nx \, dx$$

Immediately we see that the constant $(n = 0)$, given by $\frac{1}{2}a_0 = 1/2\pi \int_0^{2\pi} f(x)dx$, is just the average of the function over the interval 2π. It is, therefore, the steady or 'd.c.' level on which the alternating sine and cosine components of the series are superimposed, and the constant can be varied by moving the function with respect to the x axis. When a periodic function is symmetric about the x axis its average value, that is, its steady or d.c. base level, $\frac{1}{2}a_0$, is zero, as in the square wave system of Figure 11.1. If we raise the square waves so that they stand as pulses of height $2h$ on the x axis, the value of $\frac{1}{2}a_0$ is $h\pi$ (average value over 2π). The values of a_n represent twice the average value of the product $f(x) \cos nx$ over the interval 2π; b_n can be interpreted in a similar way.

We see also that the series representation of the function is the sum of cosine terms which are even functions $[\cos x = \cos(-x)]$ and of sine terms which are odd functions $[\sin x = -\sin(-x)]$. Now every function $f(x) = \frac{1}{2}[f(x) + f(-x)] + \frac{1}{2}[f(x) - f(-x)]$, in which the first bracket is even and the second bracket is odd. Thus, the cosine part of a Fourier series represents the even part of the function and the sine terms represent the odd part of the function. Taking the argument one stage further, a function $f(x)$ which is an even function is represented by a Fourier series having only cosine terms; if $f(x)$ is odd it will have only sine terms in its Fourier representation. Whether a function is completely even or completely odd can often be determined by the position of the y axis. Our square wave of Figure 11.1 is an odd function

$[f(x) = -f(-x)]$; it has no constant and is represented by $f(x) = 4h/\pi(\sin x + 1/3\sin 3x + 1/5\sin 5x$, etc., but if we now move the y axis a half period to the right as in Figure 11.2, then $f(x) = f(-x)$, an even function, and the square wave is represented by

$$f(x) = \frac{4h}{\pi}\left(\cos x - \frac{1}{3}\cos 3x + \frac{1}{5}\cos 5x - \frac{1}{7}\cos 7x + \cdots\right)$$

If we take the first three or four terms of the series representing the square wave of Figure 11.1 and add them together, the result is Figure 11.3. The fundamental, or first harmonic, has the frequency of the square wave and the higher frequencies build up the squareness of the wave. The highest frequencies are responsible for the sharpness of the vertical sides of the waves; this type of square wave is commonly used to test the frequency response of amplifiers. An amplifier with a square wave input effectively 'Fourier analyses' the input and responds to the individual frequency components. It then puts them together again at its output, and if a perfect square wave emerges from the amplifier it proves that the amplifier can handle the whole range of the frequency components equally well. Loss of sharpness at the edges of the waves shows that the amplifier response is limited at the higher frequency range.

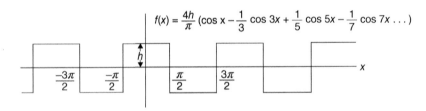

Figure 11.2   *The wave of Figure 11.1 is now symmetric about the y axis and becomes a cosine series (even function).*

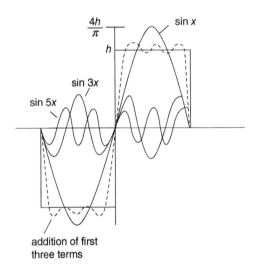

Figure 11.3   *Addition of the first three terms of the Fourier series for the square wave of Figure 11.1 shows that the higher frequencies are responsible for sharpening the edges of the pulse.*

11.1.1   Worked Example of Fourier Series

Consider the square wave of height h in Figure 11.1. The value of the function is given by

$$f(x) = h \quad \text{for} \quad 0 < x < \pi$$

and

$$f(x) = -h \quad \text{for} \quad \pi < x < 2\pi$$

The coefficients of the series representation are given by

$$a_n = \frac{1}{\pi}\left[h\int_0^\pi \cos nx\, dx - h\int_\pi^{2\pi} \cos nx\, dx\right] = 0$$

because

$$\int_0^\pi \cos nx\, dx = \int_\pi^{2\pi} \cos nx\, dx = 0$$

and

$$b_n = \frac{1}{\pi}\left[h\int_0^\pi \sin nx\, dx - h\int_\pi^{2\pi} \sin nx\, dx\right]$$

$$= \frac{h}{n\pi}\left[[\cos nx]_\pi^0 + [\cos nx]_\pi^{2\pi}\right]$$

$$= \frac{h}{n\pi}\left[(1 - \cos n\pi) + (1 - \cos n\pi)\right]$$

giving $b_n = 0$ for n even and $b_n = 4h/n\pi$ for n odd. Thus, the Fourier series representation of the square wave is given by

$$f(x) = \frac{4h}{\pi}\left(\sin x + \frac{\sin 3x}{3} + \frac{\sin 5x}{5} + \frac{\sin 7x}{7} + \cdots\right) \tag{11.2}$$

11.1.2   Fourier Series for any Interval

Although we have discussed the Fourier representation in terms of a periodic function its application is much more fundamental, for any section or interval of a well-behaved function may be chosen and expressed in terms of a Fourier series. This series will accurately represent the function only within the chosen interval. If applied outside that interval it will not follow the function but will periodically repeat the value of the function within the chosen interval. If we represent this interval by a Fourier cosine series the repetition will be that of an even function; if the representation is a Fourier sine series an odd function repetition will follow.

Suppose now that we are interested in the behaviour of a function over only one-half of its full interval and have no interest in its representation outside this restricted region. In Figure 11.4a the function $f(x)$ is

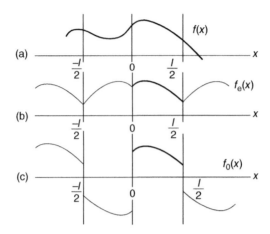

Figure 11.4   *A Fourier series may represent a function over a selected half-interval. The general function in (a) is represented in the half-interval $0 < x < l/2$ by f_e, an even function cosine series in (b), and by f_o, an odd function sine series in (c). These representations are valid only in the specified half-interval. Their behaviour outside that half-interval is purely repetitive and departs from the original function.*

shown over its full space interval $-l/2$ to $+l/2$, but $f(x)$ can be represented completely in the interval 0 to $+l/2$ by either a cosine function (which will repeat itself each half-interval as an even function) or it can be represented completely by a sine function, in which case it will repeat itself each half-interval as an odd function. Neither representation will match $f(x)$ outside the region 0 to $+l/2$, but in the half-interval 0 to $+l/2$ we can write

$$f(x) = f_e(x) = f_o(x)$$

where the subscripts e and o are the even (cosine) or odd (sine) Fourier representations, respectively.

The arguments of sines and cosines must, of course, be phase angles, and so far the variable x has been measured in radians. Now, however, the interval is specified as a distance and the variable becomes $2\pi x/l$, so that each time x changes by l the phase angle changes by 2π.

Thus

$$f_e(x) = \frac{a_0}{2} + \sum_{n=1}^{\infty} a_n \cos \frac{2\pi nx}{l}$$

where

$$a_n = \frac{1}{\frac{1}{2}\text{interval}} \int_{-l/2}^{l/2} f(x) \cos \frac{2\pi nx}{l} dx$$

$$= \frac{2}{l} \left[\int_{-l/2}^{0} f_e(x) \cos \frac{2\pi nx}{l} dx + \int_{0}^{l/2} f_e(x) \cos \frac{2\pi nx}{l} dx \right]$$

$$= \frac{4}{l} \int_{0}^{l/2} f(x) \cos \frac{2\pi nx}{l} dx$$

because

$$f(x) = f_e(x) \quad \text{from} \quad x = 0 \text{ to } l/2$$

and

$$f(x) = f(-x) = f_e(x) \quad \text{from} \quad x = 0 \text{ to } -l/2$$

Similarly we can represent $f(x)$ by the sine series

$$f(x) = f_o(x) = \sum_{n=1}^{\infty} b_n \sin \frac{2\pi nx}{l}$$

in the range $x = 0$ to $l/2$ with

$$b_n = \frac{1}{\frac{1}{2}\text{interval}} \int_{-l/2}^{l/2} f(x) \sin \frac{2\pi nx}{l} dx$$

$$= \frac{2}{l} \left[\int_{-l/2}^{0} f_o(x) \sin \frac{2\pi nx}{l} dx + \int_{0}^{l/2} f_o(x) \sin \frac{2\pi nx}{l} dx \right]$$

In the second integral $f_o(x) = f(x)$ in the interval 0 to $l/2$ whilst

$$\int_{-l/2}^{0} f_o(x) \sin \frac{2\pi nx}{l} dx = \int_{l/2}^{0} f_o(-x) \sin \frac{2\pi nx}{l} dx = -\int_{l/2}^{0} f_o(x) \sin \frac{2\pi nx}{l} dx$$

$$= \int_{0}^{l/2} f_o(x) \sin \frac{2\pi nx}{l} dx = \int_{0}^{l/2} f(x) \sin \frac{2\pi nx}{l} dx$$

Hence

$$b_n = \frac{4}{l} \int_{0}^{l/2} f(x) \sin \frac{2\pi nx}{l} dx$$

If we follow the behaviour of $f_e(x)$ and $f_o(x)$ outside the half-interval 0 to $l/2$ (Figure 11.4b, c) we see that they no longer represent $f(x)$.

11.2   Application of Fourier Sine Series to a Triangular Function

Figure 11.5 shows a function which we are going to describe by a sine series in the half-interval 0 to π. The function is

$$f(x) = x \quad \left(0 < x < \frac{\pi}{2}\right)$$

and

$$f(x) = \pi - x \quad \left(\frac{\pi}{2} < x < \pi\right)$$

Writing $f(x) = \sum b_n \sin nx$ gives

$$b_n = \frac{2}{\pi} \int_0^{\pi/2} x \sin nx \, dx + \frac{2}{\pi} \int_{\pi/2}^{\pi} (\pi - x) \sin nx \, dx$$

$$= \frac{4}{n^2 \pi} \sin \frac{n\pi}{2}$$

$$\left(\text{using } \int x \sin nx dx = \frac{1}{n^2} \sin nx - \frac{x}{n} \cos nx\right), n \neq 0$$

When n is even $\sin n\pi/2 = 0$, so that only terms with odd values of n are present and

$$f(x) = \frac{4}{\pi} \left(\frac{\sin x}{1^2} - \frac{\sin 3x}{3^2} + \frac{\sin 5x}{5^2} - \frac{\sin 7x}{7^2} + \cdots\right)$$

Note that at $x = \pi/2$, $f(x) = \pi/2$, giving

$$\frac{\pi^2}{8} = \frac{1}{1^2} + \frac{1}{3^2} + \frac{1}{5^2} + = \sum_{n=0}^{\infty} \frac{1}{(2n+1)^2}$$

We shall use this result a little later.

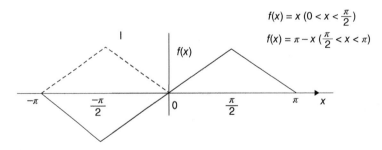

$$f(x) = x \ (0 < x < \tfrac{\pi}{2})$$
$$f(x) = \pi - x \ (\tfrac{\pi}{2} < x < \pi)$$

Figure 11.5   *Function representing a plucked string and defined over a limited interval. When the string vibrates all the permitted harmonics contribute to the initial configuration.*

Note that the solid line in the interval 0 to $-\pi$ in Figure 11.5 is the Fourier sine representation for $f(x)$ repeated outside the interval 0 to π whilst the dotted line would result if we had represented $f(x)$ in the interval 0 to π by an even cosine series.

11.3 Application to the Energy in the Normal Modes of a Vibrating String

If we take a string of length l with fixed ends and pluck its centre a distance d we have the configuration of the half interval 0 to π of Figure 11.5 which we represented as a Fourier sine series. Releasing the string will set up its normal mode or standing wave vibrations, each of which we have shown in section 5.11 to have the displacement

$$y_n = (A_n \cos \omega_n t + B_n \sin \omega_n t) \sin \frac{\omega_n x}{c} \tag{5.4}$$

where $\omega_n = n\pi c/l$ is the normal mode frequency.

The total displacement, which represents the shape of the plucked string at $t = 0$, is given by summing the normal modes

$$y = \sum y_n = \sum (A_n \cos \omega_n t + B_n \sin \omega_n t) \sin \frac{\omega_n x}{c}$$

Note that this sum resembles a Fourier series where the fixed ends of the string, $y = 0$ at $x = 0$ and $x = l$ allow only the sine terms in x in the series expansion. If the string remains plucked at rest only the terms in x with appropriate coefficients are required to describe it, but its vibrational motion after release has a time dependence which is expressed in each harmonic coefficient as

$$A_n \cos \omega_n t + B_n \sin \omega_n t$$

The significance of these coefficients emerges when we consider the initial or boundary conditions in time.

Let us write the total displacement of the string at time $t = 0$ as

$$y_0(x) = \sum y_n(x) = \sum (A_n \cos \omega_n t + B_n \sin \omega_n t) \sin \frac{\omega_n x}{c}$$
$$= \sum A_n \sin \frac{\omega_n x}{c} \quad \text{at} \quad t = 0$$

Similarly we write the velocity of the string at time $t = 0$ as

$$v_0(x) = \frac{\partial}{\partial t} y_0(x) = \sum \dot{y}_n(x)$$
$$= \sum (-\omega_n A_n \sin \omega_n t + \omega_n B_n \cos \omega_n t) \sin \frac{\omega_n x}{c}$$
$$= \sum \omega_n B_n \sin \frac{\omega_n x}{c} \quad \text{at} \quad t = 0$$

Both $y_0(x)$ and $v_0(x)$ are thus expressed as Fourier sine series, but if the string is at rest at $t = 0$, then $v_0(x) = 0$ and all the B_n coefficients are zero, leaving only the A_n's. If the displacement of the string $y_0(x) = 0$ at time $t = 0$ whilst the string is moving, then all the A_n's are zero and the Fourier coefficients are the $\omega_n B_n$'s.

We can solve for both A_n and $\omega_n B_n$ in the usual way for if

$$y_0(x) = \sum A_n \sin \frac{\omega_n x}{c}$$

and

$$v_0(x) = \sum \omega_n B_n \sin \frac{\omega_n x}{c}$$

for a string of length l then

$$A_n = \frac{2}{l} \int_0^l y_0(x) \sin \frac{\omega_n x}{c} dx$$

and

$$\omega_n B_n = \frac{2}{l} \int_0^l v_0(x) \sin \frac{\omega_n x}{c} dx$$

If the plucked string of mass m (linear density ρ) is released from rest at $t = 0$ ($v_0(x) = 0$) the energy in each of its normal modes of vibration, given at the end of section 5.13 as

$$E_n = \frac{1}{4} m \omega_n^2 (A_n^2 + B_n^2)$$

is simply

$$E_n = \frac{1}{4} m \omega_n^2 A_n^2$$

because all B_n's are zero.

The total vibrational energy of the released string will be the sum $\sum E_n$ over all the modes present in the vibration.

Let us now solve the problem of the plucked string released from rest. The configuration of Figure 11.5 (string length l, centre plucked a distance d) is given by

$$y_0(x) = \frac{2dx}{l} \qquad 0 \le x \le \frac{l}{2}$$
$$= \frac{2d(l-x)}{l} \qquad \frac{l}{2} \le x \le l$$

so

$$A_n = \frac{2}{l} \left[\int_0^{1/2} \frac{2dx}{l} \sin \frac{\omega_n x}{c} dx + \int_{1/2}^l \frac{2d(l-x)}{l} \sin \frac{\omega_n x}{c} dx \right]$$
$$= \frac{8d}{n^2 \pi^2} \sin \frac{n\pi}{2} \left(\text{for } \omega_n = \frac{n\pi c}{l} \right)$$

We see at once that $A_n = 0$ for n even (when the sine term is zero) so that all even harmonic modes are missing. The physical explanation for this is that the even harmonics would require a node at the centre of the string which is always moving after release.

The displacement of our plucked string is therefore given by the addition of all the permitted (odd) modes as

$$y_0(x) = \sum_{n\,\text{odd}} y_n(x) = \sum_{n\,\text{odd}} A_n \sin \frac{\omega_n x}{c}$$

where

$$A_n = \frac{8d}{n^2\pi^2} \sin \frac{n\pi}{2}$$

The energy of the nth mode of oscillation is

$$E_n = \frac{1}{4}m\omega_n^2 A_n^2 = \frac{64 d^2 m \omega_n^2}{4(n^2\pi^2)^2}$$

and the total vibrational energy of the string is given by

$$E = \sum_{n\,\text{odd}} E_n = \frac{16 d^2 m}{\pi^4} \sum_{n\,\text{odd}} \frac{\omega_n^2}{n^4} = \frac{16 d^2 c^2 m}{\pi^2 l^2} \sum_{n\,\text{odd}} \frac{1}{n^2} \tag{11.3}$$

for

$$\omega_n = \frac{n\pi c}{l}$$

But we saw in the last section that

$$\sum_{n\,\text{odd}} \frac{1}{n^2} = \frac{\pi^2}{8}$$

so

$$E = \sum E_n = \frac{2 m c^2 d^2}{l^2} = \frac{2 T d^2}{l} \tag{11.4}$$

where $T = \rho c^2$ is the constant tension in the string.

This vibrational energy, in the absence of dissipation, must be equal to the potential energy of the plucked string before release and the reader should prove this by calculating the work done in plucking the centre of the string a small distance d, where $d \ll l$.

To summarize, our plucked string can be represented as a sine series of Fourier components, each giving an allowed normal mode of vibration when it is released. The concept of normal modes allows the energies of each mode to be added to give the total energy of vibration which must equal the potential energy of the plucked string before release. The energy of the nth mode is proportional to n^{-2} and therefore decreases with increasing frequency. Even modes are forbidden by the initial boundary conditions.

The boundary conditions determine which modes are allowed. If the string were struck by a hammer those harmonics having a node at the point of impact would be absent, as in the case of the plucked string. Pianos are commonly designed with the hammer striking a point one seventh of the way along the string, thus eliminating the seventh harmonic which combines to produce discordant effects.

11.4 Fourier Series Analysis of a Rectangular Velocity Pulse on a String

Let us now consider a problem similar to that of the last section except that now the displacement $y_0(x)$ of the string is zero at time $t = 0$ whilst the velocity $v_0(x)$ is non-zero. A string of length l, fixed at both ends, is struck by a mallet of width a about its centre point. At the moment of impact the displacement

$$y_0(x) = 0$$

but the velocity

$$v_0(x) = \frac{\partial y_0(x)}{\partial t} = 0 \quad \text{for} \quad \left|x - \frac{l}{2}\right| \geq \frac{a}{2}$$

$$= v \quad \text{for} \quad \left|x - \frac{l}{2}\right| < \frac{a}{2}$$

This situation is shown in Figure 11.6.
The Fourier series is given by

$$v_0(x) = \sum_n \dot{y}_n = \sum_n \omega_n B_n \sin \frac{\omega_n x}{c}$$

where

$$\omega_n B_n = \frac{2}{l} \int_{+l/2-a/2}^{l/2+a/2} v \sin \frac{\omega_n x}{c} dx$$

$$= \frac{4v}{n\pi} \sin \frac{n\pi}{2} \sin \frac{n\pi a}{2l}$$

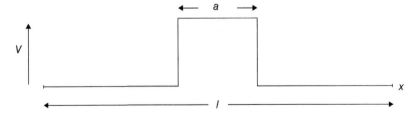

Figure 11.6   *Velocity distribution at time $t = 0$ of a string length l, fixed at both ends and struck about its centre point by a mallet of width a. Displacement $y_0(x) = 0$; velocity $v_0(x) = v$ for $|x - l/2| < a/2$ and zero outside this region.*

Again we see that $\omega_n B_n = 0$ for n even $(\sin n\pi/2 = 0)$ because the centre point of the string is never stationary, as is required in an even harmonic.

Thus

$$v_0(x) = \sum_{n \, \text{odd}} \frac{4v}{n\pi} \sin \frac{n\pi a}{2l}$$

The energy per mode of oscillation

$$\begin{aligned} E_n &= \frac{1}{4} m \omega_n^2 (A_n^2 + B_n^2) \\ &= \frac{1}{4} m \omega_n^2 B_n^2 \quad (\text{All } A_n\text{'s} = 0) \\ &= \frac{1}{4} m \frac{16 v^2}{n^2 \pi^2} \sin^2 \frac{n\pi a}{2l} \\ &= \frac{4 m v^2}{n^2 \pi^2} \sin^2 \frac{n\pi a}{2l} \end{aligned}$$

Now

$$n = \frac{\omega_n}{\omega_1} = \frac{\omega_n l}{\pi c}$$

for the fundamental frequency

$$\omega_1 = \frac{\pi c}{l}$$

So

$$E_n = \frac{4 m v^2 c^2}{l^2 \omega_n^2} \sin^2 \frac{\omega_n a}{2c}$$

Again we see, since $\omega_n \propto n$ that the energy of the nth mode $\propto n^{-2}$ and decreases with increasing harmonic frequency. We may show this by rewriting

$$\begin{aligned} E_n(\omega) &= \frac{m v^2 a^2}{l^2} \frac{\sin^2(\omega_n a/2c)}{(\omega_n a/2c)^2} \\ &= \frac{m v^2 a^2}{l^2} \frac{\sin^2 \alpha}{\alpha^2} \end{aligned}$$

where

$$\alpha = \omega_n a/2c$$

and plotting this expression as an energy-frequency spectrum in Figure 11.7.

The familiar curve of $\sin^2 \alpha/\alpha^2$ again appears as the envelope of the energy values for each ω_n.

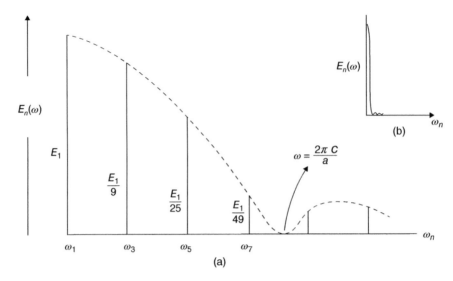

Figure 11.7   (a) Distribution of the energy in the harmonics ω_n of the string of Figure 11.6. The spectrum $E_n(\omega) \propto$ $\sin^2 \alpha/\alpha^2$ where $\alpha = \omega_n a/2c$. Most of the energy in the string is contained in the frequency range $\Delta\omega \approx 2\pi c/a$, and for $a = \Delta x$ (the spatial width of the pulse), $\Delta x/c = \Delta t$ and $\Delta\omega\Delta t \approx 2\pi$ (Bandwidth Theorem). Note that the values of $E_n(\omega)$ for ω_3, ω_5, ω_7, etc. are magnified for clarity. (b) The true shape of the pulse.

If the energy at ω_1 is E_1 then $E_3 = E_1/9$ and $E_5 = E_1/25$ so the major portion of the energy in the velocity pulse is to be found in the low frequencies. The first zero of the envelope $\sin^2 \alpha/\alpha^2$ occurs when

$$\alpha = \frac{\omega a}{2c} = \pi$$

so the width of the central frequency pulse containing most of the energy is given by

$$\omega \approx \frac{2\pi c}{a}$$

This range of energy-bearing harmonics is known as the 'spectral width' of the pulse written

$$\Delta\omega \approx \frac{2\pi c}{a}$$

The 'spatial width' a of the pulse may be written as Δx so we have

$$\Delta x \Delta\omega \approx 2\pi c$$

Reducing the width Δx of the mallet will increase the range of frequencies $\Delta\omega$ required to take up the energy in the rectangular velocity pulse. Now c is the velocity of waves on the string so a wave travels a distance Δx along the string in a time

$$\Delta t = \Delta x/c$$

which defines the duration of the pulse giving

$$\Delta\omega\Delta t \approx 2\pi$$

or

$$\Delta\nu\Delta t \approx 1$$

the Bandwidth Theorem we first met in section 6.2.
Note that the harmonics have frequencies

$$\omega_n = \frac{n\pi c}{l}$$

so $\pi c/l$ is the harmonic interval. When the length l of the string becomes very long and $l \to \infty$ so that the pulse is isolated and non-periodic, the harmonic interval becomes so small that it becomes differential and the Fourier series summation becomes the Fourier Integral.

11.5 Three-Phase Full Wave Rectification

One of the most common ways of producing direct current with low ripple from an alternating current source is three-phase full wave rectification. Figure 11.8a shows the waveform produced by the distributing circuit of Figure 11.8b. Each of the three leads I_+, I_0, I_- produces the same a.c. signal of amplitude I_0 with respect to a neutral point with $I_+ = I_0 \cos(\omega t + \pi/3)$, $I_0 = I_0 \cos \omega t$ and $I_- = I_0 \cos(\omega t - \pi/3)$. The resulting wave is almost fully rectified. The sum of the three components

$$I = I_0 \cos(\omega t + \pi/3) + I_0 \cos \omega t + I_0 \cos(\omega t - \pi/3) = 2I_0 \cos \omega t$$

because $I_+ = I_- = \frac{1}{2}I_0 \cos \omega t$
The value is repeated 6 times per cosine cycle with the Fourier series coefficients equal to

$$a_n = 2I_0 \left(\frac{12}{\pi}\right) \int_0^{\frac{\pi}{6}} \cos \omega t \cos n\omega t = \frac{I_0}{\pi} \frac{12}{n^2 - 1}$$

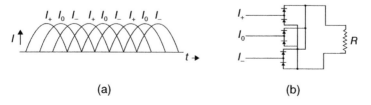

(a) (b)

Figure 11.8 *(a) Waveform of rectified full wave, three-phase a.c. signal. (b) Circuit for producing waveform in (a).*

with alternate $+$ve and $-$ve signs of n giving

$$I = \frac{12I_0}{\pi} \left[\frac{1}{2} + \frac{1}{35} \cos 6\omega t - \frac{1}{143} \cos 12\omega t + \cdots \right]$$

with 6th, 12th and 18th harmonics of the fundamental. Note that the second term of the series is only $\approx 6\%$ of the constant d/c term and the following terms are even smaller. The ripple on the d.c. term is very low.

11.6 The Spectrum of a Fourier Series

The Fourier series can always be represented as a frequency spectrum. In Figure 11.9a a the relative amplitudes of the frequency components of the square wave of Figure 11.1 are plotted, each sine term giving a single spectral line. In a similar manner, the distribution of energy with frequency may be displayed for the plucked string of the earlier section. The frequency of the rth mode of vibration is given by $\omega_r = r\pi c/l$, and the energy in each mode varies inversely with r^2, where r is odd. The spectrum of energy distribution is therefore given by Figure 11.9b.

Suppose now that the length of this string is halved but that the total energy remains constant. The frequency of the fundamental is now increased to $\omega_r' = 2r\pi c/l$ and the frequency interval between consecutive spectral lines is doubled (Figure 11.9c). Again, the smaller the region in which a given amount of energy is concentrated the wider the frequency spectrum required to represent it.

Frequently, a Fourier series is expressed in its complex or exponential form

$$f(t) = \sum_{n=-\infty}^{\infty} d_n e^{in\omega t}$$

where $2\,d_n = a_n - ib_n(n \geq 0)$ and $2\,d_n = a_{-n} + ib_{-n}(n < 0)$.

Because

$$\cos n\omega t = \frac{1}{2}\left(e^{in\omega t} + e^{-in\omega t} \right)$$

and

$$\sin n\omega t = \frac{1}{2i}\left(e^{in\omega t} - e^{-in\omega t} \right)$$

a frequency spectrum in the complex plane produces two spectral lines for each frequency component $n\omega$, one at $+n\omega$ and the other at $-n\omega$. Figure 11.9d shows the cosine representation, which lies wholly in the real plane, and Figure 11.9e shows the sine representation, which is wholly imaginary. The amplitudes of the lines in the positive and negative frequency ranges are, of course, complex conjugates, and the modulus of their product gives the square of the true amplitude. The concept of a negative frequency is seen to arise because the $e^{-in\omega t}$ term increases its phase in the opposite sense to that of the positive term $e^{in\omega t}$. The negative amplitude of the negative frequency in the sine representation indicates that it is in anti-phase with respect to that of the positive term.

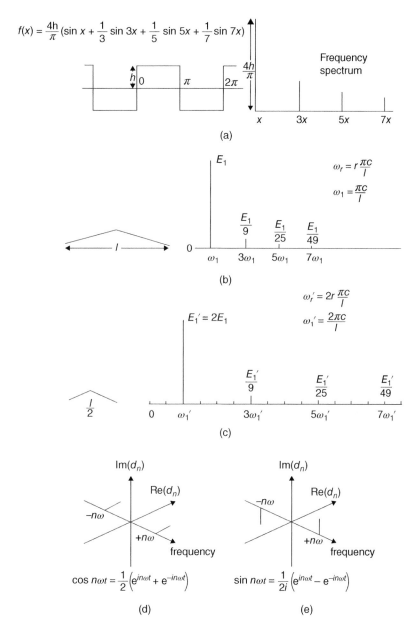

Figure 11.9   *(a) Fourier sine series of a square wave represented as a frequency spectrum; (b) energy spectrum of a plucked string of length l; and (c) the energy spectrum of a plucked string of length l/2 with the same total energy as (b), demonstrating the Bandwidth Theorem that the greater the concentration of the energy in space or time the wider its frequency spectrum. Complex exponential frequency spectrum of (d) $\cos \omega t$ and (e) $\sin \omega t$.*

Problem 11.1.   After inspection of the two waveforms in the diagram what can you say about the values of the constant, absence or presence of sine terms, cosine terms, odd or even harmonics, and range of harmonics required in their Fourier series representation? (Do not use any mathematics.)

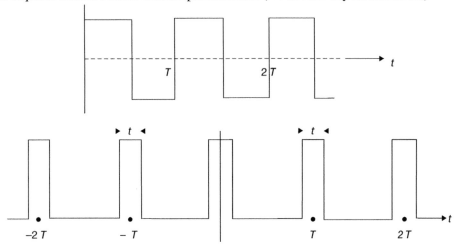

Problem 11.2.   Show that if a periodic waveform is such that each half-cycle is identical except in sign with the previous one, its Fourier spectrum contains no even order frequency components. Examine the result physically.

Problem 11.3.   Show that advancing the phase of the series in Figure 11.1 produces the series in Figure 11.2.

Problem 11.4.   A half-wave rectifier removes the negative half-cycles of a pure sinusoidal wave $y = h \sin x$. Show that the Fourier series is given by

$$y = \frac{h}{\pi}\left(1 + \frac{\pi}{1\cdot 2}\sin x - \frac{2}{1\cdot 3}\cos 2x - \frac{2}{3\cdot 5}\cos 4x - \frac{2}{5\cdot 7}\cos 6x \ldots\right)$$

Problem 11.5.   A full-wave rectifier merely inverts the negative half-cycle in Problem 11.4. Show that this doubles the output and removes the undesirable modulating ripple of the first harmonic.

Problem 11.6.   Can you suggest a method of obtaining the full wave rectifier series by using only the half wave rectifier series (Problem 11.4 and 11.5)? Hint – use Problem 11.3 as a model. The calculation is optional.

Problem 11.7.

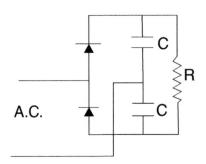

The accompanying circuit denotes the potential of an applied full wave rectifier. Each capacitor is charged once a cycle to a potential $V/2$ where V is the potential appearing across R. The charge flows from the capacitor through R for a time t approximately equal to the period of the a/c cycle. Use an expression for the current to show that the percentage voltage fluctuation across the load $\Delta V/V \approx 2t/RC$ where t is the a.c. period and RC is the relaxation time of the circuit.

Problem 11.8. Use equations 11.3 and 11.4 with $T = \rho c^2$ to show that the lowest three excited modes contain $\approx 93.5\%$ of the total energy.

Problem 11.9. Show that $f(x) = x^2$ may be represented in the interval $\pm \pi$ by

$$f(x) = \frac{\pi^2}{3} + \sum (-1)^n \frac{4}{n^2} \cos nx$$

Problem 11.10. Use the square wave sine series of unit height $f(x) = 4/\pi(\sin x + \frac{1}{3}\sin 3x + \frac{1}{5}\sin 5x)$ to show that

$$1 - \frac{1}{3} + \frac{1}{5} - \frac{1}{7} = \pi/4$$

Problem 11.11. An infinite train of pulses of unit height, with pulse duration 2τ and a period between pulses of T, is expressed as

$$
\begin{aligned}
f(t) &= 0 \quad \text{for} \quad -\frac{1}{2}T < t < -\tau \\
&= 1 \quad \text{for} \quad -\tau < t < \tau \\
&= 0 \quad \text{for} \quad \tau < t < \frac{1}{2}T
\end{aligned}
$$

and

$$f(t + T) = f(t)$$

Show that this is an even function with the cosine coefficients given by

$$a_n = \frac{2}{n\pi} \sin \frac{2\pi}{T} n\tau$$

Problem 11.12. Show, in Problem 11.11, that as τ becomes very small the values of $a_n \to 4\tau/T$ and are independent of n, so that the spectrum consists of an infinite set of lines of constant height and spacing. The representation now has the same form in both time and frequency; such a function is called 'self reciprocal'. What is the physical significance of the fact that as $\tau \to 0$, $a_n \to 0$?

Problem 11.13.   The pulses of Problems 11.11 and 11.12 now have amplitude $1/2\tau$ with unit area under each pulse. Show that as $\tau \to 0$ the infinite series of pulses is given by

$$f(t) = \frac{1}{T} + \frac{2}{T} \sum_{n=1}^{\infty} \cos 2\pi nt/T$$

Under these conditions the amplitude of the original pulses becomes infinite, the energy per pulse remains finite and for an infinity of pulses in the train the total energy in the waveform is also infinite. The amplitude of the individual components in the frequency representation is finite, representing finite energy, but again, an infinity of components gives an infinite energy.

12

Waves in Optics (1) Interference

12.1   Light. Waves or Rays?

Light exhibits a dual nature. In practice, its passage through optical instruments such as telescopes and microscopes is most easily shown by geometrical ray diagrams but the fine detail of the images formed by these instruments is governed by diffraction which, together with interference, requires light to propagate as waves. This chapter explains the basis of wavefront propagation and we shall consider the effects of interference.

The electromagnetic wave nature of light was convincingly settled by Clerk–Maxwell in 1864 but as early as 1690 Huygens was trying to reconcile waves and rays. He proposed that light be represented as a wavefront, each point on this front acting as a source of secondary wavelets whose envelope became the new position of the wavefront, shown in Figure 12.1(a). Light propagation was seen as the progressive development of such a process. In this way, reflection and refraction at a plane boundary separating two optical media could be explained as shown in Figure 12.1(b) and (c).

Huygens' theory was explicit only on those contributions to the new wavefront directly ahead of each point source of secondary waves. No statement was made about propagation in the backward direction nor about contributions in the oblique forward direction. Each of these difficulties is resolved in the more rigorous development of the theory by Kirchhoff which uses the fact that light waves are oscillatory.

The way in which rays may represent the propagation of wavefronts is shown in Figure 12.2 where spherically diverging, plane and spherically converging wavefronts are moving from left to right. All parts of the wavefront (a surface of constant phase) take the same time to travel from the source and all points on the wavefront are the same optical distance from the source. This optical distance must take account of the changes of refractive index met by the wavefront as it propagates. If the physical path length is measured as x in a medium of refractive index n then the optical path length in the medium is the product nx. In travelling from one point to another light chooses a unique optical path which may always be defined in terms of Fermat's Principle.

Introduction to Vibrations and Waves, First Edition. H. J. Pain and Patricia Rankin.
© 2015 John Wiley & Sons, Ltd. Published 2015 by John Wiley & Sons, Ltd.
Companion website: http://booksupport.wiley.com

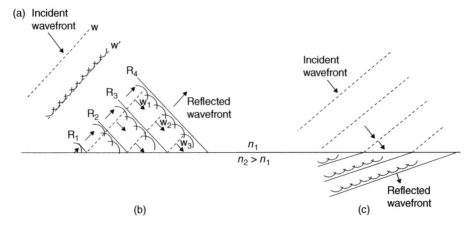

Figure 12.1   *(a) Incident plane wavefront* W *propagates via Huygens wavelets to* W′. *(b) At the plane boundary between the media (refractive index* $n_2 > n_1$*) the incident wavefront* W_1 *has a reflected section* R_1. *Increasing sections* R_2 *and* R_3 *are reflected until the whole wavefront is reflected as* R_4. *(c) An increasing section of the incident wavefront is refracted. Incident wavefronts are shown dashed, and reflected and refracted wavefronts as a continuous line.*

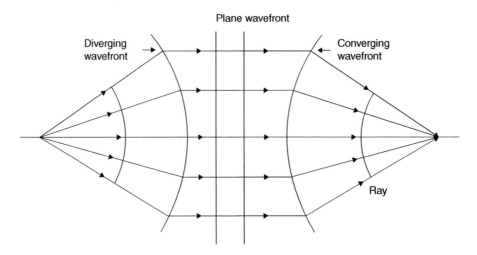

Figure 12.2   *Ray representation of spherically diverging, plane and spherically converging wavefronts.*

12.2   Fermat's Principle

Fermat's Principle states that the optical path length has a stationary value; its first order variation or first derivative in a Taylor series expansion is zero. This means that when an optical path lies wholly within a medium of constant refractive index the path is a straight line, the shortest distance between its end points, and the light travels between these points in the minimum possible time. When the medium has a varying refractive index or the path crosses the boundary between media of different refractive indices the direction of the path always adjusts itself so that the time taken between its end points is a minimum. Fermat's Principle is therefore sometimes known as the Principle of Least Time. Figure 12.3

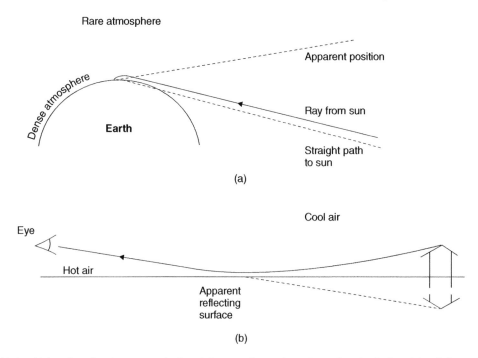

Figure 12.3   *Light takes the shortest optical path in a medium of varying refractive index. (a) A light ray from the sun bends towards the earth in order to shorten its path in the denser atmosphere. The sun remains visible after it has passed below the horizon. (b) A light ray avoids the denser atmosphere and the road immediately below warm air produces an apparent reflection.*

shows examples of light paths in a medium of varying refractive index. As examples of light meeting a boundary between two media we use Fermat's Principle to derive the laws of reflection and refraction.

12.3   The Laws of Reflection

In Figure 12.4a Fermat's Principle requires that the optical path length OSI should be a minimum where O is the object, S lies on the plane reflecting surface and I is the point on the reflected ray at which the image of O is viewed. The plane OSI must be perpendicular to the reflecting surface for, if reflection takes place at any other point S′ on the reflecting surface where OSS′ and ISS′ are right angles then evidently OS′ > OS and IS′ > IS, giving OS′I > OSI.

The laws of reflection also require, in Figure 12.4a, that the angle of incidence i equals the angle of reflection r. If the coordinates of O, S and I are those shown and the velocity of light propagation is c then the time taken to traverse OS is

$$t = (x^2 + y^2)^{1/2}/c$$

and the time taken to traverse SI is

$$t' = [(X - x)^2 + y^2]^{1/2}/c$$

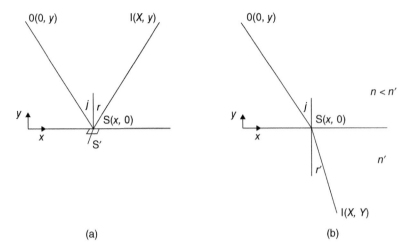

Figure 12.4   *The time for light to follow the path OSI is a minimum (a) in reflection, when OSI forms a plane perpendicular to the reflecting surface and $\hat{\imath} = \hat{r}$; and (b) in refraction, when $n \sin i = n' \sin r'$ (Snell's Law).*

so that the total time taken to travel the path OSI is

$$T = t + t'$$

The position of S is now varied along the x axis and we seek, via Fermat's Principle of Least Time, that value of x which minimizes T, so that

$$\frac{\mathrm{d}T}{\mathrm{d}x} = \frac{x}{c(x^2 + y^2)^{1/2}} - \frac{X - x}{c[(X - x)^2 + y^2]^{1/2}} = 0$$

But

$$\frac{x}{(x^2 + y^2)^{1/2}} = \sin i$$

and

$$\frac{X - x}{[(X - x)^2 + y^2]^{1/2}} = \sin r$$

Hence

$$\sin i = \sin r$$

and

$$\hat{\imath} = \hat{r}$$

12.4   The Law of Refraction

Exactly similar arguments lead to Snell's Law.

Here we express it as

$$n \sin i = n' \sin r'$$

where i is the angle of incidence in the medium of refractive index n and r' is the angle of refraction in the medium of refractive index $n'(n' > n)$. In Figure 12.4b a plane boundary separates the media and light from O $(0, y)$ is refracted at S$(x, 0)$ and viewed at I (X, Y) on the refracted ray. If v and v' are respectively the velocities of light propagation in the media n and n' then OS is traversed in the time

$$t = (x^2 + y^2)^{1/2}/v$$

and SI is traversed in the time

$$t' = [(X - x)^2 + Y^2]^{1/2}/v'$$

The total time to travel from O to I is $T = t + t'$ and we vary the position of S along the x axis which lies on the plane boundary between n and n', seeking that value of x which minimizes T. So

$$\frac{\mathrm{d}T}{\mathrm{d}x} = \frac{1}{v}\frac{x}{(x^2 + y^2)^{1/2}} - \frac{1}{v'}\frac{(X - x)}{[(X - x)^2 + Y^2]^{1/2}} = 0$$

where

$$\frac{x}{(x^2 + y^2)^{1/2}} = \sin i$$

and

$$\frac{(X - x)}{[(X - x)^2 + Y^2]^{1/2}} = \sin r'$$

But

$$\frac{1}{v} = \frac{n}{c}$$

and

$$\frac{1}{v'} = \frac{n'}{c}$$

Hence

$$n \sin i = n' \sin r'$$

12.5   Interference and Diffraction

All waves display the phenomena of interference and diffraction which arise from the superposition of more than one wave. At each point of observation within the interference or diffraction pattern the phase difference between any two component waves of the same frequency will depend on the different paths they have followed and the resulting amplitude may be greater or less than that of any single component. Although we speak of separate waves the waves contributing to the interference and diffraction pattern must ultimately derive from the same single source. This avoids random phase effects from separate sources and guarantees coherence. However, even a single source has a finite size and spatial coherence of the light from different parts of the source imposes certain restrictions if interference effects are to be observed. This is discussed in section 12.14.1, subsection on spatial coherence. In this chapter we shall consider the effects of interference. Chapter 13 will discuss diffraction.

12.6   Interference

Interference effects may be classified in two ways:

1. Division of amplitude
2. Division of wavefront

1. Division of amplitude. Here a beam of light or ray is reflected and transmitted at a boundary between media of different refractive indices. The incident, reflected and transmitted components form separate waves and follow different optical paths. They interfere when they are recombined.
2. Division of wavefront. Here the wavefront from a single source passes simultaneously through two or more apertures each of which contributes a wave at the point of superposition. Diffraction also occurs at each aperture.

The difference between interference and diffraction is merely one of scale: in optical diffraction from a narrow slit (or source) the aperture is of the order of the wavelength of the diffracted light. According to Huygens' Principle every point on the wavefront in the plane of the slit may be considered as a source of secondary wavelets and the further development of the diffracted wave system may be obtained by superposing these wavelets.

In the interference pattern arising from two or more such narrow slits each slit may be seen as the source of a single wave so the number of superposed components in the final interference pattern equals the number of slits (or sources). This suggests that the complete pattern for more than one slit will display both interference and diffraction effects and we shall see that this is indeed the case.

12.7   Division of Amplitude

First of all we consider interference effects produced by division of amplitude. In Figure 12.5 a ray of monochromatic light of wavelength λ in air is incident at an angle i on a plane parallel slab of material thickness t and refractive index $n > 1$. It suffers partial reflection and transmission at the upper surface, some of the transmitted light is reflected at the lower surface and emerges parallel to the first reflection with a phase difference determined by the extra optical path it has travelled in the material. These parallel

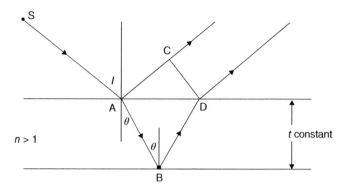

Figure 12.5 *Fringes of constant inclination. Interference fringes formed at infinity by division of amplitude when the material thickness t is constant. The mth order bright fringe is a circle centred at S and occurs for the constant θ value in $2nt\cos\theta = (m + \frac{1}{2})\lambda$.*

beams meet and interfere at infinity but they may be brought to focus by a lens. Their optical path difference is seen to be

$$
\begin{aligned}
n(\mathrm{AB} + \mathrm{BD}) - \mathrm{AC} &= 2n\mathrm{AB} - \mathrm{AC} \\
&= 2nt/\cos\theta - 2t\tan\theta\sin i \\
&= \frac{2nt}{\cos\theta}(1 - \sin^2\theta) = 2nt\cos\theta
\end{aligned}
$$

(because $\sin i = n\sin\theta$).

    This path difference introduces a phase difference

$$
\delta = \frac{2\pi}{\lambda} 2nt\cos\theta
$$

but an additional phase change of π rad occurs at the upper surface.

    The phase difference δ between the two interfering beams is achieved by writing the beam amplitudes as

$$
y_1 = a(\sin\omega t + \delta/2) \quad \text{and} \quad y_2 = a\sin(\omega t - \delta/2)
$$

with a resultant amplitude

$$
\begin{aligned}
R &= a[\sin(\omega t + \delta/2) + \sin(\omega t - \delta/2) \\
&= 2a\sin\omega t\cos\delta/2
\end{aligned}
$$

and an intensity

$$
I = R^2 = 4a^2\sin^2\omega t\cos^2\delta/2
$$

Figure 12.6 shows the familiar $\cos^2\delta/2$ intensity fringe pattern for the spatial part of I.

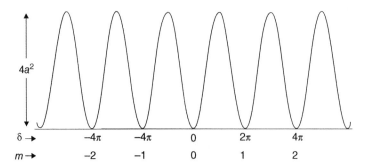

Figure 12.6 *Interference fringes of cos² intensity produced by the division of amplitude in Figure 12.5. The phase difference δ = 2πnt cos θ/λ and m is the order of interference.*

Thus, if $2nt \cos \theta = m\lambda$ (m an integer) the two beams are anti-phase and cancel to give zero intensity, a minimum of interference. If $2nt \cos \theta = (m + \frac{1}{2})\lambda$ the amplitudes will reinforce to give an interference maximum.

Since t is constant the locus of each interference fringe is determined by a constant value of θ which depends on a constant angle i. This gives a circular fringe centred on S. An extended source produces a range of constant θ values at one viewing position so the complete pattern is obviously a set of concentric circular fringes centred on S and formed at infinity. They are fringes of equal inclination and are called Haidinger fringes. They are observed to high orders of interference, that is values of m, so that t may be relatively large.

When the thickness t is not constant and the faces of the slab form a wedge, Figure 12.7a, the interfering rays are not parallel but meet at points (real or virtual) near the wedge. The resulting interference fringes are localized near the wedge and are almost parallel to the thin end of the wedge. When observations are made at or near the normal to the wedge $\cos \theta \sim 1$ and changes slowly in this region so that $2nt \cos \theta \approx 2nt$. The condition for bright fringes then becomes

$$2nt = (m + \tfrac{1}{2})\lambda$$

and each fringe locates a particular value of the thickness t of the wedge and this defines the patterns as fringes of equal thickness. As the value of m increases to $m + 1$ the thickness of the wedge increases by $\lambda/2n$ so the fringes allow measurements to be made to within a fraction of a wavelength and are of great practical importance.

The spectral colours of a thin film of oil floating on water are fringes of constant thickness. Each frequency component of white light produces an interference fringe at that film thickness appropriate to its own particular wavelength.

In the laboratory the most familiar fringes of constant thickness are Newton's Rings.

Worked Example

A thin air wedge is formed by separating one end of a pair of flat glass plates one of which rests on the other. The plates are 50 cm long and the end separation is 0.5 mm. Light of $\lambda = 600$ nm falls almost vertically on the glass. If $x = 0$ when the wedge thickness is $t = 0$, show that the width of a fringe is $\Delta x = \lambda/2\alpha$ where α is the wedge angle in radians. Calculate the number of fringes per cm.

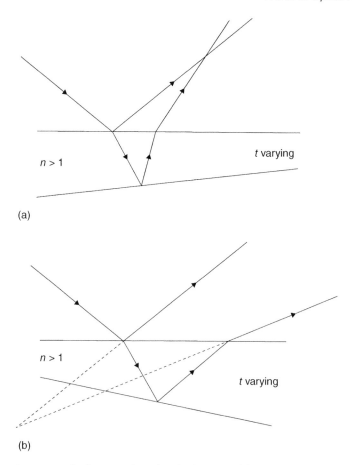

Figure 12.7 *Fringes of constant thickness. When the thickness t of the material is not constant the fringes are localized where the interfering beams meet (a) in a real position and (b) in a virtual position. These fringes are almost parallel to the line where t = 0 and each fringe defines a locus of constant t.*

Solution

The text gives $2t = (m + \frac{1}{2})\lambda/2\alpha$ where $t = x\alpha$ so $\Delta x = x_{m+1} - x_m = \lambda/2\alpha$

$$\alpha = .5 \times 10^{-3}/50 \times 10^{-2} \text{ i.e. } \alpha = 10^{-3} \text{ radians}$$
$$\Delta x = 600 \times 10^{-9}/2 \times 10^{-3} = 3 \times 10^{-4} \text{ m}$$
$$\therefore \text{ No. of fringes per cm } = 10^{-2}/3 \times 10^{-4} = 33.3.$$

12.8 Newton's Rings

Here the wedge of varying thickness is the air gap between two spherical surfaces of different curvature. A constant value of t yields a circular fringe and the pattern is one of concentric fringes alternately dark and bright. The simplest example, Figure 12.8, is a *plano-convex* lens resting on a plane reflecting surface where the system is illuminated from above using a partially reflecting glass plate tilted at 45°. Each downward ray is partially reflected at each surface of the lens and at the plane surface. Interference takes place between the light beams reflected at each surface of the air gap. At the lower (air to glass) surface of the gap there is a π rad phase change upon reflection and the centre of the interference fringe pattern, at the point of contact, is dark. Moving out from the centre, successive rings are light and dark as the air gap thickness increases in units of $\lambda/2$. If R is the radius of curvature of the spherical face of the lens, the thickness t of the air gap at a radius r from the centre is given approximately by $t \approx r^2/2R$. In the mth order of interference a bright ring requires

$$2t = (m + \frac{1}{2})\lambda = r^2/R$$

and because $t \propto r^2$ the fringes become more crowded with increasing r. Rings may be observed with very simple equipment and good quality apparatus can produce fringes for $m > 100$.

Worked Example

We have just seen that the mth bright Newton's ring appears at $2t = (m + \frac{1}{2})\lambda = r^2/R$ where R is the radius of curvature of the plano-convex lens and t is the thickness of the film under the lens at radius r

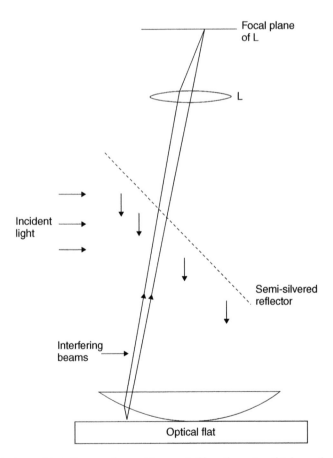

Figure 12.8 *Newton's rings of interference formed by an air film of varying thickness between the lens and the optical flat. The fringes are circular, each fringe defining a constant value of the air film thickness.*

(Figure 12.8). If θ is the angle between the light shining $r = 0$ and $r = r$ where the thickness of the film is t then we may write $t = R(1 - \cos\theta) \approx R(\theta^2/2! + \theta^4/4!) \approx r^2/2R$ where $r^2/2R$ is a parabola which is the first approximation to a circle for small r. This approximation is good enough if the difference between the parabola and the circle is $\ll \lambda$. If $R = 1$ m and $\lambda = 500$ nm $= 5 \times 10^{-7}$ m, then $\theta \ll 5.8 \times 10^{-2}$ radians, i.e. $\approx 3°$. If $r = 1$ cm, $t = 1/200$ cm and $2t \approx m\lambda$. Writing $\lambda = 5 \times 10^{-5}$ cm we have $m = 10^{-2}/5 \times 10^{-5} \approx 200$ rings. $\therefore$ the first order $t = r^2/2R$ is justified.

12.9 Michelson's Spectral Interferometer

This instrument can produce both types of interference fringes, that is, circular fringes of equal inclination at infinity and localized fringes of equal thickness. At the end of the nineteenth century it was one of the most important instruments for measuring the structure of spectral lines.

As shown in Figure 12.9 it consists of two identical plane parallel glass plates G_1 and G_2 and two highly reflecting plane mirrors M_1 and M_2. G_1 has a partially silvered back face, G_2 does not. In the

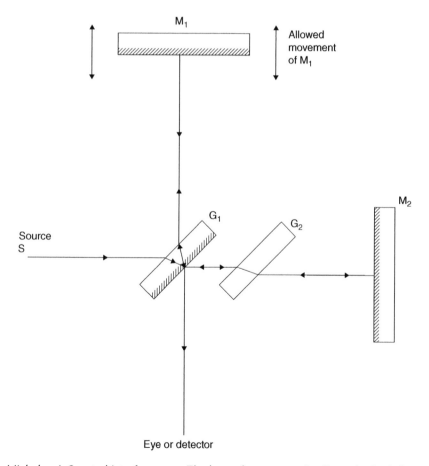

Figure 12.9 *Michelson's Spectral Interferometer. The beam from source S splits at the back face of G_1, and the two parts are reflected at mirrors M_1 and M_2 to recombine and interfere at the eye or detector. G_2 is not necessary with monochromatic light but is required to produce fringes when S is a white light source.*

figure G_1 and G_2 are parallel and M_1 and M_2 are perpendicular. Slow, accurately monitored motion of M_1 is allowed in the direction of the arrows but the mounting of M_2 is fixed although the angle of the mirror plane may be tilted so that M_1 and M_2 are no longer perpendicular.

The incident beam from an extended source divides at the back face of G_1. A part of it is reflected back through G_1 to M_1 where it is returned through G_1 into the eye or detector. The remainder of the incident beam reaches M_2 via G_2 and returns through G_2 to be reflected at the back face of G_1 into the eye or detector where it interferes with the beam from the M_1 arm of the interferometer. The presence of G_2 assures that each of the two interfering beams has the same optical path in glass. This condition is not essential for fringes with monochromatic light but it is required with a white light source where dispersion in glass becomes important.

An observer at the detector looking into G_1 will see M_1, a reflected image of $M_2(M_2'$, say) and the images S_1 and S_2' of the source provided by M_1 and M_2. This may be represented by the linear configuration of Figure 12.10 which shows how interference takes place and what type of fringes are produced.

When the optical paths in the interferometer arms are equal and M_1 and M_2 are perpendicular the planes of M_1 and the image M_2' are coincident. However a small optical path difference t between the arms becomes a difference of $2t$ between the mirrored images of the source as shown in Figure 12.10. The divided ray from a single point P on the extended source is reflected at M_1 and M_2 (shown as M_2') but these reflections appear to come from P_1 and P_2' in the image planes of the mirrors. The path difference between the rays from P_1 and P_2' is evidently $2t \cos \theta$. When $2t \cos \theta = m\lambda$ a maximum of interference occurs and for constant θ the interference fringe is a circle. The extended source produces a range of constant θ values and a pattern of concentric circular fringes of constant inclination.

If the path difference t is very small and the plane of M_2 is now tilted, a wedge is formed and straight localized fringes may be observed at the narrowest part of the wedge. As the wedge thickens the fringes begin to curve because the path difference becomes more strongly dependent upon the angle of observation. These curved fringes are always convex towards the thin end of the wedge.

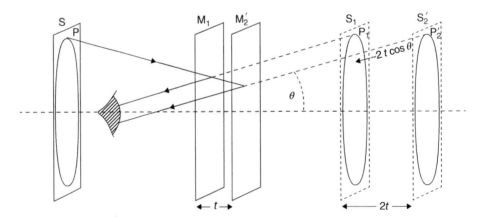

Figure 12.10   *Linear configuration to show fringe formation by a Michelson interferometer. A ray from point P on the extended source S reflects at M_1, and appears to come from P_1 in the reflected plane S_1. The ray is reflected from M_2 (shown here as M_2') and appears to come from P_2' in the reflected plane S_2'. The path difference at the detector between the interfering beams is effectively $2t \cos \theta$ where t is the difference between the path lengths from the source S to the separate mirrors M_1 and M_2.*

12.10 The Structure of Spectral Lines

The discussion on spatial coherence (section 12.14.1, subsection Spacial Coherence) will show that two close identical sources emitting the same wavelength λ produce interference fringe systems slightly displaced from each other (Figure 12.20).

The same effect is produced by a single source, such as sodium, emitting two wavelengths, λ and $\lambda - \Delta\lambda$ so that the maxima and minima of the $\cos^2$ fringes for λ are slightly displaced from those for $\lambda - \Delta\lambda$. This displacement increases with the order of interference m until a value of m is reached when the maximum for λ coincides with a minimum for $\lambda - \Delta\lambda$ and the fringes disappear as their visibility is reduced to zero.

Worked Example

In 1862, Fizeau, using a sodium source to produce Newton's Rings, found that the fringes disappeared at the order $m = 490$ but returned to maximum visibility at $m = 980$. He correctly identified the presence of two components in the spectral line.

The visibility

$$(I_{\max} - I_{\min})/(I_{\max} + I_{\min})$$

equals zero when

$$m\lambda = (m + \tfrac{1}{2})(\lambda - \Delta\lambda)$$

and for $\lambda = 0.5893$ µm and $m = 490$ we have $\Delta\lambda = 0.0006$ µm (6 Å), which are the accepted values for the D lines of the sodium doublet.

Using his spectral interferometer, Michelson extended this work between the years 1890 and 1900, plotting the visibility of various fringe systems and building a mechanical harmonic analyser into which he fed different component frequencies in an attempt to reproduce his visibility curves. The sodium doublet with angular frequency components ω and $\omega + \Delta\omega$ produced a visibility curve similar to that of Figure 12.20 and was easy to interpret. More complicated visibility patterns were not easy to reproduce and the modern method of Fourier transform spectroscopy reverses the procedure by extracting the frequency components from the observed pattern.

Michelson did however confirm that the cadmium red line, $\lambda = 0.6438$ µm, was highly monochromatic. The visibility had still to reach a minimum when the path difference in his interferometer arms was 0.2 m.

Worked Example

Michelson was unlucky. Another 4.06 cm and his fringes would have disappeared. The coherence length of a wavetrain is given by $\Delta l = \lambda^2/\Delta\lambda$, calculated as follows:

$$c = \nu\lambda \;\therefore\; \Delta\nu = c\left|\frac{\Delta\nu}{\nu^2}\right| = \lambda^2\left(\frac{\Delta\nu}{c}\right) \quad \text{where } \Delta\nu = 1/\Delta t \text{(Bandwidth Theorem)}.$$

$$\therefore \Delta\lambda = \lambda^2/c\Delta t \text{ and } \Delta l = c\Delta t = \lambda^2/\Delta\lambda$$

Cadmium light has $\lambda = 643.847$ nm with a line width of $\Delta\lambda = 0.0013$ nm.

$$\therefore \Delta l = (6 \cdot 4.3 \cdot 847 \times 10^{-9})^2/0.0013 \times 10^{-9} = 31.89 \,\text{cm}$$

When the Optical Path Difference $2t = \Delta l$ the fringes disappear, so at $t = 15.94$ cm the fringe visibility is zero. Michelson reached $t = 20$ cm.

12.11   Fabry–Pérot Interferometer

The interference fringes produced by division of amplitude which we have discussed so far have been observed as reflected light and have been produced by only two interfering beams. We now consider fringes which are observed in transmission and which require multiple reflections. They are fringes of constant inclination formed in a pattern of concentric circles by the Fabry–Pérot interferometer. The fringes are particularly narrow and sharply defined so that a beam consisting of two wavelengths λ and $\lambda - \Delta\lambda$ forms two patterns of rings which are easily separated for small $\Delta\lambda$. This instrument therefore has an extremely high resolving power. The main component of the interferometer is an etalon (Figure 12.11) which consists of two plane parallel glass plates with identical highly reflecting inner surfaces S_1 and S_2 which are separated by a distance d.

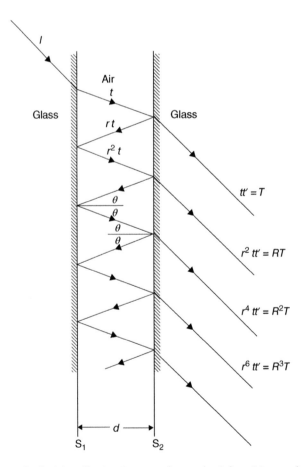

Figure 12.11 S_1 and S_2 are the highly reflecting inner surfaces of a Fabry–Pérot etalon with a constant air gap thickness d. Multiple reflections produce parallel interfering beams with amplitudes T, RT, R^2T, etc. each beam having a phase difference $\delta = 4\pi d \cos\theta / \lambda$ with respect to its neighbour.

Suppose a monochromatic beam of unit amplitude, angular frequency ω and wavelength (in air) of λ strikes the surface S_1 as shown. A fraction t of this beam is transmitted in passing from glass to air. At S_2 a further fraction t' is transmitted in passing from air to glass to give an emerging beam of amplitude $tt' = T$. The reflection coefficient at the air $-S_1$ and air$-S_2$ surfaces is r so each subsequent emerging beam is parallel but has an amplitude factor $r^2 = R$ with respect to its predecessor. Other reflection and transmission losses are common to all beams and do not affect the analysis. Each emerging beam has a phase lag $\delta = 4\pi d \cos\theta/\lambda$ with respect to its predecessor and these parallel beams interfere when they are brought to focus via a lens.

The vector sum of the transmitted interfering amplitudes together with their appropriate phases may be written

$$A = Te^{i\omega t} + TRe^{i(\omega t-\delta)} + TR^2 e^{i(\omega t-2\delta)} \cdots$$
$$= Te^{i\omega t}[1 + Re^{-i\delta} + R^2 e^{-i2\delta} \cdots]$$

which is an infinite geometric progression with the sum

$$A = Te^{i\omega t}/(1 - Re^{-i\delta})$$

This has a complex conjugate

$$A^* = Te^{-i\omega t}/(1 - Re^{i\delta})$$

If the incident unit intensity is I_0 the fraction of this intensity in the transmitted beam may be written

$$\frac{I_t}{I_0} = \frac{|AA^*|}{I_0} = \frac{T^2}{(1 - Re^{-i\delta})(1 - Re^{i\delta})} = \frac{T^2}{(1 + R^2 - 2R\cos\delta)}$$

(See section 2.1 Complex Numbers (vi).)
or, with

$$\cos\delta = 1 - 2\sin^2\delta/2$$

as

$$\frac{I_t}{I_0} = \frac{T^2}{(1 - R)^2 + 4R\sin^2\delta/2} = \frac{T^2}{(1 - R)^2} \frac{1}{1 + [4R\sin^2\delta/2/(1 - R)^2]}$$

But the factor $T^2/(1 - R)^2$ is a constant, written C, so

$$\frac{I_t}{I_0} = C \cdot \frac{1}{1 + [4R\sin^2\delta/2/(1 - R)^2]}$$

Writing $CI_0 = I_{max}$, the graph of I_t versus δ in Figure 12.12 shows that as the reflection coefficient of the inner surfaces is increased, the interference fringes become narrow and more sharply defined. Values of $R > 0.9$ may be reached using the special techniques of multilayer dielectric coating. In one of these techniques a glass plate is coated with alternate layers of high and low refractive index materials so that each boundary presents a large change of refractive index and hence a large reflection. If the optical thickness of each layer is $\lambda/4$ the emerging beams are all in phase and the reflected intensity is high.

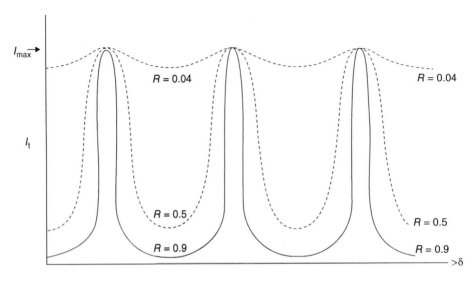

Figure 12.12 *Observed intensity of fringes produced by a Fabry–Pérot interferometer. Transmitted intensity I_t versus δ. $R = r^2$ where r is the reflection coefficient of the inner surfaces of the etalon. As R increases the fringes become narrower and more sharply defined.*

Worked Example

If the reflection coefficient of intensity of a Fabry–Pérot etalon is 60%, what is the ratio of intensity $I_{max} = CI_0$ to that half way between two maxima?

Solution

The text gives

$$I_t = I_{max} \frac{1}{1 + \left[\frac{4R \sin^2 \delta/2}{(1-R)^2} \right]} \text{ where } I_{max} = CI_0$$

The phase difference between two maxima is $\delta = 2\pi/2 = \pi$ so in the equation for I_t/I_{max}, $\sin^2 \delta/2 = \sin^2 \pi/2 = 1$ and

$$I_t = I_{max} \frac{1}{1 + 4R/(1-R)^2} = I_{max} \frac{1}{1 + (4 \cdot 0.6)/.4^2} = \frac{I_{max}}{1 + 15}$$

∴ Intensity at half way between two maxima is $I_{max}/16$.

12.12   Resolving Power of the Fabry–Pérot Interferometer

Figure 12.12 shows that a value of $R = 0.9$ produces such narrow and sharply defined fringes that if the incident beam has two components λ and $\lambda - \Delta\lambda$ the two sets of fringes should be easily separated. The criterion for separation depends on the shape of the fringes: the diffraction grating of section 13.16,

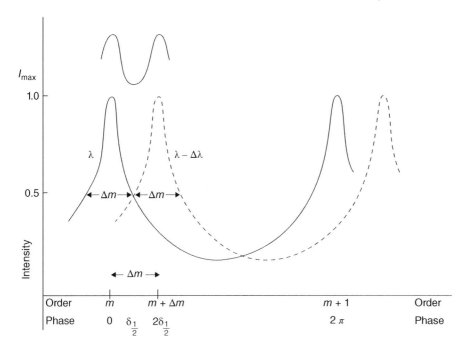

Figure 12.13 *Fabry–Pérot interference fringes for two wavelength λ and $\lambda - \Delta\lambda$ are resolved at order m when they cross at half their maximum intensity. Moving from order m to m + 1 changes the phase δ by 2π rad and the full 'half-value' width of each maximum is given by $\Delta m = 2\delta_{1/2}$ which is also the separation between the maxima of λ and $\lambda - \Delta\lambda$ when the fringes are just resolved.*

Resolving Power of Diffraction Grating, uses the Rayleigh criterion, but the fringes here are so sharp that they are resolved at a much smaller separation than that required by Rayleigh.

Here the fringes of the two wavelengths may be resolved when they cross at half their maximum intensities; that is, at $I_t = I_{\max}/2$ in Figure 12.13.

Using the expression

$$I_t = I_{\max} \cdot \frac{1}{1 + \frac{4R \sin^2 \delta/2}{(1-R)^2}}$$

we see that $I_t = I_{\max}$ when $\delta = 0$ and $I_t = I_{\max}/2$ when the factor

$$4R \sin^2 \delta/2/(1 - R)^2 = 1$$

The fringes are so narrow that they are visible only for very small values of δ so we may replace $\sin \delta/2$ by $\delta/2$ in the expression

$$4R \sin^2 \delta/2/(1 - R)^2 = 1$$

to give the value

$$\delta_{1/2} = \frac{1 - R}{R^{1/2}}$$

as the phase departure from the maximum, $\delta = 0$, which produces the intensity $I_t = I_{max}/2$ for wavelength λ. Our criterion for resolution means, therefore, that the maximum intensity for $\lambda - \Delta\lambda$ is removed an extra amount $\delta_{1/2}$ along the phase axis of Figure 12.13. This axis also shows the order of interference m at which the wavelengths are resolved, together with the order $m + 1$ which represents a phase shift of $\delta = 2\pi$ along the phase axis.

12.12.1   Resolving Power

In the mth order of interference we have

$$2d \cos \theta = m\lambda$$

and for fringes of equal inclination (θ constant), logarithmic differentiation gives

$$\lambda/\Delta\lambda_{min} = -m/\Delta m$$

where $\Delta\lambda_{min}$ is the minimum resovable difference between two wavelengths.

Now $\Delta m = 1$ represents a phase change of $\delta = 2\pi$ and the phase difference of $2.\delta_{1/2}$ which just resolves the two wavelengths corresponds to a change of order

$$\Delta m = 2.\delta_{1/2}/2\pi$$

Thus, the resolving power, defined as

$$\text{R.P.} = \frac{\lambda}{\Delta\lambda_{min}} = \left|\frac{m}{\Delta m}\right| = \frac{m\pi}{\delta_{1/2}} = \frac{m\pi R^{1/2}}{(1 - R)} = \frac{\pi R^{1/2}}{1 - R}\frac{2d}{\lambda}$$

where $\Delta\lambda_{min}$ is the minimum resolvable difference between two wavelengths.

12.12.2   Finesse

The equivalent expression for the resolving power in the mth order for a diffracting grating of N lines (interfering beams) is shown in section 13.6, Resolving Power of Diffraction Grating, to be

$$\frac{\lambda}{\Delta\lambda_{min}} = mN$$

so we may express

$$N' = \pi R^{1/2}/(1 - R)$$

as the effective number of interfering beams in the Fabry–Pérot interferometer.

This quantity N' is called the finesse of the etalon and is a measure of its quality. We see that

$$N' = \frac{2\pi}{2\delta_{1/2}} = \frac{1}{\Delta m} = \frac{\text{separation between orders } m \text{ and } m+1}{\text{'half value' width of } m\text{th order}}$$

Thus, using one wavelength only, the ratio of the separation between successive fringes to the narrowness of each fringe measures the quality of the etalon. A typical value of $N' \sim 30$.

Worked Example

If the reflection coefficient intensity of a Fabry-Pérot etalon is 90% and the etalon plate separation $d = 10$ mm, calculate the resolving power $\lambda/\Delta\lambda_{\min}$ for length of $\lambda = 600$ nm.

Solution

$$\lambda/\Delta\lambda_{\min} = \pi R^{1/2} 2d/(1-R)\lambda = \frac{\pi(.9)^{1/2} 20 \cdot 10^{-2} \cdot 10^9}{0.1 \times 600}$$
$$\approx \frac{10 \cdot \pi \cdot 20 \cdot 10^7}{600} \approx \frac{2\pi}{6} \cdot 10^7 \approx 10^7$$

12.12.3   Free Spectral Range

There is a limit to the wavelength difference $\Delta\lambda$ which can be resolved with the Fabry–Pérot interferometer. This limit is reached when $\Delta\lambda$ is such that the circular fringe for λ in the mth order coincides with that for $\lambda - \Delta\lambda$ in the $m + 1$th order. The pattern then loses its unique definition and this value of $\Delta\lambda$ is called the free spectral range.

From the preceding section we have the expression

$$\frac{\lambda}{\Delta\lambda} = -\frac{m}{\Delta m}$$

and in the limit when $\Delta\lambda$ represents the free spectral range then

$$\Delta m = 1$$

and

$$\Delta\lambda_{\text{fsr}} = -\lambda/m$$

where the subscript fsr indicates free spectral range. But $m\lambda = 2d$ when $\theta \simeq 0$ so the free spectral range

$$\Delta\lambda_{\text{fsr}} = -\lambda^2/2d$$

But

$$\frac{\lambda^2}{2d} = N'\Delta\lambda_{\min}$$

so

$$\frac{\lambda_{\text{fsr}}}{\lambda_{\text{min}}} = \frac{\pi R^{1/2}}{1 - R} = \text{Finesse}$$

The aim therefore is to have λ_{fsr} as large as possible and λ_{min} as small as possible.

12.12.4 The Laser Cavity

The laser cavity is in effect an extended Fabry–Pérot etalon. Mirrors coated with multi-dielectric films can produce reflection coefficients $R \approx 0.99$ and the amplified stimulated emission in the laser produces a beam which is continuously reflected between the mirror ends of the cavity. The high value of R allows the amplitudes of the beam in opposing directions to be taken as equal, so a standing wave system is generated (Figure 12.14) to form a longitudinal mode in the cavity.

The superposed amplitudes after a return journey from one mirror to the other and back are written for a wave number k and a frequency $\omega = 2\pi\nu$ as

$$E = A_1\left(e^{i(\omega t - kx)} - e^{i(\omega t + kx)}\right)$$
$$= A_1(e^{-ikx} - e^{ikx})e^{i\omega t} = -2iA_1 \sin kx\, e^{i\omega t}$$

of which the real part is $E = 2A_1 \sin kx \sin \omega t$.

If the cavity length is L, one round trip between the mirrors creates a phase change of

$$\phi = -2Lk + 2\alpha = -\frac{4\pi L}{c}\nu + 2\alpha$$

where α is the phase change on reflection at each mirror.

For this standing wave mode to be maintained, the phase change must be a multiple of 2π, so for m an integer

$$\phi = m2\pi = \frac{4\pi L}{c}\nu - 2\alpha$$

or

$$\nu = \frac{mc}{2L} + \frac{\alpha c}{2\pi L}$$

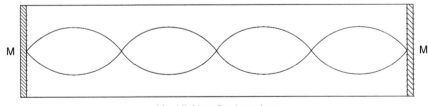

M = Highly reflecting mirror

Figure 12.14 *A longitudinal mode in a laser cavity which behaves as an extended Fabry–Pérot etalon with highly reflecting mirrors at each end. The standing wave system acquires an extra $\lambda/2$ for unit change in the mode number m. A typical output is shown in Figure 12.15.*

When m changes to $m + 1$, the phase change of 2π corresponds to an extra wavelength λ for the return journey; that is, an extra $\lambda/2$ in the standing wave mode. A series of longitudinal modes can therefore exist with frequency intervals $\Delta\nu = c/2L$ determined by a unit change in m.

The intensity profile for each mode and the separation $\Delta\nu$ is best seen by reference to Figure 12.13, where $\phi \equiv \delta$ is given by the unit change in the order of interference from m to $m + 1$.

The intensity profile for each cavity mode is that of Figure 12.13, where the full width at half maximum intensity is given by the phase change

$$2\delta_{1/2} = \frac{2(1 - R)}{R^{1/2}}$$

where R is the reflection coefficient. This corresponds to a full width intensity change over a frequency $d\nu$ generated by the phase change

$$d\phi = \frac{4\pi L}{c}d\nu \text{ in the expression for } \phi \text{ above}$$

The width at half maximum intensity for each longitudinal mode is therefore given by

$$\frac{4\pi L}{c}d\nu = \frac{2(1 - R)}{R^{1/2}}$$

or

$$d\nu = \frac{(1 - R)c}{R^{1/2}2\pi L}$$

For a laser 1 m long with $R = 0.99$, the longitudinal modes are separated by frequency intervals

$$\Delta\nu = \frac{c}{2L} = 1.5 \times 10^8 \text{ Hz}$$

Each mode intensity profile has a full width at half maximum of

$$d\nu = 10^{-2}\frac{c}{2\pi} \approx 4.5 \times 10^5 \text{Hz}$$

Worked Example

For a He–Ne laser the mean frequency of the output at 632.8 nm is 4.74×10^{14} Hz. The pattern for $\Delta\nu$ and $d\nu$ is shown in Figure 12.15, where the intensities are reduced under the dotted envelope as the frequency difference from the mean is increased.

The finesse of the laser cavity is given by

$$\frac{\Delta\nu}{d\nu} = \frac{1.5 \times 10^8}{4.5 \times 10^5} \approx 300$$

for the example quoted.

The intensity of each longitudinal mode is, of course, amplified by each passage of the stimulated emission. Radiation allowed from out of one end represents the laser output but the amplification process is dominant and the laser produces a continuous beam.

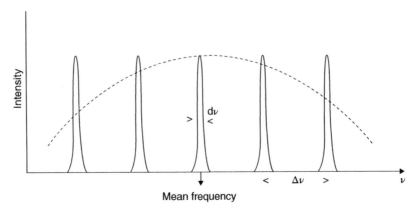

Figure 12.15   *Output of a laser cavity. A series of longitudinal modes separated by frequency intervals $\Delta\nu = c/2L$, where c is the velocity of light and L is the cavity length. The modes are centred about the mean output frequency and are modulated under the dotted envelope. For a He–Ne Laser 1 m long the separation $\Delta\nu$ between the modes ≈ 300 full widths of a mode intensity profile at half its maximum value.*

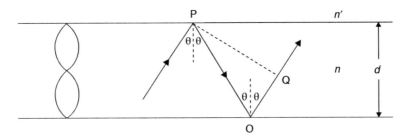

Figure 12.16   *A thin dielectric film or fibre acts as an optical wave guide. The reflection angle θ must satisfy the relation $n\sin\theta \geq n'$, where n' is the refractive index of the coating over the film of refractive index n. Propagating modes have standing wave systems across the film and constructive interference occurs on the standing wave axis where the amplitude is a maximum. Point Q (with added phase) would be midway between the cavity walls to coincide with a maximum amplitude of $m = 1$ for standing wave. Destructive interference occurs at the nodes.*

12.12.5   Total Internal Reflection

Snell's Law of Refraction gives $n\sin\theta = n'\sin\phi$. When light travels from a dielectric into air $n > n'$ and $\phi > \theta$. Eventually at some critical angle θ_c, $\phi = 90°$ and $\sin\theta = n'/n < 1$. For glass into air $n' = 1/1.5$ and $\theta_c = 42°$. Beyond θ_c the light is totally internally reflected with $\theta_i = \theta_r$.

12.12.6   The Thin Film Optical Wave Guide

Figure 12.16 shows a thin film of width d and refractive index n along which light of frequency ν and wave number k is guided by multiple internal reflections. The extent of the wave guide is infinite in the direction normal to the page. The internal reflection angle θ must satisfy

$$n\,\sin\theta \geq n'$$

where n' is the refractive index of the medium bounding the thin film surfaces. Each reflected ray is normal to a number of wavefronts of constant phase separated by λ, where $k = 2\pi/\lambda$ and constructive interference is necessary for any mode to propagate. Reflections may take place at any pair of points P and O along the film and we examine the condition for constructive interference by considering the phase difference along the path POQ, taking into account a phase difference α introduced by reflection at each of P and O.

Now

$$PO = \cos\theta/d$$

and

$$OQ = PO \cos 2\theta$$

so with

$$\cos 2\theta = 2\cos^2\theta - 1$$

we have

$$PO + OQ = 2d\cos\theta$$

giving a phase difference

$$\Delta\phi = \phi_Q - \phi_P = -\frac{2\pi\nu}{c}\,(n\,2d\cos\theta) + 2\alpha$$

Constructive interference requires

$$\Delta\phi = m\,2\pi$$

where m is an integer, so we write

$$m\,2\pi = \frac{2\pi\nu}{c}n\,2d\cos\theta - 2\pi\Delta m$$

where

$$\Delta m = 2\alpha/2\pi$$

represents the phase change on reflection.

Radiation will therefore propagate only when

$$\cos\theta = \frac{c(m + \Delta m)}{\nu 2nd}$$

for $m = 0, 1, 2, 3.$

The condition $n \sin \theta \geq n'$ restricts the values of the frequency ν which can propagate. If $\theta = \theta_m$ for mode m and

$$\cos \theta_m = (1 - \sin^2 \theta_m)^{1/2}$$

then

$$n \sin \theta_m \geq n'$$

becomes

$$\cos \theta_m \leq \left[1 - \left(\frac{n'}{n} \right)^2 \right]^{1/2}$$

and ν must satisfy

$$\nu \geq \frac{c(m + \Delta m)}{2d(n^2 - n'^2)^{1/2}}$$

The mode $m = 0$ is the mode below which ν will not propagate, while Δm is a constant for a given wave guide. Each mode is represented by a standing wave system across the wave guide normal to the direction of propagation. Constructive interference occurs on the axis of this wave system where the amplitude is a maximum and destructive interference occurs at the nodes.

12.13   Division of Wavefront

12.13.1   Interference between Waves from Two Slits or Sources

In Figure 12.17, let S_1 and S_2 be two equal sources separated by a distance f, each generating a wave of angular frequency ω and amplitude a. At a point P sufficiently distant from S_1 and S_2 only plane wavefronts arrive with displacements

$$y_1 = a \sin(\omega t - kx_1) \quad \text{from } S_1$$

and

$$y_2 = a \sin(\omega t - kx_2) \quad \text{from } S_2$$

so that the phase difference between the two signals at P is given by

$$\delta = k(x_2 - x_1) = \frac{2\pi}{\lambda}(x_2 - x_1)$$

This phase difference δ, which arises from the path difference $x_2 - x_1$, depends only on x_1, x_2 and the wavelength λ and not on any variation in the source behaviour. This requires that there shall be no sudden changes of phase in the signal generated at either source – such sources are called coherent.

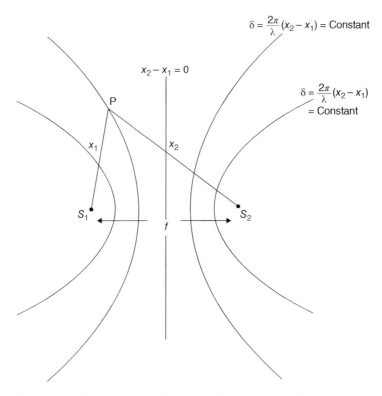

$$\delta = \frac{2\pi}{\lambda}(x_2 - x_1) = \text{Constant}$$

$x_2 - x_1 = 0$

$$\delta = \frac{2\pi}{\lambda}(x_2 - x_1)$$
$$= \text{Constant}$$

Figure 12.17   *Interference at P between waves from equal sources S_1 and S_2, separation f, depends only on the path difference $x_2 - x_1$. Loci of points with constant phase difference $\delta = (2\pi/\lambda)(x_2 - x_1)$ are the family of hyperbolas with S_1 and S_2 as foci.*

The superposition of displacements at P gives a resultant

$$R = y_1 + y_2 = a[\sin(\omega t - kx_1) + \sin(\omega t - kx_2)]$$

Writing $X \equiv (x_1 + x_2)/2$ as the average distance from the two sources to point P we obtain

$$kx_1 = kX - \delta/2 \quad \text{and} \quad kx_2 = kX + \delta/2$$

to give

$$R = a[\sin(\omega t - kX + \delta/2) + \sin(\omega t - kX - \delta/2)]$$
$$= 2a\sin(\omega t - kX)\cos\delta/2$$

and an intensity

$$I = R^2 = 4a^2\sin^2(\omega t - kX)\cos^2\delta/2$$

When

$$\cos \frac{\delta}{2} = \pm 1$$

the spatial intensity is a maximum,

$$I = 4a^2$$

and the component displacements reinforce each other to give constructive interference. This occurs when

$$\frac{\delta}{2} = \frac{\pi}{\lambda}(x_2 - x_1) = n\pi$$

that is, when the path difference

$$x_2 - x_1 = n\lambda$$

When

$$\cos \frac{\delta}{2} = 0$$

the intensity is zero and the components cancel to give destructive interference. This requires that

$$\frac{\delta}{2} = (2n + 1)\frac{\pi}{2} = \frac{\pi}{\lambda}(x_2 - x_1)$$

or, the path difference

$$x_2 - x_1 = \left(n + \frac{1}{2}\right)\lambda$$

The loci or sets of points for which $x_2 - x_1$ (or δ) is constant are shown in Figure 12.17 to form hyperbolas about the foci S_1 and S_2 (in three dimensions the loci would be the hyperbolic surfaces of revolution).

12.14   Interference from Two Equal Sources of Separation f

12.14.1   Separation $f \gg \lambda$. Young's Slit Experiment

One of the best-known methods for producing optical interference effects is the Young's slit experiment. Here the two coherent sources, Figure 12.18, are two identical slits S_1 and S_2 illuminated by a monochromatic wave system from a single source equidistant from S_1 and S_2. The observation point P lies on a screen which is set at a distance l from the plane of the slits.

The intensity at P is given by

$$I = R^2 = 4a^2 \cos^2 \frac{\delta}{2}$$

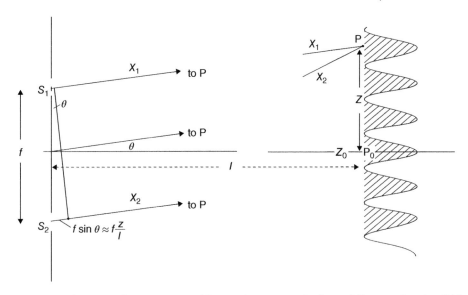

Figure 12.18   *Waves from equal sources S_1 and S_2 interfere at P with phase difference $\delta = (2\pi/\lambda)(x_2 - x_1) = (2\pi/\lambda)f \sin\theta \approx (2\pi/\lambda)f(z/l)$. The distance $l \gg z$ and f so S_1P and S_2P are effectively parallel. Interference fringes of intensity $I = I_0 \cos^2 \delta/2$ are formed in the plane PP_0.*

and the distances $PP_0 = z$ and slit separation f are both very much less than I (experimentally $\approx 10^{-3}l$). This is indicated by the break in the lines x_1 and x_2 in Figure 12.18 where S_1P and S_2P may be considered as sufficiently parallel for the path difference to be written as

$$x_2 - x_1 = f \sin\theta = f\frac{z}{l}$$

to a very close approximation.
  Thus

$$\delta = \frac{2\pi}{\lambda}(x_2 - x_1) = \frac{2\pi}{\lambda}f \sin\theta = \frac{2\pi}{\lambda}f\frac{z}{l}$$

  If

$$I = 4a^2 \cos^2 \frac{\delta}{2}$$

then

$$I = I_0 = 4a^2 \quad \text{when} \quad \cos\frac{\delta}{2} = 1$$

that is, when the path difference

$$f\frac{z}{l} = 0, \quad \pm\lambda, \pm 2\lambda, \ldots \pm n\lambda$$

and

$$I = 0 \quad \text{when} \quad \cos\frac{\delta}{2} = 0$$

that is, when

$$f\frac{z}{l} = \pm\frac{\lambda}{2}, \pm\frac{3\lambda}{2}, \pm\left(n + \frac{1}{2}\right)\lambda$$

Taking the point P_0 as $z = 0$, the variation of intensity with z on the screen P_0P will be that of Figure 12.18, a series of alternating straight bright and dark fringes parallel to the slit directions, the bright fringes having $I = 4a^2$ whenever $z = n\lambda l/f$ and the dark fringes $I = 0$, occurring when $z = \left(n + \frac{1}{2}\right)\lambda l/f$, n being called the order of interference of the fringes. The zero order $n = 0$ at the point P_0 is the central bright fringe. The distance on the screen between two bright fringes of orders n and $n + 1$ is given by

$$z_{n+1} - z_n = [(n + 1) - n]\frac{\lambda l}{f} = \frac{\lambda l}{f}$$

which is also the physical separation between two consecutive dark fringes. The spacing between the fringes is therefore constant and independent of n, and a measurement of the spacing, l and f determines λ.

The intensity distribution curve (Figure 12.19) shows that when the two wavetrains arrive at P exactly out of phase they interfere destructively and the resulting intensity or energy flux is zero. Energy conservation requires that the energy must be redistributed, and that lost at zero intensity is found in the intensity peaks. The average value of $\cos^2 \delta/2$ is $\frac{1}{2}$, and the dotted line at $I = 2a^2$ is the average intensity value over the interference system which is equal to the sum of the separate intensities from each slit.

There are two important points to remember about the intensity interference fringes when discussing diffraction phenomena; these are

- The intensity varies with $\cos^2 \delta/2$.
- The maxima occur for path differences of zero or integral numbers of the wavelength, whilst the minima represent path differences of odd numbers of the half-wavelength.

Spatial Coherence
In the preceding section nothing has been said about the size of the source producing the plane wave which falls on S_1 and S_2. If this source is an ideal point source A equidistant from S_1 and S_2, Figure 12.20,

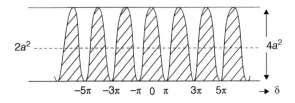

Figure 12.19   *Intensity of interference fringes is proportional to $\cos^2 \delta/2$, where δ is the phase difference between the interfering waves. The energy which is lost in destructive interference (minima) is redistributed into regions of constructive interference (maxima).*

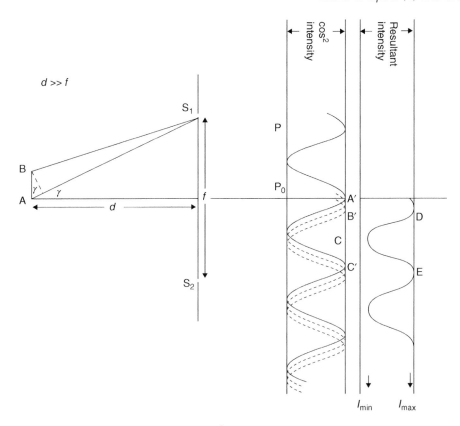

Figure 12.20 *The point source A produces the cos² interference fringes represented by the solid curve A′C′. Other points on the line source AB produce cos² fringes (the displaced broken curves B′) and the observed total intensity is the curve DE. When the points on AB extend A′B′ to C′ the fringes disappear and the field is uniformly illuminated.*

then a single set of $\cos^2$ fringes is produced. But every source has a finite size, given by AB in Figure 12.20, and each point on the line source AB produces its own set of interference fringes in the plane PP_0; the eye observing the sum of their intensities.

If the solid curve A′C′ is the intensity distribution for the point A of the source and the broken curves up to B′ represent the corresponding fringes for points along AB the resulting intensity curve is DE. Unless A′B′ extends to C the variations of DE will be seen as faint interference bands. These intensity variations were quantified by Michelson, who defined the

$$\text{Visibility} = \frac{I_{\max} - I_{\min}}{I_{\max} + I_{\min}}$$

The $\cos^2$ fringes from a point source obviously have a visibility of unity because the minimum intensity $I_{\min} = 0$.

When A′B′ (Figure 12.20) = A′C′, the point source fringe separation (or a multiple of it) of the field is uniformly illuminated, fringe visibility = 0 and the fringes disappear.

This occurs when the path difference

$$AS_2 - BS_1 \approx AB \sin \gamma = \lambda/2 \quad \text{where} \quad AS_2 = AS_1.$$

Thus, the requirement for fringes of good visibility imposes a limit on the finite size of the source. Light from points on the source must be spatially coherent in the sense that $AB \sin \gamma \ll \lambda/2$ in Figure 12.20.

But for $f \ll d$,

$$\sin \gamma \approx f/2d$$

so the coherence condition becomes

$$\sin \gamma = f/2d \ll \lambda/2AB$$

or

$$AB/d \ll \lambda/f$$

where AB/d measures the angle subtended by the source at the plane $S_1 S_2$.

Spatial coherence therefore requires that the angle subtended by the source

$$\ll \lambda/f$$

where f is the linear size of the diffracting system. (Note also that λ/f measures $\theta(\sim z/l)$ the angular separation of the fringes in Figure 12.18.)

Worked Example

As an example of spatial coherence we may consider the production of Young's interference fringes using the sun as a source.

The sun subtends an angle of 0.018 rad at the earth and if we accept the approximation

$$\frac{AB}{d} \ll \frac{\lambda}{f} \approx \frac{\lambda}{4f}$$

with $\lambda = 0.5\,\mu\text{m}$,
we have

$$f \sim \frac{0.5}{4(0.018)} \sim 14\,\mu\text{m}$$

This small value of slit separation is required to meet the spatial coherence condition.

12.14.2   Separation $f \ll \lambda$ ($kf \ll 1$ where $k = 2\pi/\lambda$)

If there is a zero phase difference between the signals leaving the sources S_1 and S_2 of Figure 12.18 then the intensity at some distant point P may be written

$$I = 4a^2 \cos^2 \frac{\delta}{2} = 4I_{\rm s} \cos^2 \frac{kf \sin \theta}{2} \approx 4I_{\rm s},$$

where the path difference $S_2P - S_1P = f \sin \theta$ and $I_{\rm s} = a^2$ is the intensity from each source.

We note that, since $f \ll \lambda$ ($kf \ll 1$), the intensity has a very small θ dependence and the two sources may be effectively replaced by a single source of amplitude $2a$.

12.14.3   Dipole Radiation ($f \ll \lambda$)

Suppose, however, that the signals leaving the sources S_1 and S_2 are anti-phase so that their total phase difference at some distant point P is

$$\delta = (\delta_0 + kf \sin \theta)$$

where $\delta_0 = \pi$ is the phase difference introduced at source.

The intensity at P is given by

$$I = 4 I_{\rm s} \cos^2 \frac{\delta}{2} = 4 I_{\rm s} \cos^2 \left(\frac{\pi}{2} + \frac{kf \sin \theta}{2} \right)$$
$$= 4 I_{\rm s} \sin^2 \left(\frac{kf \sin \theta}{2} \right)$$
$$\approx I_{\rm s} (kf \sin \theta)^2$$

because

$$kf \ll 1$$

Two anti-phase sources of this kind constitute a dipole whose radiation intensity $I \ll I_{\rm s}$ the radiation from a single source, when $kf \ll 1$. The efficiency of radiation is seen to depend on the product kf and, for a fixed separation f the dipole is a less efficient radiator at low frequencies (small k) than at higher frequencies. Figure 12.21 shows the radiation intensity I plotted against the polar angle θ and we see that for the dipole axis normal to the direction $\theta = \pi/2$, completely destructive interference occurs only on the parallel axis $\theta = 0$ and $\theta = \pi$. There is no direction (value of θ) giving completely constructive interference. The highest value of the radiated intensity occurs along the axis $\theta = \pi/2$ and $\theta = 3\pi/2$ but even this is only

$$I = (kf)^2 I_{\rm s},$$

where

$$kf \ll 1$$

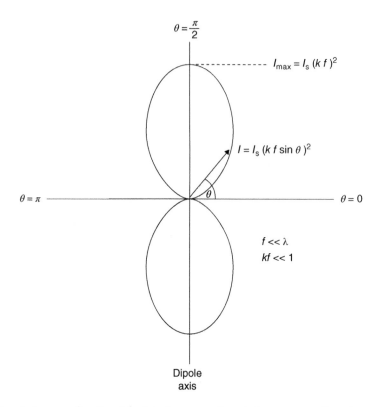

Figure 12.21   *Intensity I versus direction θ for interference pattern between waves from two equal sources, π rad out of phase (dipole) with separation f ≪ λ. This pattern is achieved by an End Fire array with dipole elements parallel to the θ = 0 to π axis, with alternate elements phased positive and negative and spaced λ/2 apart. In the direction 0 = π/2 to 3π/2 the dipole contributions are in phase, the delay in space between alternate dipoles matched by the phase difference π between them. In the θ = 0 to π directions alternate dipole contributions cancel each other.*

The directional properties of a radiating dipole are incorporated in the design of transmitting aerials. In acoustics a loudspeaker may be considered as a multi-dipole source, the face of the loudspeaker generating compression waves whilst its rear propagates rarefactions. Acoustic reflections from surrounding walls give rise to undesirable interference effects which are avoided by enclosing the speaker in a cabinet. Bass reflex or phase inverter cabinets incorporate a vent on the same side as the speaker face at an acoustic distance of half a wavelength from the rear of the speaker. The vent thus acts as a second source in phase with the speaker face and radiation is improved.

12.15   Interference from Linear Array of N Equal Sources

Figure 12.22 shows a linear array of N equal sources with constant separation f generating signals which are all in phase ($\delta_0 = 0$). At a distant point P in a direction θ from the sources the phase difference between the signals from two successive sources is given by $\delta = \frac{2\pi}{\lambda} f \sin \theta$ and the resultant at P is found by superposing the equal contribution from each source with the constant phase difference δ between successive contributions. Unlike dipole radiation there is no restriction that $f \ll \lambda$.

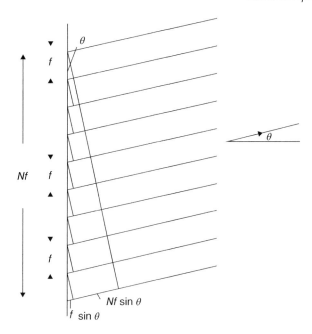

Figure 12.22   *Linear array of N equal sources separation f radiating in a direction θ to a distant point P. The resulting amplitude at P is given by R = a[sin N(δ/2)/ sin(δ/2)] where a is the amplitude from each source and δ = (2π/λ)f sin θ is the common phase difference between successive sources.*

But we find in Appendix 3 that the resultant of such a superposition is given by

$$R = a\frac{\sin(N\delta/2)}{\sin(\delta/2)}$$

where a is the signal amplitude at each source, so the intensity may be written

$$I = R^2 = a^2\frac{\sin^2(N\delta/2)}{\sin^2(\delta/2)} = I_s\frac{\sin^2(N\pi f \sin \theta/\lambda)}{\sin^2(\pi f \sin \theta/\lambda)}$$
$$= I_s\frac{\sin^2 N\beta}{\sin^2 \beta}$$

where I_s is the intensity from each source and $\beta = \pi f \sin \theta/\lambda$.
If we take the case of $N = 2$, then

$$I = I_s\frac{\sin^2 2\beta}{\sin^2 \beta} = 4I_s \cos^2 \beta = 4I_s \cos^2 \frac{\delta}{2}$$

which gives us the Young's Slit Interference pattern.
   We can follow the intensity pattern for N sources by considering the behaviour of the term $\sin^2 N\beta/\sin^2 \beta$.

We see that when

$$\beta = \frac{\pi}{\lambda} \sin\theta = 0 \pm \pi \pm 2\pi, \text{etc.}$$

i.e. when

$$f \sin\theta = 0, \pm\lambda, \pm2\lambda \ldots \pm n\lambda$$

constructive interference of the order n takes place, and

$$\frac{\sin^2 N\beta}{\sin^2 \beta} \rightarrow \frac{N^2\beta^2}{\beta^2} \rightarrow N^2$$

giving

$$I = N^2 I_{\mathrm{s}}$$

that is, a very strong intensity at the Principal Maximum condition of

$$f \sin\theta = n\lambda$$

We can display the behaviour of the $\sin^2 N\beta / \sin^2 \beta$ term as follows

Numerator $\sin^2 N\beta$ is zero for $N\beta \rightarrow 0\pi \ldots N\pi \ldots 2N\pi$
$$\downarrow \qquad \downarrow \qquad \downarrow$$
Denominator $\sin^2 \beta$ is zero for $\beta \rightarrow 0 \ldots \pi \ldots 2\pi$

The coincidence of zeros for both numerator and denominator determine the Principal Maxima with the factor N^2 in the intensity, i.e. whenever $f \sin\theta = n\lambda$.

Between these principal maxima are $N-1$ points of zero intensity which occur whenever the numerator $\sin^2 N\beta = 0$ but where $\sin^2 \beta$ remains finite.

These occur when

$$f \sin\theta = \frac{\lambda}{N}, \frac{2\lambda}{N} \ldots (n-1)\frac{\lambda}{N}$$

The $N-2$ subsidiary maxima which occur between the principal maxima have much lower intensities because none of them contains the factor N^2. Figure 12.23 shows the intensity curves for $N = 2, 4, 8$ and $N \rightarrow \infty$.

Two scales are given on the horizontal axis. One shows how the maxima occur at the order of interference $n = f \sin\theta / \lambda$. The other, using units of $\sin\theta$ as the ordinate displays two features. It shows that the separation between the principal maxima in units of $\sin\theta$ is λ/f and that the width of half the base of the principal maxima in these units is λ/Nf (the same value as the width of the base of subsidiary maxima). As N increases not only does the principal intensity increase as N^2 but the width of the principal maximum becomes very small.

As N becomes very large, the interference pattern becomes highly directional, very sharply defined peaks of high intensity occurring whenever $\sin\theta$ changes by λ/f.

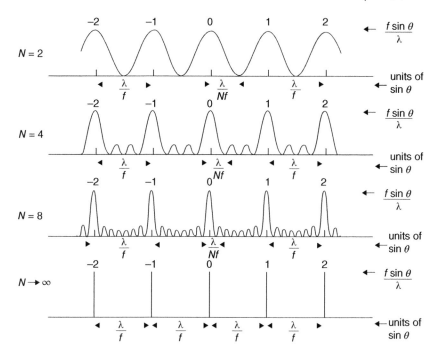

Figure 12.23   *Intensity of interference patterns from linear arrays of N equal sources of separation f. The horizontal axis in units of f sin θ/λ gives the spectral order n of interference. The axis in units of sin θ shows that the separation between principal maxima is given by sin θ = λ/f and the half-width of the principal maximum is given by sin θ = λ/Nf. Such an array is known as a Broadside Array.*

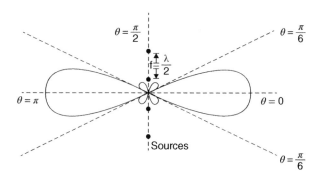

Figure 12.24   *Polar plot of the intensity of the interference pattern from a linear array of four sources with common separation f = λ/2. Note that the half-width of the principal maximum is θ = π/6 satisfying the relation sin θ = λ/Nf and that the separation between principal maxima satisfies the relation that the change in sin θ = λ/f. Such an array is known as a Broadside Array.*

The directional properties of such a broadside linear array are widely used in both transmitting and receiving aerials and the polar plot for $N = 4$ (Figure 12.24) displays these features. For N large, such an array, used as a receiver, forms the basis of a radio telescope where the receivers (sources) are set at a constant (but adjustable) separation f and tuned to receive a fixed wavelength. Each receiver takes the form of a parabolic reflector, the axes of which are kept parallel as the reflectors are oriented in different directions. The angular separation between the directions of incidence for which the received signal is a maximum is given by $\sin \theta = \lambda/f$.

Problem 12.1.   Suppose that Newton's Rings are formed by the system of Figure 12.4 except that the plano-convex lens now rests centrally in a concave surface of radius of curvature R_1 and not on an optical flat. Show that the radius r_n of the nth dark ring is given by

$$r_n^2 = R_1 R_2 n\lambda/(R_1 - R_2)$$

where R_2 is the radius of curvature of the lens and $R_1 > R_2$ (note that R_1 and R_2 have the same sign).

Problem 12.2.   A system of Newton's Rings is formed by an air film between a plano-convex lens and resting on a flat plate of glass. The radius of the 40th ring is 2.6 cm. The air film is replaced by an oil film with a refractive index of 1.69. What happens to the ring system and what is the new radius of the 40th ring?

Problem 12.3.   Young's fringes are produced by light from a slit source which falls on two narrow slits 1 mm apart and 100 mm from the slit source. The fringes are observed on a screen 1 m away. The source is white light filtered so that only the wave band from 480 to 520 nm is used. (a) What is the separation of the fringes? (b) Approximately how many fringes are clearly visible? (c) How wide can the source slit be made without seriously reducing the fringe visibility?

Problem 12.4.   Two identical radio masts transmit at a frequency of 1500 kilocycles per second and are 400 m apart. Show that the intensity of the Interference pattern between these radiators is given by $I = 2I_0[1 + \cos(4\pi \sin \theta)]$, where I_0 is the radiated intensity of each. Plot this intensity distribution on a polar diagram in which the masts lie on the $90° - 270°$ axis to show that there are two major cones of radiation in opposite directions along this axis and 6 minor cones at $0°$, $30°$, $150°$, $180°$, $210°$ and $330°$.

Problem 12.5.   (a) Two equal sources radiate a wavelength λ and are separated a distance $\lambda/2$. There is a phase difference $\delta_0 = \pi$ between the signals at source. If the intensity of each source is I_s, show that the intensity of the radiation pattern is given by

$$I = 4I_s \sin^2 \left(\frac{\pi}{2} \sin \theta\right)$$

where the sources lie on the axis $\pm\pi/2$.
   Plot I versus θ.
(b) If the sources in (a) are now $\lambda/4$ apart and $\delta_0 = \pi/2$ show that

$$I = 4I_s \left[\cos^2 \frac{\pi}{4}(1 + \sin \theta)\right]$$

Plot I versus θ.

Problem 12.6.   The minimum wavelength difference that a Fabry-Pérot spectrometer with reflectivity $R = 90\%$ can resolve is $\Delta\lambda_{min} = 2.947 \times 10^{-11}$ m. Calculate the value of the free spectral range $\Delta\lambda_{fsr}$ and the number of interfering beams which contribute to the fringe intensity.

Problem 12.7.   (a) A large number of identical radiators is arranged in rows and columns to form a lattice of which the unit cell is a square of side d. Show that all the radiation from the lattice in the direction θ will be in phase at a large distance if $\tan \theta = m/n$, where m and n are integers.

(b) If the lattice of section (a) consists of atoms in a crystal where the rows are parallel to the crystal face, show that radiation of wavelength λ incident on the crystal face at a grazing angle of θ is scattered to give interference maxima when $2d \sin \theta = n\lambda$ (Bragg reflection).

Problem 12.8.   Show that the separation of equal sources in a linear array producing a principal maximum along the line of the sources ($\theta = \pm\pi/2$) is equal to the wavelength being radiated. Such a pattern is called 'end fire'. Determine the positions (values of θ) of the secondary maxima for $N = 4$ and plot the angular distribution of the intensity.

Problem 12.9.   The first multiple radio astronomical interferometer was equivalent to a linear array of $N = 32$ sources (receivers) with a separation $f = 7$ m working at a wavelength $\lambda = 0.21$ m. Show that the angular width of the central maximum is 6 min of arc and that the angular separation between successive principal maxima is $1°42'$.

Problem 12.10.   Two transmitters radiate signals at the same frequency in a Broadside Array. Their phase difference $\delta_0 = 0$ in the bracket $(\delta_0 + kf \sin \theta)$ (section 12.14.3). The central beam is normal to the axis joining the transmitters. The effect of changing δ_0 is to vary $\sin \theta$ and rotate the central beam. If $f/\lambda = 3$ and a phase change of $\delta_0 = \pi/3$ show that the main beam is rotated by $\theta = 3°09'$.

13

Waves in Optics (2) Diffraction

13.1 Diffraction

Diffraction is classified as Fraunhofer or Fresnel. In Fraunhofer diffraction the pattern is formed at such a distance from the diffracting system that the waves generating the pattern may be considered as plane. A Fresnel diffraction pattern is formed so close to the diffracting system that the waves generating the pattern still retain their curved characteristics.

13.1.1 Fraunhofer Diffraction

The single narrow slit. Earlier in the last chapter it was stated that the difference between interference and diffraction is merely one of scale and not of physical behaviour.

Suppose we contract the scale of the N equal sources separation f of Figure 12.22 until the separation between the first and the last source, originally Nf, becomes equal to a distance d where d is now assumed to be the width of a narrow slit on which falls a monochromatic wavefront of wavelength λ where $d \sim \lambda$. Each of the large number N equal sources may now be considered as the origin of secondary wavelets generated across the plane of the slit on the basis of Huygens' Principle to form a system of waves diffracted in all directions.

When these diffracted waves are focused on a screen as shown in Figure 13.1 the intensity distribution of the diffracted waves may be found in terms of the aperture of the slit, the wavelength λ and the angle of diffraction θ. In Figure 13.1 a plane light wave falls normally on the slit aperture of width d and the waves diffracted at an angle θ are brought to focus at a point P on the screen PP_0. The point P is sufficiently distant from the slit for all wavefronts reaching it to be plane and we limit our discussion to Fraunhofer Diffraction.

Finding the amplitude of the light at P is the simple problem of superposing all the small contributions from the N equal sources in the plane of the slit, taking into account the phase differences which arise from the variation in path length from P to these different sources. We have already solved this

Introduction to Vibrations and Waves, First Edition. H. J. Pain and Patricia Rankin.
© 2015 John Wiley & Sons, Ltd. Published 2015 by John Wiley & Sons, Ltd.
Companion website: http://booksupport.wiley.com

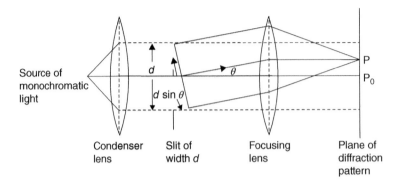

Figure 13.1   *A monochromatic wave normally incident on a narrow slit of width d is diffracted through an angle θ and the light in this direction is focused at a point P. The amplitude at P is the superposition of all the secondary waves in the plane of the slit with their appropriate phases. The extreme phase difference from contributing waves at opposite edges of the slit is $\phi = 2\pi d \sin\theta/\lambda = 2\alpha$.*

problem several times. Here we reapply the result already used in Appendix 3, that the intensity at P is given by

$$I = I_{\mathrm{s}}\frac{\sin^2 N\beta}{\sin^2 \beta} \quad \text{where} \quad N\beta = \frac{\pi}{\lambda}Nf\sin\theta$$

is half the phase difference between the contributions from the first and last sources. But now $Nf = d$ the slit width, and if we replace β by α where $\alpha = (\pi/\lambda)d\sin\theta$ is now half the phase difference between the contributions from the opposite edges of the slit, the intensity of the diffracted light at P is given by

$$I = I_{\mathrm{s}} = \frac{\sin^2(\pi/\lambda)d\sin\theta}{\sin^2(\pi/\lambda N)d\sin\theta} = I_{\mathrm{s}}\frac{\sin^2\alpha}{\sin^2(\alpha/N)}$$

For N large

$$\sin^2\frac{\alpha}{N} \to \left(\frac{\alpha}{N}\right)^2$$

and we have

$$I = N^2 I_{\mathrm{s}}\frac{\sin^2\alpha}{\alpha^2} = I_0\frac{\sin^2\alpha}{\alpha^2}$$

Plotting $I = I_0(\sin^2\alpha/\alpha^2)$ with $\alpha = (\pi/\lambda)d\sin\theta$ in Figure 13.2 we see that its pattern is symmetrical about the value

$$\alpha = \theta = 0$$

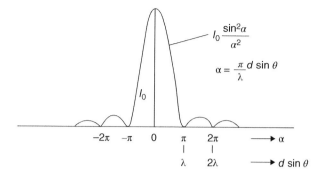

Figure 13.2 *Diffraction pattern from a single narrow slit of width d has an intensity I $= I_0 \sin^2 \alpha/\alpha^2$ where $\alpha = \pi d \sin\theta/\lambda$.*

where $I = I_0$ because $\sin\alpha/\alpha \to 1$ as $\alpha \to 0$. The intensity $I = 0$ whenever $\sin\alpha = 0$ that is, whenever α is a multiple of π or

$$\alpha = \frac{\pi}{\lambda}d\sin\theta = \pm\pi \pm 2\pi \pm 3\pi, \text{etc.}$$

giving

$$d\sin\theta = \pm\lambda \pm 2\lambda \pm 3\lambda, \text{etc.}$$

This condition for diffraction minima is the same as that for interference maxima between two slits of separation d, and this is important when we consider the problem of light transmission through more than one slit.

The intensity distribution maxima occur whenever the factor $\sin^2\alpha/\alpha^2$ has a maximum; that is, when

$$\frac{d}{d\alpha}\left(\frac{\sin\alpha}{\alpha}\right)^2 = \frac{d}{d\alpha}\left(\frac{\sin\alpha}{\alpha}\right) = 0$$

or

$$\frac{\cos\alpha}{\alpha} - \frac{\sin\alpha}{\alpha^2} = 0$$

This occurs whenever $\alpha = \tan\alpha$, and Figure 13.3 shows that the roots of this equation are closely approximated by $\alpha = \pm 3\pi/2, \pm 5\pi/2$, etc. (see problem at end of chapter on exact values).

Table 13.1 shows the relative intensities of the subsidiary maxima with respect to the principal maximum I_0.

The rapid decrease in intensity as we move from the centre of the pattern explains why only the first two or three subsidiary maxima are normally visible.

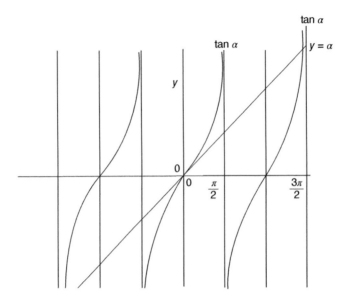

Figure 13.3   *Position of principal and subsidiary maxima of single slit diffraction pattern is given by the intersections of $y = \alpha$ and $y = \tan\alpha$.*

Table 13.1   *Relative intensities of subsidiary maxima in Figure 13.2 where I_0 is the principal maximum intensity, the slit width is d and $a = n/\lambda d \sin\theta$. These are the origins of a, b and c in Figure 13.9.*

α	$\dfrac{\sin^2\alpha}{\alpha^2}$	$\dfrac{I_0\sin^2\alpha}{\alpha^2}$
0	1	I_0
$\dfrac{3\pi}{2}$	$\dfrac{4}{9\pi^2}$	$\dfrac{I_0}{22.2}$
$\dfrac{5\pi}{2}$	$\dfrac{4}{25\pi^2}$	$\dfrac{I_0}{61.7}$
$\dfrac{7\pi}{2}$	$\dfrac{4}{49\pi^2}$	$\dfrac{I_0}{121}$

13.2   Scale of the Intensity Distribution

The width of the principal maximum is governed by the condition $d \sin\theta = \pm\lambda$. A constant wavelength λ means that a decrease in the slit width d will increase the value of $\sin\theta$ and will widen the principal maximum and the separation between subsidiary maxima. The narrower the slit the wider the diffraction pattern; that is, the narrower the pulse in x-space the greater the region in k- or wave number space required to represent it.

13.3   Intensity Distribution for Interference with Diffraction from N Identical Slits

The extension of the analysis from the example of one slit to that of N equal slits of width d and common spacing f, Figure 13.4, is straightforward.

To obtain the expression for the intensity at a point P of diffracted light from a single slit we considered the contributions from the multiple equal sources across the plane of the slit.

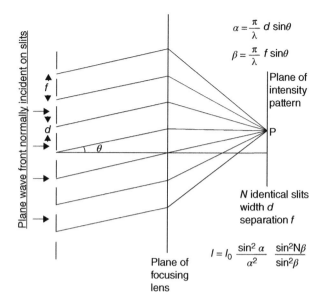

$$\alpha = \frac{\pi}{\lambda}\, d\sin\theta$$

$$\beta = \frac{\pi}{\lambda}\, f\sin\theta$$

Plane of
intensity
pattern

P

N identical slits
width *d*
separation *f*

$$I = I_0 \frac{\sin^2 \alpha}{\alpha^2}\, \frac{\sin^2 N\beta}{\sin^2\beta}$$

Plane of
focusing
lens

Plane wave front normally incident on slits

Figure 13.4 *Intensity distribution for diffraction by N equal slits is* $I = I_0 \frac{\sin^2\alpha}{\alpha^2}\frac{\sin^2 N\beta}{\sin^2\beta}$ *the product of the diffraction intensity for one slit,* $I_0 \sin^2\alpha/\alpha^2$ *and the interference intensity between N sources* $\sin^2 N\beta/\sin^2\beta$, *where* $\alpha = (\pi/\lambda)d\sin\theta$ *and* $\beta = (\pi/\lambda)f\sin\theta$.

We obtained the result

$$I = I_0 \frac{\sin^2\alpha}{\alpha^2}$$

by contracting the original linear array of *N* sources of spacing *f* in Appendix 3. If we expand the system again to recover the linear array, where each source is now a slit giving us the diffraction contribution

$$I_s = I_0 \frac{\sin^2\alpha}{\alpha^2}$$

we need only insert this value at I_s in the original expression for the interference intensity,

$$I = I_s \frac{\sin^2 N\beta}{\sin^2\beta}$$

in section 12.15, Interference from Linear Array of *N* Equal Sources, where

$$\beta = \frac{\pi}{\lambda} f\sin\theta$$

to obtain, for the intensity at P in Figure 13.4, the value

$$I = I_0 \frac{\sin^2\alpha}{\alpha^2} \frac{\sin^2 N\beta}{\sin^2\beta},$$

where

$$\alpha = \frac{\pi}{\lambda} d \sin \theta$$

Note that this expression combines the diffraction term $\sin^2 \alpha / \alpha^2$ for each slit (source) and the interference term $\sin^2 N\beta / \sin^2 \beta$ from N sources (which confirms what we expected from the opening paragraphs on interference). The diffraction pattern for any number of slits will always have an envelope

$$\frac{\sin^2 \alpha}{\alpha^2} \quad \text{(single slit diffraction)}$$

modifying the intensity of the multiple slit (source) interference pattern

$$\frac{\sin^2 N\beta}{\sin^2 \beta}$$

13.4   Fraunhofer Diffraction for Two Equal Slits ($N = 2$)

When $N = 2$ the factor

$$\frac{\sin^2 N\beta}{\sin^2 \beta} = 4 \cos^2 \beta$$

so that the intensity

$$I = 4I_0 \frac{\sin^2 \alpha}{\alpha^2} \cos^2 \beta$$

the factor 4 arising from N^2 whilst the $\cos^2 \beta$ term is familiar from the double source interference discussion. The intensity distribution for $N = 2, f = 2d$, is shown in Figure 13.5. The intensity is zero at the diffraction minima when $d \sin \theta = n\lambda$. It is also zero at the interference minima when $f \sin \theta = (n + \frac{1}{2})\lambda$.

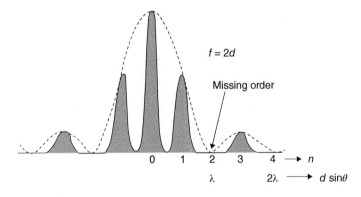

Figure 13.5   *Diffraction pattern for two equal slits, showing interference fringes modified by the envelope of a single slit diffraction pattern. Whenever diffraction minima coincide with interference maxima a fringe is suppressed to give a 'missing order' of interference.*

At some value of θ an interference maximum occurs for $f \sin \theta = n\lambda$ at the same position as a diffraction minimum occurs for $d \sin \theta = m\lambda$.

In this case the diffraction minimum suppresses the interference maximum and the order n of interference is called a missing order.

The value of n depends upon the ratio of the slit spacing to the slit width for

$$\frac{n\lambda}{m\lambda} = \frac{f \sin \theta}{d \sin \theta}$$

i.e.

$$\frac{n}{m} = \frac{f}{d} = \frac{\beta}{\alpha}$$

Thus, if

$$\frac{f}{d} = 2$$

the missing orders will be $n = 2, 4, 6, 8$, etc. for $m = 1, 2, 3, 4$, etc.
The ratio

$$\frac{f}{d} = \frac{\beta}{\alpha}$$

governs the scale of the diffraction pattern since this determines the number of interference fringes between diffraction minima and the scale of the diffraction envelope is governed by α.

13.5   Transmission Diffraction Grating (N Large)

A large number N of equivalent slits forms a transmission diffraction grating where the common separation f between successive slits is called the grating space.

Again, in the expression for the intensity

$$I = I_0 \frac{\sin^2 \alpha}{\alpha^2} \frac{\sin^2 N\beta}{\sin^2 \beta}$$

the pattern lies under the single slit diffraction term (Figure 13.6).

$$\frac{\sin^2 \alpha}{\alpha^2}$$

The principal interference maxima occur at

$$f \sin \theta = n\lambda$$

having the factor N^2 in their intensity and these are observed as spectral lines of order n. We see, however, that the intensities of the spectral lines of a given wavelength decrease with increasing spectral order because of the modifying $\sin^2 \alpha/\alpha^2$ envelope.

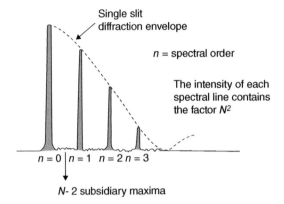

Figure 13.6   *Spectral line of a given wavelength produced by a diffraction grating loses intensity with increasing order n as it is modified by the single slit diffraction envelope. At the principal maxima each spectral line has an intensity factor N² where N is the number of lines in the grating.*

13.6   Resolving Power of Diffraction Grating

The importance of the diffraction grating as an optical instrument lies in its ability to resolve the spectral lines of two wavelengths which are too close to be separated by the naked eye. If these two wavelengths are λ and $\lambda + d\lambda$ where $d\lambda/\lambda$ is very small the Resolving Power for any optical instrument is given by the ratio $\lambda/d\lambda$.

Two such lines are just resolved, according to Rayleigh's Criterion, when the maximum of one falls upon the first minimum of the other. If the lines are closer than this their separate intensities cannot be distinguished.

If we recall that the spectral lines are the principal maxima of the interference pattern from many slits we may display Rayleigh's Criterion in Figure 13.7 where the nth order spectral lines of the two wavelengths are plotted on an axis measured in units of $\sin\theta$. We have already seen in Figure 12.23 that the half width of the spectral lines (principal maxima) measured in such units is given by λ/Nf where N is now the number of grating lines (slits) and f is the grating space. In Figure 13.7 the nth order of wavelength λ occurs when

$$f \sin\theta = n\lambda$$

whilst the nth order for $\lambda + d\lambda$ satisfies the condition

$$f[\sin\theta + \Delta(\sin\theta)] = n(\lambda + \mathrm{d}\lambda)$$

so that

$$f\Delta(\sin\theta) = n\,\mathrm{d}\lambda$$

Rayleigh's Criterion requires that the fractional change

$$\Delta(\sin\theta) = \frac{\lambda}{Nf}$$

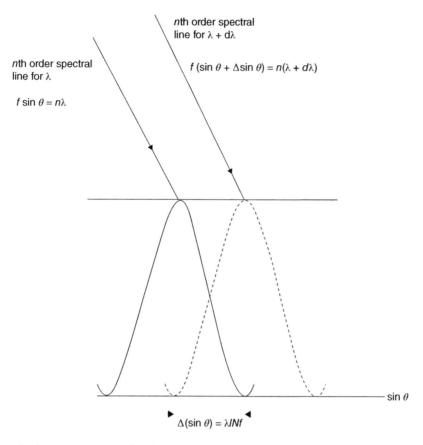

Figure 13.7   *Rayleigh's criterion states that the two wavelengths λ and $\lambda + d\lambda$ are just resolved in the nth spectral order when the maximum of one line falls upon the first minimum of the other as shown. This separation, in units of $\sin\theta$, is given by λ/Nf where N is the number of diffraction lines in the grating and f is the grating space. This leads to the result that the resolving power of the grating $\lambda/d\lambda = nN$.*

so that

$$f\Delta(\sin\theta) = n\, d\lambda = \frac{\lambda}{N}$$

Hence the Resolving Power of the diffraction grating in the *n*th order is given by

$$\frac{\lambda}{d\lambda} = Nn$$

Note that the Resolving Power increases with the number of grating lines N and the spectral order n. A limitation is placed on the useful range of n by the decrease of intensity with increasing n due to the modifying diffraction envelope

$$\frac{\sin^2\alpha}{\alpha^2} \quad \text{(Figure 13.6)}$$

13.7   Resolving Power in Terms of the Bandwidth Theorem

A spectral line in the nth order is formed when $f \sin \theta = n\lambda$ where $f \sin \theta$ is the path difference between light coming from two successive slits in the grating. The extreme path difference between light coming from opposite ends of the grating of N lines is therefore gives by

$$Nf \sin \theta = Nn\lambda$$

and the time difference between signals travelling these extreme paths is

$$\Delta t = \frac{Nn\lambda}{c}$$

where c is the velocity of light.

The light frequency $\nu = c/\lambda$ has a resolvable differential change

$$|\Delta \nu| = c\frac{|\Delta \lambda|}{\lambda^2} = \frac{c}{Nn\lambda}$$

because $\Delta \lambda / \lambda = 1/Nn$ (from the inverse of the Resolving Power).

Hence

$$\Delta \nu = \frac{c}{Nn\lambda} = \frac{1}{\Delta t}$$

or $\Delta \nu \Delta t = 1$ (the Bandwidth Theorem).

Thus, the frequency difference which can be resolved is the inverse of the time difference between signals following the extreme paths

$$(\Delta \nu \Delta t = 1 \quad \text{is equivalent of course to } \Delta \omega \Delta t = 2\pi)$$

If we now write the extreme path difference as

$$Nn\lambda = \Delta x$$

we have, from the inverse of the Resolving Power, that

$$\frac{\Delta \lambda}{\lambda} = \frac{1}{Nn}$$

so

$$\frac{|\Delta \lambda|}{\lambda^2} = \Delta \left(\frac{1}{\lambda} \right) = \frac{\Delta k}{2\pi} = \frac{1}{Nn\lambda} = \frac{1}{\Delta x}$$

where the wave number $k = 2\pi/\lambda$.

Hence we also have

$$\Delta x \Delta k = 2\pi$$

where Δk is a measure of the resolvable wavelength difference expressed in terms of the difference Δx between the extreme paths.

In section 3.5 we discussed the quality factor Q of an oscillatory system. Note that the resolving power may be considered as the Q of an instrument such as the diffraction grating or a Fabry–Pérot cavity for

$$\frac{\lambda}{\Delta\lambda} = \left|\frac{\nu}{\Delta\nu}\right| = \frac{\omega}{\Delta\omega} = Q$$

13.8   Fraunhofer Diffraction from a Rectangular Aperture

In Figure 13.8 a plane wavefront is diffracted as it passes through the rectangular aperture of dimensions a in the x direction and b in the y direction. The vector $\mathbf{k}$, which is normal to the diffracted wavefront, has direction cosines l and m with respect to the x and y axes respectively. This wavefront is brought to a focus at point P and the amplitude at P is the superposition of the contributions from all points (x, y) in the aperture with their appropriate phases.

A typical point (x, y) in the aperture is denoted by the vector $\mathbf{r'}$; the phase difference between the contribution from this point and that from O, the central point of the aperture, is $2\pi/\lambda$ (path difference). The path difference is the projection of vector $\mathbf{r'}$ upon the vector $\mathbf{k}$ so the phase difference is $\mathbf{r'} \cdot \mathbf{k} = (2\pi/\lambda)(lx + my)$ where $(lx + my)$ is the projection of $\mathbf{r'}$ upon $\mathbf{k}$. We write $2\pi l/\lambda = k_x$ and $2\pi m/\lambda = k_y$.

First of all we sum the simple harmonic contributions to P from a strip parallel to the x axis of length a and width dy divided into areas d$\bar{x}$dy each of which makes the same contribution to the final image at P. The explanation of the notation d$\bar{x}$ is the following. This strip passes through O.

A differential change of the cosine l as x moves from $-a/2$ to $+a/2$ appears in the mathematical technique of the Fourier Transform which is beyond the scope of this book. However a constant change of phase with ldx may be achieved by varying the length dx and associating each phase change δ with the average value d$\bar{x}$ where nd$\bar{x} = a$ (the length of the aperture). There is no loss of accuracy in the final result which depends only on the phases at $x = \pm a/2$ both of which have equal values of the cosine l and are precisely defined by $2\pi lx/\lambda$ with $x = \pm a/2$. The first of the n harmonic contributions to the sum from the strip comes from the edge at $x = -a/2$ which has a phase difference of $-2\pi la/2\lambda$ from the

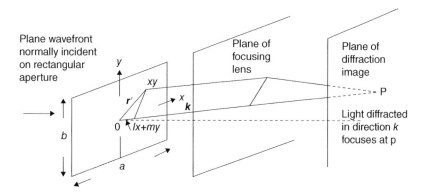

Figure 13.8 *Plane waves of monochromatic light incident normally on a rectangular aperture are diffracted in a direction **k**. All light in this direction is brought to focus at P in the image plane. The amplitude at P is the superposition of contributions from all the typical points, x, y in the aperture plane with their appropriate phase relationships.*

contribution of the centre point O which is zero. The phase difference between O ($x = 0$) and the other end of the strip $x = \pm a/2$ is $2\pi la/2\lambda$ so the phase difference between the first and last contributions is $2\pi la/\lambda = n\delta$ (with n large).

Applying the geometry of Appendix 3 to Figure 13.8, we have the resultant sum of all contributions as $R = d\bar{x}\sin(n\delta/2)/\sin\delta/2$ with a phase angle $\frac{n\delta}{2} = \alpha$ as half the difference in phase between the first and last contributions to the sum. When δ is very small we may write with the text in the final section of Appendix 3,

$$R_x = d\bar{x}\frac{\sin(n\delta/2)}{\sin\delta/2} = d\bar{x}\frac{\sin\alpha}{\alpha/n} = nd\bar{x}\frac{\sin\alpha}{\alpha} = a\frac{\sin\alpha}{\alpha} \quad \text{where } \alpha = \pi la/\lambda.$$

This is the result we expect from adding harmonic contributions with a small constant phase difference between neighbours. Repeating the process with a strip of length b and width dx parallel to the y axis and passing through O, we find the resultant to be $R_y = b\frac{\sin\beta}{\beta}$ where $\beta = \pi mb/\lambda$.

This gives a conbined resultant amplitude $R_x R_y = ab\frac{\sin\alpha}{\alpha}\frac{\sin\beta}{\beta}$ where the product ab is the area of the aperture each unit $dxdy$ of which contributes the same radiation to the intensity of the image at P resulting in an intensity $(R_x R_y)^2 = I = I_0\frac{\sin^2\alpha}{\alpha^2}\frac{\sin^2\beta}{\beta^2}$. The intensities relative to I_0 depend upon the product of the two diffractive terms $\frac{\sin^2\alpha}{\alpha^2}$ and $\frac{\sin^2\beta}{\beta^2}$. The diffraction pattern from such an aperture together with a plan showing the relative intensities is given in Figure 13.9 (see Table 13.1). Note: a and b in Table 13.1 are those of Figure 13.9 and are not the aperture dimensions. See Appendix 7 for a more formal derivation.

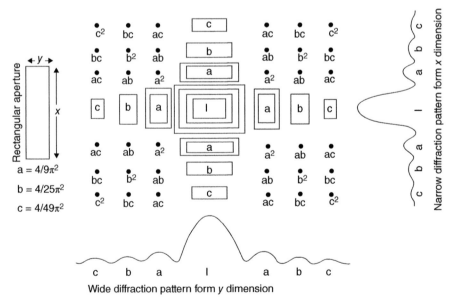

Figure 13.9 *The distribution of intensity in the diffraction pattern from a rectangular aperture is seen as the product of two single-slit diffraction patterns, a wide diffraction pattern from the narrow dimension of the slit and a narrow diffraction pattern from the wide dimension of the slit. This 'rotates' the diffraction pattern through 90° with respect to the aperture.* **Note that a and b in Figure 13.9 are not the length and width of the aperture.** *For the origin of the values of a, b and c see Table 13.1.*

13.9   Fraunhofer Diffraction from a Circular Aperture

The Fraunhofer diffraction pattern from a circular aperture is always circular as in Figures 13.10a and 13.10b. It has a bright central disc surrounded by concentric rings alternately dark and bright with decreasing intensity. Figure 13.10a is the general pattern. Figure 13.10b is a special case called an Airy disc which results from even the smallest point source. It limits the resolving power of optical instruments and modifies Rayleigh's criterion.

A graph of the intensity of these patterns is shown in Figure 13.11 with a series of minima defining the radial centres r' of the dark rings. Figure 13.12 analyses the method by which the pattern is produced.

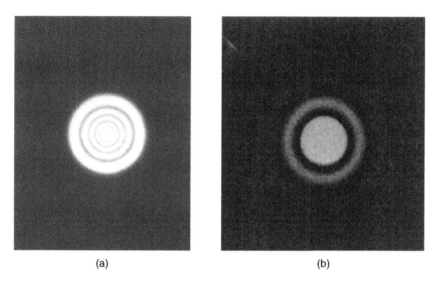

(a)   (b)

Figure 13.10   *(a) Fraunhofer diffraction pattern from a circular aperture. (b) Airy disc Fraunhofer diffraction pattern from the small circular point source.*

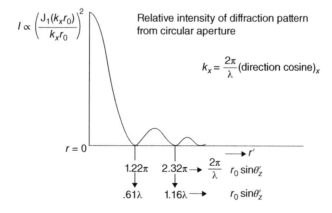

Figure 13.11   *Intensity of the diffraction pattern from a circular aperture of radius r_0 versus r', the radius of the pattern. The intensity is proportional to $[J_1(k_x r_0)/k_x r_0]^2$, where J_1 is Bessel's function of order 1. The pattern consists of a central circular principal maximum surrounded by a series of concentric rings of minima and subsidiary maxima of rapidly diminishing intensity.*

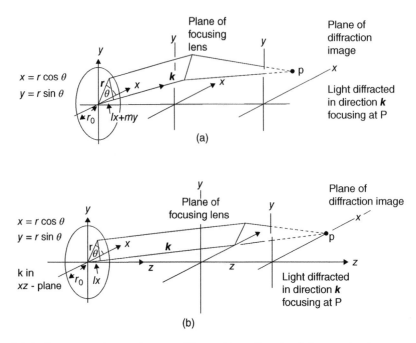

Figure 13.12 (a) A plane monochromatic wave diffracted in a direction **k** from a circular aperture is focused at a point P in the image plane. Contributions from all points x, y in the aperture superpose at P with appropriate phase relationships. (b) The direction **k** of (a) is chosen to lie wholly in the xz-plane to simplify the analysis. No generality is lost because of circular symmetry. The variation of the amplitude of diffracted light along any one radius determines the complete pattern.

The principle of pattern formation is the same as that of the rectangular aperture. A plane wave illuminates the aperture uniformly. The aperture is divided into concentric circular areas each having its own constant phase difference with respect to the centre $r = 0$. These areas contribute to the final image according to their size and their respective phase differences. A concentric ring is defined by its radius r from the centre of the aperture, its width dr, its length $2\pi r$ and its area $2\pi r dr$. The areas are integrated from $r = 0$ to the radius r_0 of the aperture. The mathematics is more complex than in the case of the rectangular aperture and the result is described in terms of a Bessel function, the y axis of Figure 13.11, a power series which satisfies circular boundary conditions in r and θ in the same way that sines and cosines satisfy the boundary conditions of a rectangular aperture in Cartesian coordinates. The appropriate function here is written

$$J_1(x') = \frac{x'}{2} - \frac{x'^3}{2^2 \cdot 4} + \frac{x'^5}{2^2 \cdot 4^2 \cdot 6} - \frac{x'^7}{2^2 \cdot 4^2 \cdot 6^2 \cdot 8}$$

where $x' = k_x r_0 = 2\pi l r_0/\lambda = 2\pi r_0 \sin\theta'_z/\lambda$ where θ'_z is the angle between the vector **k** and the z axis and defines the angle of diffraction. The intensity of the pattern is proportional to $[\frac{J_1(k_x r_0)}{k_x r_0}]^2$ and Figure 13.11 shows that the first minimum occurs at $r_0 \sin\theta'_z = 0.61\lambda$ and that the next minimum is at $r_0 \sin\theta'_z = 1.16\lambda$. These minima locate the centres of the dark bands. Reducing the radius r_0 of the aperture increases the values of θ'_z for the minima in accordance with $\Delta x \Delta k \approx 2\pi$ in the Bandwidth Theorem. Figure 13.12 shows that the analysis is simplified by choosing the direction of the vector **k**(z) of the diffracted light to lie wholly in the xz plane and to consider only the projection of a point $r(x,y)$ on **k**(z) where r is the

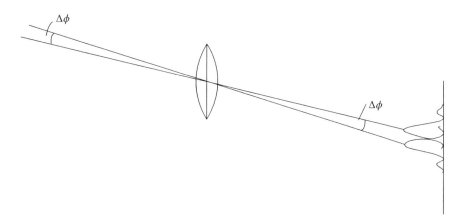

Figure 13.13   *Two stars with angular separation $\Delta\phi$ form separate Airy disc images when viewed through a telescope. Rayleigh's criterion (Figure 13.7) states that these images are resolved when the central maximum of one falls upon the first minimum of the other.*

radius of the ring. Because of circular symmetry the variation of the amplitudes of diffracted light along any one radius determines the complete diffraction pattern.

13.10   The Airy Disc and Resolving Power

When the two components of a double star with an angular separation of $\Delta\phi$ are viewed through a telescope with an objective lens of focal length l and diameter d their images will appear as two Airy discs separated by the angle $\Delta\phi$. The two diffraction patterns will be resolved if $\Delta\phi$ is much wider than the angular width of a disc but not if it is much less. Lord Rayleigh's criterion (Figure 13.7) gives the critical angle $\Delta\phi$ for resolution as that when the maximum of one disc falls on the first minimum of the other, Figure 13.13. Figure 13.11 then gives

$$\Delta\phi = \frac{0.61\lambda}{r_0} = \frac{1.22\lambda}{d}$$
$$(\Delta\phi = \Theta'_z \text{ in Figure 13.11})$$

where λ is the radiated wavelength.

This condition is known as diffraction-limited resolution. A poor quality lens will introduce aberrations and will not meet this criterion.

13.11   The Michelson Stellar Interferometer

In the discussion on Spatial Coherence (section 12.14.1) we saw that the relative displacement of the interference fringes from separate sources 1 and 2 led to a partial loss of the visibility of the fringes defined as

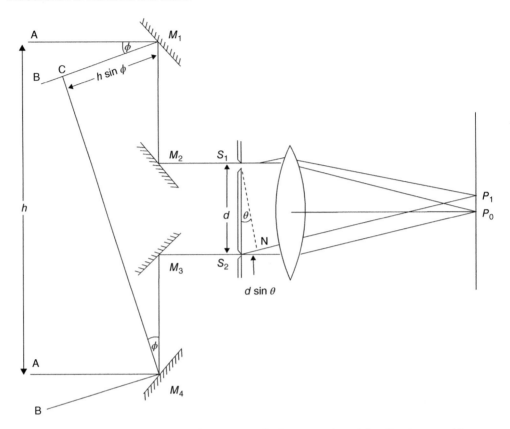

Figure 13.14 *In the Michelson stellar interferometer Light from stars A and B strike the movable outer mirrors M_1 and M_4 to be reflected via fixed mirrors M_2 and M_3 through two slits S_1 and S_2 and a lens to form interference fringes. Light from Star A forms its zero order fringe at P_0 and its first-order fringe at P_1 when $S_2N = d \sin \theta = \lambda_0$. The minimum separation h of M_1M_4 is found for light from B to reduce the fringe visibility to zero, that is, when the path difference $h = \sin\phi = \lambda_0/2$. The angles are so small that θ and ϕ replace their sines. Note that the fringe separation depends on d, but the fringe visibility is governed by h.*

$$V = \frac{I_{max} - I_{min}}{I_{max} + I_{min}}$$

and eventually when the displacement was equal to half a fringe width $V = 0$ and there was a complete loss of contrast.

Michelson's Stellar Interferometer (1920) used this to measure the angular separation between the two components of a double star or, alternatively, the angular width of a single star.

Initially, we take the simplest case to illustrate the principle and then discuss the practical problems which arise. We assume in the first instance that light from the stars is monochromatic with a wavelength λ_0. Michelson used four mirrors, M_1, M_2, M_3 and M_4, mounted on a girder with two slits S_1 and S_2 in front of the lens of an astronomical telescope, Figure 13.14. The slits were perpendicular to the line joining the two stars. The separation h of the outer pair of mirrors ($\sim$metres) was increased until the fringes observed in the focal plane of the objective just disappeared. Assuming zero path difference between $M_1M_2 P_0$ and $M_4M_3 P_0$ the light from star A will form its zero order fringe maximum at P_0 and its first-order fringe maximum at P_1, due to a path difference $S_2N = d \sin \theta = \lambda_0$ so the fringe spacing is determined by d, the separation between the inner mirrors M_2 and M_3.

The condition for fringe disappearance is that rays from star B will form a first-order maximum fringe midway between P_0 and P_1, that is, when

$$CM_1M_2S_1P_0 - M_4M_3S_2P_0 = CM_1 = h\sin\phi = \lambda_0/2$$

The condition for fringe disappearance is therefore determined by h while the angular size of the fringes depends on d so there is an effective magnification of h/d over a fringe system produced by the slits alone.

The angles θ and ϕ are small and the minimum value of h is found which produces $V = 0$ so that the fringes disappear at

$$h\phi = \lambda_0/2 \quad \text{or} \quad h = \frac{\lambda}{2\phi}$$

Measurement of h thus determines the double-star angular separation.

Several assumptions have been made in this simple case presentation. First, that the intensities of the light radiated by the stars are equal and that they are coherent sources. In fact, even if the sources are incoherent their radiation is essentially coherent at the interferometer. Second, the radiation is not monochromatic and only a few fringes around the zero order were visible so λ_0 must be taken as a mean wavelength. Finally, the introduction of a lens into the system inevitably creates Airy discs and the visibility must be expressed in terms of the Airy disc intensity distribution. This results in

$$V = 2\left(\frac{J_1(u)}{u}\right)$$

where

$$u = \pi h\phi/\lambda_0$$

If this visibility is plotted against $h\phi/\lambda_0$ its first zero occurs at 1.22 so the fringes disappear when $h = 1.22\,\lambda_0/\phi$.

Worked Example

In fact, Michelson first used his interferometer in 1920 to measure the angular diameter of the star Betelgeuse, the colour of which is orange. His astronomical telescope was the 2.54 m (100 in.) telescope of the Mt. Wilson Observatory. A mean wavelength $\lambda_0 = 570 \times 10^{-9}$ m was used and the fringes vanished when $h = 3.07$ m to give an angular diameter $\phi = 22.6 \times 10^{-8}$ radians or 0.047 arc seconds. The distance of Betelgeuse from the Earth was known and its diameter was calculated to be about 384×10^6 km, roughly 280 times that of the Sun. This magnitude is greater than that of the orbital diameter of Mars around the Sun.

13.12   Fresnel Diffraction

13.12.1   The Straight Edge and Slit

Our discussion of Fraunhofer diffraction considered a plane wave normally incident upon a slit in a plane screen so that waves at each point in the plane of the slit were in phase. Each point in the plane became the source of a new wavefront and the superposition of these wavefronts generated a diffraction pattern.

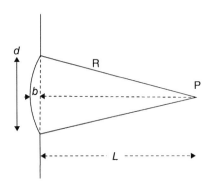

Figure 13.15   *Spherical wave at slit aperture width d converges at point P. Fresnel diffraction requires $b = d^2/8L \ll \lambda$ i.e. $L \gg d^2/8\lambda$. L is the Rayleigh distance. Beyond L the diffraction merges from Fresnel to Fraunhofer.*

At a sufficient distance from the slit the superposed wavefronts were plane and this defined the condition for Fraunhofer diffraction. Its pattern follows from summing the contributions from these waves together with their relative phases and in Appendix 3 we see that these form an arc of constant length. When the contributions are all in phase the arc is a straight line but as the relative phases increase the arcs curve to form closed circles of decreasing radii. The length of the chord joining the ends of the arc measure the resulting amplitude of the superposition and the square of that length measures the light intensity within the pattern.

Nearer the slit where the superposed wavefronts are not yet plane but retain their curved character the diffraction pattern is that of Fresnel. There is no sharp division between Fresnel and Fraunhofer diffraction, the pattern changes continuously from Fresnel to Fraunhofer as the distance from the slit increases.

Figure 13.15 shows a spherical wavefront at a slit aperture of width d. The wavefront converges at a point P located at a distance L from the aperture. If R is the radius of curvature of the wavefront the sagitta $b = (d/2)^2/2R$ and the phase difference from the centre of the sagitta to the centre of the slit is $\phi = \frac{2\pi}{\lambda}\frac{b^2}{2L}$ where $b/L \ll 1$. Now $b = (\frac{d}{2})^2\frac{1}{2L} = \frac{d^2}{8L}$ when $d^2/L^2 \ll 1$ and $L \approx R$.

Fraunhofer diffraction (plane wavefront) requires the curvature at the slit to be sufficiently large that $b \ll \lambda$ so that $L \gg d^2/8\lambda$. Below the value of L the diffraction is Fresnel; beyond L the diffraction merges into Fraunhofer. L is called the Rayleigh distance.

The Fresnel pattern is determined by a procedure exactly similar to that in Fraunhofer diffraction, an arc of constant length is obtained but now it convolutes around the arms of a pair of joined spirals, Figure 13.19, and not around closed circles.

An understanding of Fresnel diffraction is most easily gained by first considering, not the slit, but a straight edge formed by covering the lower half of the incident plane wavefront with an infinite plane screen. The undisturbed upper half of the wavefront will contribute one half of the total spiral pattern, that part in the first quadrant.

The Fresnel diffraction pattern from a straight edge, Figure 13.16, has several significant features. In the first place light is found beyond the geometric shadow; this confirms its wave nature and requires a Huygens wavelet to contribute to points not directly ahead of it. Also, near the edge there are fringes of intensity greater and less than that of the normal undisturbed intensity (taken here as unity). On this scale the intensity at the geometric shadow is exactly 0.25.

To explain the origin of this pattern we consider the point O at the straight edge of Figure 13.17 and the point P directly ahead of O. The line OP defines the geometric shadow. Below O the screen cuts off

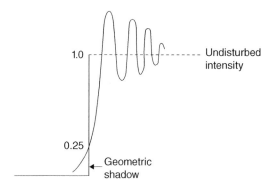

Figure 13.16 *Fresnel diffraction pattern from a straight edge. Light is found within the geometric shadow and fringes of varying intensity form the observed pattern. The intensity at the geometric shadow is 0.25 of that due to the undisturbed wavefront.*

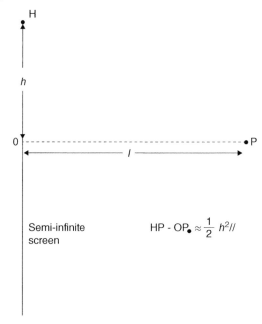

Figure 13.17 *Line OP normal to the straight edge defines the geometric shadow. The wavefront at height h above 0 makes a contribution to the disturbance at P which has a phase lag of $\pi h^2 / \lambda l$ with respect to that from 0. The total disturbance at P is the vector sum (amplitude and phase) of all contributions from the wavefront section above 0.*

the wavefront. The phase difference between the contributions to the disturbance at P from O and from a point H, height h above O is given by

$$\Delta(h) = \frac{2\pi}{\lambda}(\text{HP} - \text{OP}) \simeq \frac{2\pi}{\lambda}\frac{1}{2}\frac{h^2}{l}$$

(taking h as a sagitta and $l \approx$ radius of curvature)
where OP $= l$ and higher powers of h^2/l^2 are neglected.

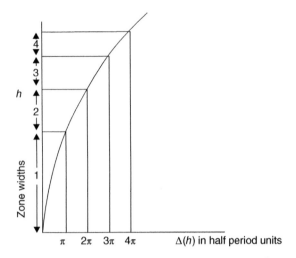

Figure 13.18   *Variation of the width of each half period zone with height h above the straight edge.*

We now divide the wavefront above O into strips which are parallel to the infinite straight edge and we call these strips 'half period zones'. This name derives from the fact that the width of each strip is chosen so that the contributions to the disturbance at P from the lower and upper edges of a given strip differ in phase by π radians.

Since the phase $\Delta(h) \propto h^2$ we shall not expect these strips or half period zones to be of equal width and Figure 13.18 shows how the width of each strip decreases as h increases. The total contribution from a strip will depend upon its area; that is, upon its width. The amplitude and phase of the contribution at P from a narrow strip of width $\mathrm{d}h$ at a height h above O may be written as $(\mathrm{d}h)e^{i\Delta}$ where $\Delta = \pi h^2 / \lambda l$.

This contribution may be resolved into two perpendicular components

$$\mathrm{d}x = \mathrm{d}h \cos\Delta \quad \text{and} \quad \mathrm{d}y = \mathrm{d}h \sin\Delta$$

If we now plot the vector sum of these contributions the total disturbance at P from that section of the wavefront measured from O to a height h will have the component values $x = \int \mathrm{d}x$ and $y = \int \mathrm{d}y$. These integrals are usually expressed in terms of the dimensionless variable $u = h(2/\lambda l)^{1/2}$, the physical significance of which we shall see shortly.

We then have $\Delta = \pi u^2 / 2$ and $\mathrm{d}h = (\lambda l/2)^{1/2} \mathrm{d}u$ and the integrals become

$$x = \int \mathrm{d}x = \int_0^u \cos(\pi u^2 / 2)\mathrm{d}u$$

and

$$y = \int \mathrm{d}y = \int_0^u \sin(\pi u^2 / 2)\mathrm{d}u$$

These integrals are called Fresnel's Integrals and the locus of the coordinates x and y with variation of u (that is, of h) is the spiral in the first quadrant of Figure 13.19. The complete figure is known as Cornu's spiral.

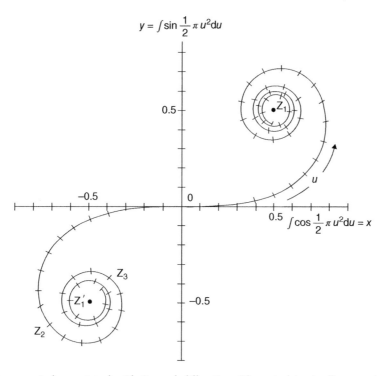

Figure 13.19 *Cornu spiral associated with Fresnel diffraction. The spiral in the first quadrant represents the contribution from the upper half of an infinite plane wavefront above an infinite straight edge. The third quadrant spiral results from the downward withdrawal of the straight edge. The width of the wavefront contributing to the diffraction pattern is correlated with the length u along the spiral. The upper half of the wavefront above the straight edge contributes an intensity $(OZ_1)^2$ which is the square of the length of the chord from the origin to the spiral eye. This intensity is 0.25 of the intensity $(Z_1Z_1')^2$ due to the whole wavefront.*

As h, the width of the contributing wavefront above the straight edge, increases, we measure the increasing length u from 0 along the curve of the spiral in the first quadrant unit, as h and $u \to \infty$ we reach Z_1 the centre of the spiral eye with coordinate $x = \frac{1}{2}$, $y = \frac{1}{2}$.

The tangent to the spiral curve is

$$\frac{dy}{dx} = \tan \frac{\pi u^2}{2}$$

and this is zero when the phase

$$\Delta(h) = \pi h^2 / \lambda l = \pi u^2 / 2 = m\pi$$

where m is an integer so that $u = \sqrt{(2m)}$ relates u, the distance measured along the spiral to m the number of half period zones contributing to the disturbance at P. The total intensity at P due to all the half period zones above the straight edge is given by the square of the length of the 'chord' OZ_1. This is the intensity at the geometric shadow.

Suppose now that we keep P fixed as we slowly withdraw the screen vertically downwards from O. This begins to reveal contributions to P from the lower part of the wavefront; that is, the part which contributes to the Cornu spiral in the third quadrant. The length u now includes not only the whole of the upper spiral arm but an increasing part of the lower spiral until, when u has extended to Z_2 the 'chord' $Z_1 Z_2$ has its maximum value and this corresponds to the fringe of maximum intensity nearest the straight edge. Further withdrawal of the screen lengthens u to the position Z_3 which corresponds to the first minimum of the fringe pattern and the convolutions of an increasing length u around the spiral eye will produce further intensity oscillations of decreasing magnitude until, with the final removal of the screen, u is now the total length of the spiral and the square of the 'chord' length $Z_1 Z_1'$ gives the undisturbed intensity of unit value. But $Z_1 Z_1' = 2Z_1 O$ so that the undisturbed intensity $(Z_1 Z_1')^2$ is a factor of four greater than $(Z_1 O)^2$ the intensity at the geometric shadow.

The Fresnel Diffraction pattern from a slit may now be seen as that due to a fixed height h of the wavefront equal to that of the slit width. This defines a fixed length u of the spiral between the end points of which the 'chord' is drawn and its length measured and squared to give the intensity. At a given distance from the slit the intensity at a point P in the diffraction pattern will correlate with the precise location of the fixed length u along the spiral. At the centre of the pattern P is symmetric with respect to the upper and lower edges of the slit and the fixed length u is centred about O (Figure 13.20). As P moves across the pattern towards the geometric shadow the length u moves around the convolutions of the spiral. In the geometric shadow this length is located entirely within the first or third quadrant of the spiral and the magnitude of the 'chord' between its ends is less than OZ_1. When the slit is wide enough to produce the central minimum of the diffraction pattern in Figure 13.21 the length u is centred at O with its ends at Z_3 and Z_4 in Figure 13.20.

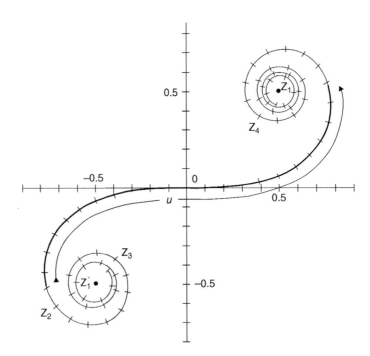

Figure 13.20 *The slit width h defines a fixed length u of the spiral. The intensity at a point P in the diffraction pattern is correlated with the precise location of u on the spiral. When P is at the centre of the pattern u is centred on 0 and moves along the spiral as P moves towards the geometric shadow. Within the geometric shadow the chord joining the ends of u is less than OZ_1.*

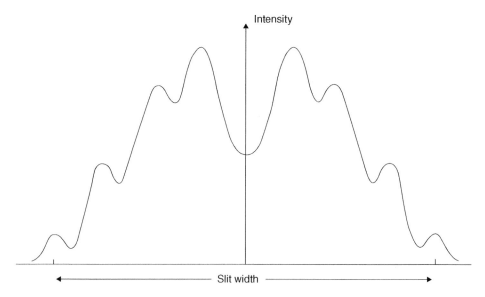

Figure 13.21   *Fresnel diffraction pattern from a slit which is wide enough for the spiral length u to be centred at 0 and to end on points Z_3 and Z_4 of Figure 13.20. This produces the intensity minimum at the centre of the pattern.*

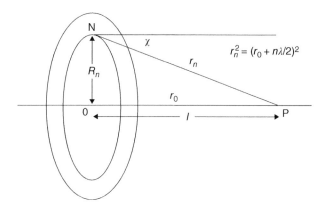

Figure 13.22   *Fresnel diffraction from a circular aperture. The mean radius R_n(ON) defines the half period zone with a phase lag of $n\pi$ at P with respect to the contribution from the central zone. The obliquity angle χ which reduces the zone contribution at P increases with n.*

13.12.2   Circular Aperture (Fresnel Diffraction)

In this case the half period zones become annuli of decreasing width. If R_n is the mean radius of the half period zone whose phase lag is $n\pi$ with respect to the contribution from the central ring the path difference in Figure 13.22 is given by

$$\text{NP} - \text{OP} = \Delta = n\lambda/2$$

Worked Example

So

$$R_n^2 = r_n^2 - r_0^2 = \left(r_0 + n\frac{\lambda}{2}\right)^2 - r_0^2 = n\lambda r_0 + \frac{1}{4}n^2\lambda^2$$

which for $\lambda \ll r_0$ gives $R_n^2 = n\lambda r_0$. Note too that the area of the aperture enclosed by $R_n^2 = \pi R_n^2 = n\pi\lambda r_0$ which shows that all Fresnel half period zones have the same area equal to $\pi\lambda r_0$. **This is a most important result.**

For a circular aperture radius a the area $\pi a^2 = n\pi\lambda r_0$ where n is the number of zones observed at $P = r_0$. Varying r_0 changes the value of n. If P moves towards the screen r_0 is reduced and n the number of zones increases. This reduces the value of R_n of a given zone as more zones crowd into observation from the pattern circumference. If P retreats from the screen, r_0 increases, n decreases and zones increase in size from the centre to the circumference until only the central spot remains after which there is darkness. Note that if r_0 gives a central bright spot then $r_0 + \lambda/2$ will give a central dark spot.

Each zone contributes equally to the disturbance at P except for a factor arising from the rigorous Kirchhoff theory which, until now, we have been able to ignore. This is the so-called obliquity factor $\cos\chi$ where χ is shown in the figure. This factor is negligible for small values of n but its effect is to reduce a zone contribution as n increases. A large circular aperture with many zones produces, in the limit, an undisturbed normal intensity on the axis and from Figure 13.24 where we show the magnitude and phase from successive half zones we see that the sum of these vectors which represents the amplitude produced by an undisturbed wave is only half of that from the innermost zone.

The spiral in Figure 13.23 explains the origin of the half period zones in the Fresnel diffraction pattern from a circular aperture Figure 13.22. The total contribution from an annular half zone is a semicircle

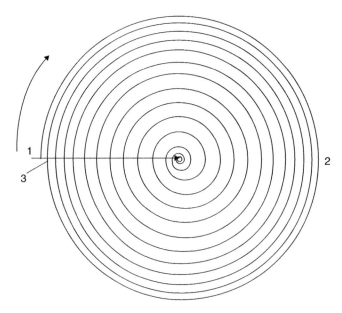

Figure 13.23   *The spiral phasor diagram for a spherical wavefront. Diameter 1 to 2 is the amplitude of the first half zone. Diameter 2 to 3 is the amplitude of the second amplitude half zone.*

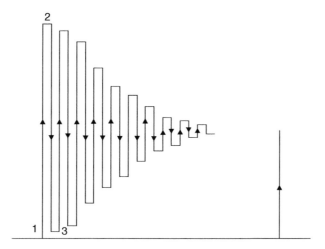

Figure 13.24   *The vector contributions from successive zones in the circular aperture. The amplitude produced by an undisturbed wave is seen to be only half of that from the central zone. Removing the contributions from alternate zones leaves the remainder in phase and produces a very high intensity. This is the principle of the zone plate of Figure 13.25.*

resulting from a number of small wave amplitude vectors each with a phase difference with respect to its predecessor. The total phase difference increases from 0 to π between the first and last vectors on the semicircle (representing a wavelength difference of $\lambda/2$). The diameter of the semicircle is the amplitude of a given half zone with amplitudes decreasing with distance from the central zone due to the obliquity factor and the $1/r$ term in the amplitude of a spherical wave which decreases with r, the distance travelled. Each semicircle with its vector diameter is known as a phasor. A tight spiral results in the amplitude of a bright ring being almost cancelled by that of the following anti-phase dark ring (Figure 13.24). The amplitude of the central bright spot in Figure 13.24 is the diameter of the semicircular circumference traced by the clockwise rotation of the outer end of the arrow between positions 1 and 2. The amplitude is twice that of the unobstructed amplitude of the incident wave at the right of Figure 13.24. Continued rotation between positions 2 and 3 produces the slightly reduced amplitude anti-phase half zone of the first dark ring. The numbers 1, 2, 3 appear in Figure 13.23 and 13.24 for correlation.

13.13   Zone Plate

It is evident that if alternate zones transmit no light then the contributions from the remaining zones would all be in phase and combine to produce a high intensity at P similar to the focusing effect of a lens. Such circular 'zone plates' are made by blacking out the appropriate areas of a glass slide, Figure 13.25. A further refinement increases the intensity still more. If the alternate zone areas are not blacked out but become areas where the optical thickness of the glass is reduced, via etching, by $\lambda/2$ the light transmitted through these zones is advanced in phase by π rad so that the contributions from all the zones are now in phase.

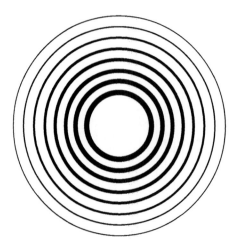

Figure 13.25   *Zone plate produced by removing alternate half zones from a circular aperture to leave the remaining contributions in phase.*

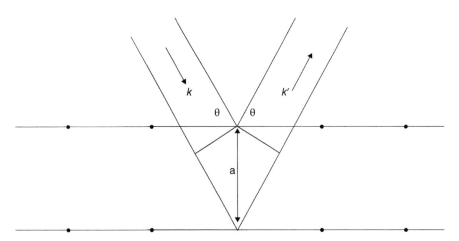

Figure 13.26   *Elastic Bragg reflection occurs when electron waves are scattered by atoms in planes separated by a distance a. Principal maxima are formed when $2a \sin \theta = n\lambda$.*

13.14   Electron Diffraction and Brillouin Zones

Atoms in crystals are arranged in regular three-dimensional patterns of unit cells, the simplest of which is the cubic cell where atoms are located at each corner of a cube. The resulting atomic planes diffract X rays (electromagnetic waves) with wavelengths $\lambda \sim 3 \times 10^{-10}$ m, a process known as Bragg scattering. Diffracted waves superpose to give maxima when $2a \sin \theta = n\lambda$ (Figure 13.26; see also Figure 12.5). Electrons with energy of ≈ 16 eV have a de Broglie wavelength $\lambda \approx 3 \times 10^{-10}$ m and display the same phenomenon. Internal scattering within the crystal occurs because many charged valence electrons have left atomic sites occupied by positively charged ions with which the negative electrons interact as they move under the influence of external applied voltages. Only 'free' electrons escape such interactions. When $\theta = \pi/2$ in Figure 13.26 the scattering takes place along a linear lattice and when the

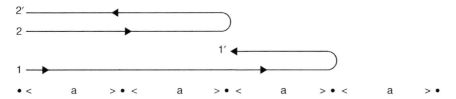

Figure 13.27   *When $\theta = \pi/2$ in Figure 13.26 Bragg scattering by electron–ion interactions gives principal maxima when electron waves are reflected from ions separated by multiples of a. The condition $2a = n\lambda$ defines the Brillouin zone boundaries for $n = 1, 2, 3$, etc.*

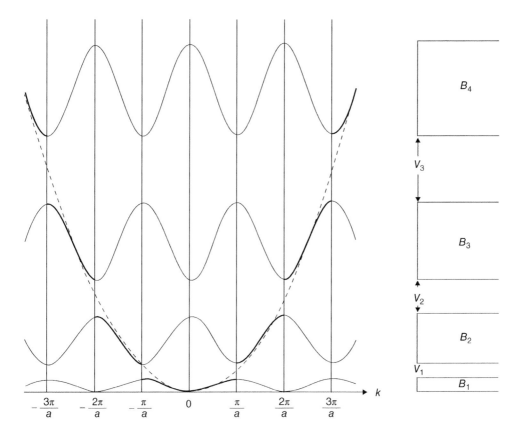

Figure 13.28   *Allowed electron energies versus k. The dotted parabola defines the free electron energy $E = \hbar^2 k^2/2m$ and the allowed energy bands are the Brillouin zones B_i. V_1, V_2, V_3 are the energy gaps between the zones. The cosine curves joining the zone boundaries are justified by Figure 6.6, i.e. all relevant information is contained in the region $\frac{-\pi}{a} \leq k \leq \frac{\pi}{a}$.*

interaction is strong enough the waves are totally reflected. This can result in phase differences of $\pm n\lambda = \pm 2n\pi/k = 2a$, i.e. $k = \pm n\pi/a$ (Figure 13.27). At exact values of k there is a gap between the allowed bands of energy in which the electrons move. These bands of energy are called Brillouin zones (see Figure 13.28 and Figure 5.15). If an electron gains sufficient energy from an applied voltage the gap between two bands may be small enough for the electron to jump across the gap into the higher energy band. This defines a conductor. If the energy gap is too great for this to happen for even the highest applied voltages the material is an insulator. Figure 13.28 shows the Brillouin zones and the dotted parabola defining the free electron energy $E = p^2/2m = \hbar^2 k^2/2m$ where $\hbar = h/2\pi$.

Problem 13.1.   (a) Atomic H_z in the atmosphere radiates a wavelength $\lambda = 21$ cm. An English land-based telescope has a diameter of 76 metres. What is its diffraction-limited angular resolution in degrees? (b) The Hubble space telescope has a diameter of 2.4 m. Determine its diffraction limited angular resolution at a wavelength of 550 nm.

Problem 13.2.   Monochromatic light is normally incident on a single slit, and the intensity of the diffracted light at an angle θ is represented in magnitude and direction by a vector I, the tip of which traces a polar diagram. Sketch several polar diagrams to show that as the ratio of slit width to the wavelength gradually increases the polar diagram becomes concentrated along the direction $\theta = 0$.

Problem 13.3.   The condition for the maxima of the intensity of light of wavelength λ diffracted by a single slit of width d is given by $\alpha = \tan \alpha$, where $\alpha = \pi d \sin \theta/\lambda$. The approximate values of α which satisfy this equation are $\alpha = 0, +3\pi/2, +5\pi/2$, etc. Writing $\alpha = 3\pi/2 - \delta, 5\pi/2 - \delta$, etc. where δ is small, show that the real solutions for α are $\alpha = 0, \pm1.43\pi, \pm2.459\pi, \pm3.471\pi$, etc.

Problem 13.4.   Prove that the intensity of the secondary maximum for a grating of three slits is $\frac{1}{9}$ of that of the principal maximum if only interference effects are considered.

Problem 13.5.   A diffraction grating has N slits and a grating space f. If $\beta = \pi f \sin \theta/\lambda$, where θ is the angle of diffraction, calculate the phase change $d\beta$ required to move the diffracted light from the principal maximum to the first minimum to show that the half width of the spectral line produced by the grating is given by $d\theta = (nN \cot \theta)^{-1}$, where n is the spectral order. (For $N = 14,000$, $n = 1$ and $\theta = 19°$, $d\theta \approx 5$s of arc.)

Problem 13.6.   (a) Dispersion is the separation of spectral lines of different wavelengths by a diffraction grating and increases with the spectral order n. Show that the dispersion of the lines when projected by a lens of focal length F on a screen is given by

$$\frac{dl}{d\lambda} = F\frac{d\theta}{d\lambda} = \frac{nF}{f}$$

for a diffraction angle θ and the nth order, where l is the linear spacing on the screen and f is the grating space.
(b) Show that the change in linear separation per unit increase in spectral order for two wavelengths $\lambda = 5 \times 10^{-7}$m and $\lambda_2 = 5.2 \times 10^{-7}$m in a system where $F = 2$m and $f = 2 \times 10^{-6}$m is 2×10^{-2}m.

Problem 13.7.   (a) A sodium doublet consists of two wavelength $\lambda_1 = 5.890 \times 10^{-7}$m and $\lambda_2 = 5.896 \times 10^{-7}$m. Show that the minimum number of lines a grating must have to resolve this doublet in the third spectral order is ≈ 328.
(b) A red spectral line of wavelength $\lambda = 6.5 \times 10^{-7}$m is observed to be a close doublet. If the two lines are just resolved in the third spectral order by a grating of 9×10^4 lines show that the doublet separation is 2.4×10^{-12}m.

Problem 13.8.   Optical instruments have circular apertures, so that the Rayleigh criterion for resolution is given $\sin \theta = 1.22\lambda/a$, where a is the diameter of the aperture.

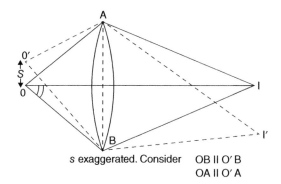

s exaggerated. Consider   OB ∥ O′B
                             OA ∥ O′A

Two points O and O′ of a specimen in the object plane of a microscope are separated by a distance *s*. The angle subtended by each at the objective aperture is 2i and their images I and I′ are just resolved. By considering the path difference between O′A and O′B show that the separation $s = 1.22\lambda/2 \sin i$.

Problem 13.9.   A screen with a small hole of 5 mm diameter is illuminated by a plane wave with a wavelength $\lambda = 5 \times 10^{-7}$ m. The first zone of diffraction pattern is a bright spot. Determine from Figures 13.23 and 13.24 the location of the darkest point of the pattern and calculate the farthest distance it can have from the screen.

Problem 13.10.   Seen from an axial point P a circular aperture uncovers the first one and a half Fresnel zones of a plane monochromatic wave. Locate this position of Figures 13.23 and 13.24 and calculate the intensity at P in terms of the unobstructed value I_0. What is the phase difference between the first and last wave vectors in this diffraction pattern?

14

Non-linear Oscillations

The oscillations discussed in this book so far have all been restricted in amplitude to those which satisfy the equation of motion where the restoring force is a linear function of the displacement. This restriction was emphasized in Chapter 1 and from time to time its limiting influence has required further discussion. We now discuss some of the consequences when this restriction is lifted.

We begin with simple examples in mechanical and solid state oscillators. Finally we discuss the development of shock waves from high amplitude sound waves.

14.1 Free Vibrations of an Anharmonic Oscillator – Large Amplitude Motion of a Simple Pendulum

In Figure 1.1 the equation of motion of the simple pendulum was written in terms of its angular displacement as

$$\frac{\mathrm{d}^2\theta}{\mathrm{d}t^2} + \omega_0^2\theta = 0$$

where $\omega_0^2 = g/l$. Here, an approximation was made by writing θ for $\sin\theta$; the equation is valid for oscillation amplitudes within this limit. When $\theta \geq 7°$ however, this validity is lost and we must consider the more complicated equation

$$\frac{\mathrm{d}^2\theta}{\mathrm{d}t^2} + \omega_0^2\sin\theta = 0$$

Multiplying this equation by $2\mathrm{d}\theta/\mathrm{d}t$ and integrating with respect to t gives $(\mathrm{d}\theta/\mathrm{d}t)^2 = 2\omega_0^2\cos\theta + A$, where A is the constant of integration. The velocity $\mathrm{d}\theta/\mathrm{d}t$ is zero at the maximum angular displacement $\theta = \theta_0$, giving $A = -2\omega_0^2\cos\theta_0$ so that

Introduction to Vibrations and Waves, First Edition. H. J. Pain and Patricia Rankin.
© 2015 John Wiley & Sons, Ltd. Published 2015 by John Wiley & Sons, Ltd.
Companion website: http://booksupport.wiley.com

$$\frac{d\theta}{dt} = \omega_0[2(\cos\theta - \cos\theta_0)]^{1/2}$$

or, upon integrating,

$$\omega_0 t = \int \frac{d\theta}{\{2[\cos\theta - \cos\theta_0]\}^{1/2}}$$

If $\theta = 0$ at time $t = 0$ and T is the new period of oscillation, then $\theta = \theta_0$ at $t = T/4$, and using half-angles we obtain

$$\omega_0\frac{T}{4} = \int_0^{\theta_0} \frac{d\theta}{2[\sin^2\theta_0/2 - \sin^2\theta/2]^{1/2}}$$

If we now express θ as a fraction of θ_0 by writing $\sin\theta/2 = \sin(\theta_0/2)\sin\phi$, where, of course, $-1 < \sin\phi < 1$, we have

$$\frac{1}{2}(\cos\theta/2)\delta\theta = (\sin\theta_0/2)\cos\phi\delta\phi$$

giving

$$\frac{\pi}{2}\frac{T}{T_0} = \int_0^{\pi/2} \frac{d\phi}{[1 - (\sin^2\theta_0/2)\sin^2\phi]^{1/2}}$$

where $T_0 = 2\pi/\omega_0$.

Expansion and integration gives

$$T = T_0(1 + \tfrac{1}{4}\sin^2\theta_0/2 + \tfrac{9}{64}\sin^4\theta_0/2 + \cdots$$

or approximately

$$T = T_0(1 + \tfrac{1}{4}\sin^2\theta_0/2)$$

14.2   Forced Oscillations – Non-linear Restoring Force

When an oscillating force is driving an undamped oscillator the equation of motion for such a system is given by

$$m\ddot{x} + s(x) = F_0\cos\omega t$$

where $s(x)$ is a non-linear function of x, which may be expressed in polynomial form:

$$s(x) = s_1 x + s_2 x^2 + s_3 x^3 \cdots$$

where the coefficients are constant. In many practical examples $s(x) = s_1 x + s_3 x^3$, where the cubic term ensures that the restoring force $s(x)$ has the same value for positive and negative displacements, so that the vibrations are symmetric about $x = 0$. When s_1 and s_3 are both positive the restoring force for a given displacement is greater than in the linear case and, if supplied by a spring, this case defines the spring as 'hard'. If s_3 is negative the restoring force is less than in the linear case and the spring is 'soft'. In Figure 14.1 the variation of restoring force is shown with displacement for s_3 zero (linear), s_3 positive (hard) and s_3 negative (soft). We see therefore that the large amplitude vibrations of the pendulum of the previous section are soft-spring controlled because

$$\sin\theta \approx \theta - \frac{1}{3}\theta^3$$

Figure 14.2 shows a mass m attached to points D and D', a vertical distance $2a$ apart, by two light elastic strings of constant stiffness s and subjected to a horizontal driving force $F_0\cos\omega t$. At zero displacement the tension in the strings is T_0 and at a displacement x (not limited in value) the tension is $T = T_0 + s(L - a)$ where L is the stretched string length.

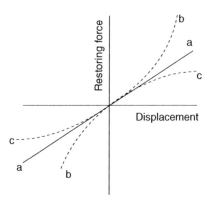

Figure 14.1 *Oscillator displacement versus restoring force for (a) linear restoring force, (b) non-linear 'hard' spring, and (c) non-linear 'soft' spring.*

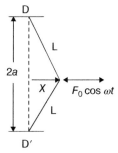

Figure 14.2 *A mass m supported by elastic strings between two points D and D' vertically separated by a distance 2a and subjected to a lateral force $F_0\cos\omega t$.*

The equation of motion (neglecting gravity) is

$$m\ddot{x} = -2T \sin\theta + F_0 \cos\omega t$$
$$= -2[T_0 + s(L-a)]\frac{x}{L} + F_0 \cos\omega t$$

Inserting the value

$$L = a \left[1 + \left(\frac{x}{a}\right)^2\right]^{1/2}$$

and expanding this expression in powers of x/a, we obtain by neglecting terms smaller than $(x/a)^3$

$$m\ddot{x} = -\frac{2T_0}{a}x - \frac{(sa - T_0)}{a^3}x^3 + F_0 \cos\omega t$$

which we may write

$$\ddot{x} + s_1 x + s_3 x^3 = \frac{F_0}{m}\cos\omega t$$

where

$$s_1 = \frac{2T_0}{ma} \quad \text{and} \quad s_3 = \frac{sa - T_0}{ma^3}$$

If s_3 is small we assume (as a first approximation) the solution $x_1 = A\cos\omega t$, which yields from the equation of motion

$$\ddot{x}_1 = -s_1 A \cos\omega t - s_3 A^3 \cos^3\omega t + \frac{F_0}{m}\cos\omega t$$

Since $\cos^3\omega t = \frac{3}{4}\cos\omega t + \frac{1}{4}\cos 3\omega t$, this becomes

$$\ddot{x}_1 = -(s_1 A + \tfrac{3}{4}s_3 A^3 - F_0/m)\cos\omega t - \tfrac{1}{4}s_3 A^3 \cos 3\omega t$$

Integrating twice, where the constants become zero from initial boundary conditions, gives as a second approximation to the equation

$$\ddot{x} + s_1 x + s_3 x^3 = \frac{F_0}{m}\cos\omega t$$

the solution

$$x_2 = \frac{1}{\omega^2}\left(s_1 A + \frac{3}{4}s_3 A^3 - \frac{F_0}{m}\right)\cos\omega t + \frac{s_3 A^3}{36\omega^2}\cos 3\omega t$$

Thus, for s_3 small we have a value of ω appropriate to a given amplitude A, and we can plot a graph of amplitude versus driving frequency. Note that we have a third harmonic. We see that for a system with a non-linear restoring force resonance does not exist in the same way as in the linear case. In the

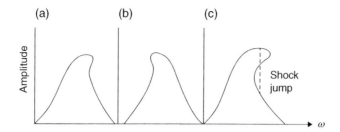

Figure 14.3 *Response curves of amplitude versus frequency for oscillators having (a) a 'hard' spring restoring force, and (b) a 'soft' spring restoring force. In the extreme case (c) the tilt of the maximum is sufficient to allow multi-valued amplitudes at a given frequency and 'shock jumps' may occur. (See Figure 14.5 for comparable behaviour in a high amplitude sound wave.).*

example above, even when no damping is present, the amplitude will not increase without limit for a driving force of a given frequency, for if ω is the natural frequency at low amplitude it is no longer the natural frequency at high amplitude. For s_3 positive (hard spring) the natural frequency increases with increasing amplitude and the amplitude versus frequency curve has a tilted maximum (Figure 14.3a). For a soft spring, s_3 is negative and the behaviour follows Figure 14.3b. It is possible for the tilt to become so pronounced (Figure 14.3c) that the amplitude is not single valued for a given ω and shock jumps in amplitude may occur at a given frequency (see the later discussion on the development of a shock front in a high amplitude acoustic wave).

14.3 Thermal Expansion of a Crystal

Chapter 1 showed that the curve of potential energy versus displacement for a harmonic oscillator was parabolic. Small departures from this curve are consistent with unharmonic oscillations. Consider the potential curve for a pair of neighbouring ions of opposite charge $\pm e$ in a crystal lattice such as that of KCl. This is shown in Figure 14.4 where r is the separation of the ions and the mutual potential energy is given by $V(r) = -e^2/4\pi\epsilon_0 r + B/r^9$ where B is a positive constant. This curve is no longer parabolic. The first term of $V(r)$ is the energy due to coulomb attraction and the second is that of a repulsive force. The value of B can be found in terms of the ion equilibrium separation r_0 because at r_0, $(\partial V/\partial r)_{r_0} = 0 = e^2/4\pi\epsilon_0 r_0^2 - 9B/r_0^{10}$ so $B = e^2 r_0^8/36\pi\epsilon_0$. X-ray diffraction from such crystals gives $r_0 = 3.15 \times 10^{-10}$ m for KCl so B may be found numerically.

To consider small displacements from the equilibrium value r_0 let us expand $V(r)$ about $r = r_0$ in a Taylor's series to give

$$V(r) = V(r_0) + x\left(\frac{\partial V}{\partial r}\right)_{r_0} + \frac{x^2}{2!}\left(\frac{\partial^2 V}{\partial r^2}\right)_{r_0} + \frac{x^3}{3!}\left(\frac{\partial^3 V}{\partial r^3}\right)_{r_0}$$

where $x = r - r_0$. At equilibrium since $(\partial V/\partial r)_{r_0} = 0$ we may write

$$V(r) - V(r_0) = V(x) = \frac{x^2}{2!}\left(\frac{\partial^2 V}{\partial r^2}\right)_{r_0} + \frac{x^3}{3!}\left(\frac{\partial^3 V}{\partial r^3}\right)_{r_0}$$

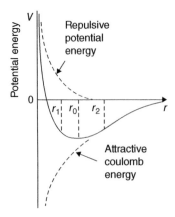

Figure 14.4   *Non-parabolic curve of mutual potential energy between oppositely charged ions in the lattice of an ionic crystal (NaCl or KCl). The combination of repulsive and attractive forces yields an equilibrium separation r_0. Very small energy increments give harmonic motion about r_0 but oscillations at higher energies are anharmonic, leading to thermal expansion of the crystal. The time averaged equilibrium position $\bar{r}$ between r_1 and r_2 is located slightly to the right of r_0.*

where the term $(\partial^2 V/\partial r^2)_{r_0}$ is the stiffness (force/distance) of a harmonically oscillatiiong chemical bond between the ions. For very small disturbances the bottom of the potential energy curve is parabolic and the ion pair oscillate symmetrically about $r = r_0$. A further increase in the ion pair energy involves the second term $x^3/3!(\partial^3 V/\partial r^3)_{r_0}$ and oscillations are no longer symmetric about r_0 because $|r_2 - r_0| > |r_1 - r_0|$ and the time average $\bar{r}$, the equilibrium position is $> r_0$. If all ion pairs acquire this amount of energy, for example by heating, the crystal expands. We may consider the force between the two ions as

$$F = -\frac{\partial V}{\partial x} = -x\left(\frac{\partial^2 V}{\partial r^2}\right)_{r_0} - \frac{x^2}{2}\left(\frac{\partial^3 V}{\partial r^3}\right)_{r_0} = -sx - \frac{x^2}{2}\left(\frac{\partial s}{\partial r}\right)_{r_0}$$

where $(\partial^2 V/\partial r^2)_{r_0} = s$ and $(\partial^3 V/\partial r^3)_{r_0}$ is its derivative. Now, for a weak spring, $\bar{r} > r_0$, $(\partial s/\partial r)_{r_0}$ is negative in the region r_0 to $\bar{r}$ to reduce the total force to zero at $\bar{r}$. We shall show shortly that $(\partial s/\partial r)_{r_0} = C(-104/r_0^4)$ where C is a constant. Writing $\bar{x} = (\bar{r} - r_0)$ we have $F = (-s\bar{x} - \frac{1}{2}(\frac{\partial s}{\partial r})_{r_0}\bar{x}^2) = 0$ because at $\bar{x}$ the oscillation has reached $\bar{r}$ the equilibrium position where every oscillating restoring force is zero as it reverses direction by $180°$. We then have, from the force bracket

$$s\bar{x} = -\frac{1}{2}\left(\frac{\partial s}{\partial r}\right)_{r_0} \bar{x}^2$$

or

$$\boxed{\bar{x} = -\frac{1}{2}\left(\frac{\partial s}{\partial r}\right)_{r_0} \bar{x}^2/s}$$

which we now evaluate.

We can now rewrite the potential energy equation using $B = e^2 r_0^8 / 36\pi\epsilon_0$ as

$$V(r) = C\left(\frac{1}{9}\frac{r_0^8}{r^9} - \frac{1}{r}\right)$$

where C is a constant $= e^2/4\pi\epsilon_0$.

$$\frac{\partial V}{\partial r} = C\left(-\frac{r_0^8}{r^{10}} + \frac{1}{r^2}\right)_{r_0} = 0 \quad \text{as we expect.}$$

$$\frac{\partial^2 V}{\partial r^2} = C\left(\frac{10 r_0^8}{r^{11}} - \frac{2}{r^3}\right)_{r_0} = C\frac{8}{r_0^3} = s$$

$$\frac{\partial^3 V}{\partial r^3} = C\left(-\frac{110 r_0^8}{r^{12}} + \frac{6}{r^4}\right)_{r_0} = C\frac{-104}{r_0^4} = \left(\frac{\partial s}{\partial r}\right)_{r_0} \quad \text{a negative quantity}$$

$$\therefore \left(\frac{\partial s}{\partial r}\right)_{r_0}/s = -\frac{13}{r_0} \quad \text{and} \quad \boxed{\bar{x} = -\frac{1}{2}\left(\frac{\partial s}{\partial r}\right)_{r_0}\bar{x^2}/s = +\frac{13}{2r_0}\bar{x^2}}$$

Now the thermal energy associated with the harmonic oscillator potential energy maximum $\frac{1}{2}s\bar{x^2} = kT$ where k is Boltzmann's constant and T is the temperature. So with $\bar{x} = 13kT/r_0 s$ and $s = C \cdot 8/r_0^3 = 2e^2/\pi\epsilon_0 r_0^3$ we have $\bar{x} = 13\pi\epsilon_0 kT r_0^2/2e^2$. The value of $r_0 = 3.15 \times 10^{-10}$ m gives a thermal expansion

$$\frac{\mathrm{d}\bar{x}}{\mathrm{d}T} = \frac{13\pi\epsilon_0 k r_0^2}{2e^2} = 9.7 \times 10^{-6}\,\text{nm} \cdot \text{K}^{-1}$$

which is about 0.5 of the experimental value.

This approach considers only one pair of ions and disregards the effect of surrounding molecules in the crystal lattice.

14.4 Non-linear Acoustic Waves and Shocks

The linearity of the longitudinal acoustic waves discussed in Chapter 7 required the assumption of a constant bulk modulus

$$B = -\frac{\mathrm{d}P}{\mathrm{d}V/V}$$

If the amplitude of the sound wave is too large this assumption is no longer valid and the wave propagation assumes a new form. A given mass of gas undergoing an adiabatic change obeys the relation

$$\frac{P}{P_0} = \left(\frac{V_0}{V}\right)^\gamma = \left[\frac{V_0}{V_0(1+\delta)}\right]^\gamma$$

in the notation of Chapter 7, so that

$$\frac{\partial P}{\partial x} = \frac{\partial p}{\partial x} = -\gamma P_0 (1 + \delta)^{-(\gamma+1)} \frac{\partial^2 \eta}{\partial x^2}$$

since $\delta = \partial \eta / \partial x$.

Since $(1 + \delta)(1 + s) = 1$, we may write

$$\frac{\partial p}{\partial x} = -\gamma P_0 (1 + s)^{\gamma+1} \frac{\partial^2 \eta}{\partial x^2}$$

and from Newton's second law we have

$$\frac{\partial p}{\partial x} = -\rho_0 \frac{\partial^2 \eta}{\partial t^2}$$

so that

$$\frac{\partial^2 \eta}{\partial t^2} = c_0^2 (1 + s)^{\gamma+1} \frac{\partial^2 \eta}{\partial x^2}, \quad \text{where} \quad c_0^2 = \frac{\gamma P_0}{\rho_0}$$

Physically this implies that the local velocity of sound, $c_0(1+s)^{(\gamma+1)/2}$, depends upon the condensation s, so that in a finite amplitude sound wave regions of higher density and pressure will have a greater sound velocity, and local disturbances in these parts of the wave will overtake those where the values of density, pressure and temperature are lower.

A single sine wave of high amplitude can be formed by a close-fitting piston in a tube which is pushed forward rapidly and then returned to its original position. Figure 14.5a shows the original shape of such a wave and Figure 14.5b shows the distortion which follows as it propagates down the tube. If the distortion continued the waveform would eventually appear as in Figure 14.5c, where analytical solutions for pressure, density and temperature would be multi-valued, as in the case of the non-linear oscillator of Figure 14.5c. Before this situation is reached, however, the waveform stabilizes into that of Figure 14.5d, where at the vertical 'shock front' the rapid changes of particle density, velocity and temperature produce the dissipating processes of diffusion, viscosity and thermal conductivity. The velocity of this 'shock front' is always greater than the velocity of sound in the gas into which it is moving, and across the 'shock front' there is always an increase in entropy. The competing effects of dissipation and non-linearity produce a stable front as long as the wave retains sufficient energy. The N-type wave of Figure 14.5d occurs naturally in explosions (in spherical dimensions) where a blast is often followed by a rarefaction.

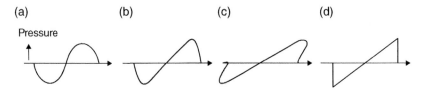

(a)   (b)   (c)   (d)

Pressure

Figure 14.5   *The local sound velocity in a high amplitude acoustic wave (a) is pressure and density dependent. The wave distorts with time (b) as the crest overtakes the lower density regions. The extreme situation of (c) is prevented by entropy-producing mechanisms and the wave stabilizes to an N-type shock wave (d) with a sharp leading edge.*

Worked Example

An explosion causes a high pressure sine wave in air shaped as in Figure 14.5a. The value of *gamma* in air is 5/3, and the value of s in the wave is 12%. If the length of the wave is 200 m and the velocity of sound in air is 330 m s^{-1}, how long does the pulse take to assume the shape of Figure 14.5d?

Solution

Velocity of the peak of the pulse $= c_0[1 + \frac{1}{2}(\gamma+1)s] = c_0(1 + \frac{4}{3} \cdot 12\%) = c_0(1.16)$. Relative velocity of peak with respect to leading edge of pulse $= .16 \cdot 330 = 52.8$ m s^{-1}. Time to overtake 50 m $= 50/52.8 = .985$ sec.

The growth of a shock front may also be seen as an extension of the Doppler effect section 7.9, where the velocity of the moving source is now greater than that of the signal. In Figure 14.6a as an aircraft moves from S to S' in a time t the air around it is displaced and the disturbance moves away with the local velocity of sound v_S. The circles show the positions at time t of the sound wavefronts generated at various points along the path of the aircraft but if the speed of the aircraft u is greater than the velocity of sound v_S regions of high density and pressure will develop, notably at the edges of the aircraft structure and along the conical surface tangent to the successive wavefronts which are generated at a speed greater than sound and which build up to a high amplitude to form a shock. The cone, whose axis is the aircraft path, has half angle α where

$$\sin \alpha = \frac{v_S}{u}$$

It is known as the 'Mach Cone' and when it reaches the ground a 'supersonic bang' is heard.

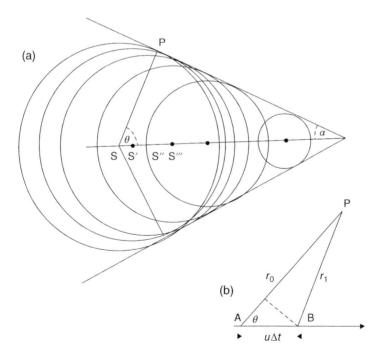

Figure 14.6 *(a) The circles are the wavefronts generated at points S along the path of the aircraft, velocity $u > v_S$ the velocity of sound. Wavefronts superpose on the surface of the Mach Cone (typical point P) of half angle $\alpha = \sin^{-1} v_S/u$ to form a shock front. (b) At point P sound waves arrive simultaneously from positions A and B along the aircraft path when $(u/v_S)\cos\theta = 1$. $(\theta + \alpha = 90°)$.*

The growth of the shock at the surface of the cone may be seen by considering the sound waves in Figure 14.6b generated at points A (time t_A) and B (time t_B) along the path of the aircraft, which travels the distance $AB = x = u\Delta t$ in the time interval $\Delta t = t_B - t_A$. The sound waves from A will travel the distance r_0 to reach the point P at a time

$$t_0 = t_A + \frac{r_0}{v_S}$$

Those from B will travel the distance r_1 to P to arrive at a time

$$t_1 = t_B + \frac{r_1}{v_S}$$

If x is small relative to r_0 and r_1, we see that

$$r_1 - r_0 \approx x \cos \theta = u\Delta t \cos \theta$$

so the time interval

$$t_1 - t_0 = t_B - t_A + \frac{(r_1 - r_0)}{v_S}$$
$$= \Delta t - \frac{u\Delta t \cos \theta}{v_S} = \Delta t \left(1 - \frac{u \cos \theta}{v_S} \right)$$

For the aircraft speed $u < v_S$, $t_1 - t_0$ is always positive and the sound waves arrive at P in the order in which they were generated.

For $u > v_S$ this time sequence depends on θ and when

$$\frac{u}{v_S} \cos \theta = 1$$

$t_1 = t_0$ and the sound waves arrive simultaneously at P to build up a shock.

Now the angles θ and α are complementary so the condition

$$\cos \theta = \frac{v_S}{u}$$

defines

$$\sin \alpha = \frac{v_S}{u}$$

so that all points P lie on the surface of the Mach Cone.

A similar situation may arise when a charged particle q emitting electromagnetic waves moves in a medium of refractive index greater than unity with a velocity v_q which may be greater than that of the phase velocity v of the electromagnetic waves in the medium ($v < c$). A Mach Cone for electromagnetic waves is formed with a half angle α where

$$\sin \alpha = \frac{v}{v_q}$$

And the resulting 'shock wave' is called Cerenkov radiation. Measuring the effective direction of propagation of the Cerenkov radiation is one way of finding the velocity of the charged particle.

14.5 Mach Number

A significant parameter in shock wave theory is the Mach number. It is a local parameter defined as the ratio of the flow velocity to the local velocity of sound. The Mach number of the shock front is therefore $M_s = u_1/c_1$, where u_1 is the velocity of the shock front propagating into a gas whose velocity of sound is c_1.

The Mach number of the gas flow behind the shock front is defined as $M_f = u/c_2$, where u is the flow velocity of the gas behind the shock front $(u < u_1)$ and c_2 is the local velocity of sound behind the shock front. There is always an increase of temperature across the shock front, so that $c_2 > c_1$ and $M_s > M_f$. The physical significance of the Mach number is seen by writing $M^2 = u^2/c^2$, which indicates the ratio of the kinetic flow energy, $\frac{1}{2}u^2 \, \text{mol}^{-1}$, to the thermal energy, $c^2 = \gamma RT \, \text{mol}^{-1}$. The higher the proportion of the total gas energy to be found as kinetic energy of flow the greater is the Mach number.

Problem 14.1. If the period of a pendulum with large amplitude oscillations is given by

$$T = T_0 \left(1 + \frac{1}{4} \sin^2 \frac{\theta_0}{2} \right)$$

where T_0 is the period for small amplitude oscillations and θ_0 is the oscillation amplitude, show that for θ_0 not exceeding $30°$, T and T_0 differ by only 2% and for $\theta_0 = 90°$ the difference is 12%.

Problem 14.2. The equation of motion of a free undamped non-linear oscillator is given by

$$m\ddot{x} = -f(x)$$

Show that for an amplitude x_0 its period

$$T_0 = 4\sqrt{\frac{m}{2}} \int_0^{x_0} \frac{dx}{[F(x_0) - F(x)]^{1/2}}, \quad \text{where} \quad F(x_0) = \int_0^{x_0} f(x)dx$$

Problem 14.3. The equation of motion of a forced undamped non-linear oscillator of unit mass is given by

$$\ddot{x} = s(x) = F_0 \cos \omega t$$

Writing $s(x) = s_1 x + s_3 x^3$, where s_1 and s_3 are constant, choose the variable $\omega t = \phi$, and for $s_3 \ll s_1$ assume a solution

$$x = \sum_{n=1}^{\infty} \left(a_n \cos \frac{n}{3}\phi + b_n \sin \frac{n}{3}\phi \right)$$

to show that all the sine terms and the even-numbered cosine terms are zero, leaving the fundamental frequency term and its third harmonic as the significant terms in the solution.

Problem 14.4.   If the mutual interionic potential in a crystal is given by

$$V = -V_0 \left[2 \left(\frac{r_0}{r} \right)^6 - \left(\frac{r_0}{r} \right)^{12} \right]$$

where r_0 is the equilibrium value of the ion separation r, show by expanding V about V_0 that the ions have small harmonic oscillations at a frequency given by $\omega^2 \approx 72 \, V_0/mr_0^2$, where m is the reduced mass.

Problem 14.5.   Repeating the worked example at the end of section 14.4 with an s value of 9%, approximately how much longer does it take to form the Figure 14.5d pulse? In practice, how would the N-type shock wave change its shape and why?

Appendix 1
The Binomial Theorem

For our purposes we may express the Binomial Theorem as $(1 + x)^n$ where n is a positive integer or fraction and $-1 < x < 1$.

Thus $n = 2$ gives $(1 + x)^2 = 1 + 2x + x^2$, $n = 3$ gives $(1 + x)^3 = 1 + 3x + 3x^2 + x^3$ and more generally

$$(1 + x)^n = 1 + nx + \frac{n(n - 1)}{2!}x^2 + \frac{(n - 1)(n - 2)(n - 3)}{3!}x^3 \cdots$$

written

$$(1 + x)^n = 1 + {}^nC_1 x + {}^nC_2 x^2 + {}^nC_3 x^3 + \cdots + {}^nC_r x^r + \cdots + x^n$$

The coefficients ${}^nC_r$ are called *combinations* and specify how many arrangements may be made by selecting the number r from n dissimilar objects when only one particular order is allowed. From the four different letters, A, B, C, D, we can form six combinations of $r = 2$, that is, AB, AC, AD, BC, BD, CD. Other arrangements, called *permutations* allow, in addition, any order to be reversed e.g. BA, CA, DA etc., so there are 12 permutations in all. A *combination* is the quotient of two *permutations* of which the numerator is written ${}^nP_r$ and the denominator is ${}^rP_r$. In forming ${}^nP_r$ we have n choices of filling the first place and $(n - 1)$ choices from which to fill the second place, that is, $n(n - 1)$ choices to fill the first two places. We proceed in this way to fill the r places in $n(n - 1)(n - 2) \cdots (n - r + 1)$ permutations. We now find the number of permutations of the r objects arranged among themselves which from the reasoning above is ${}^rP_r = r(r - 1)(r - 2) \cdots 3 \cdot 2 \cdot 1$, written $r!$ (called r factorial).

Thus

$$\frac{{}^nP_r}{{}^rP_r} = {}^nC_r = \frac{n(n - 1)(n - 2) \cdots (n - r + 1)}{r!}$$

as in $(1 + x)^n$ above.

Introduction to Vibrations and Waves, First Edition. H. J. Pain and Patricia Rankin.
© 2015 John Wiley & Sons, Ltd. Published 2015 by John Wiley & Sons, Ltd.
Companion website: http://booksupport.wiley.com

If the series is truncated before the last term the order of accuracy is that of the first term to be truncated, e.g.

$$(1+x)^{1/2} = 1 + \frac{1}{2}x - \frac{1}{8}x^2 + O(x^3)$$

where $O(x^3)$ is defined as to the order of x^3. This is often applied using $n = 1/2$ as the square root.

Appendix 2
Taylor's and the Exponential Series

Consider the function $\frac{1}{1+x}$ when $-1 < x < 1$. Actual division gives 1,

$$\frac{1}{1+x} = 1 - \frac{x}{1+x} = 1 - x + \frac{x^2}{x+1} = 1 - x + x^2 - \frac{x^3}{1+x}$$

where $1, 1-x, 1-x+x^2$ are successive approximations to $\frac{1}{1+x}$ with respective errors, $-\frac{x}{1+x}, \frac{x^2}{1+x}, -\frac{x^3}{1+x}$ which for $-1 < x < 1$ become progressively smaller. Note that for $x = 0$ the successive approximations are all equal, from $1-x$ onwards they all have the same first derivative and from $1-x+x^2$ they all have the same second derivative and so on. Consider the approximation to a function $f(x+x_0)$ where x is close to x_0 and write it as a polynomial i.e. $f(x+x_0) \approx a_0 + a_1 x + a_2 x^2 + \cdots + a_n x^n$ choosing the a's so that $f(x+x_0)$ and its first n derivatives have the same values when $x = 0$ as the polynomial and its n derivatives. This allows us to expect the polynomial to be a better approximation as the number of its terms increases. These derivatives are successively $a_1 + 2a_2 x + \cdots + na_n x^{n-1}; 2a_2 + 3 \cdot 2a_3 x + \cdots + n(n-1)a_n x^{n-2}$ etc. So the values of the polynomial and its first n derivatives are, when $x = 0, a_0, a_1, 2! \cdot a_2, 3! \cdot a_3, \cdots, n! \cdot a_n$. Equating these to $f(x + x_0)$ for $x = 0$ we have

$$f(x + x_0) = f(x_0) + xf'(x_0) + \frac{x^2}{2!}f''(x_0) + \cdots + \frac{x^n}{n!}f^n(x_0)$$

which is known as Taylor's series.

when $x = 0$ Taylor's series reduces to

$$f(x) = f(0) + xf'(0) + \frac{x^2}{2!}f''(0) \cdots \frac{x^n}{n^1}$$

Introduction to Vibrations and Waves, First Edition. H. J. Pain and Patricia Rankin.
© 2015 John Wiley & Sons, Ltd. Published 2015 by John Wiley & Sons, Ltd.
Companion website: http://booksupport.wiley.com

which is known as McLaurin's series for $f(x)$. Applying this to the exponential series e^x where all derivatives are e^x, we have for $x = 0$,

$$f(0) = f'(0) = f''(0) = \cdots = f^n(0) = e^0 = 1$$

hence,

$$e^x = 1 + x + \frac{x^2}{2!} + \frac{x^3}{3!} + \cdots + \frac{x^n}{n!}$$

which is valid for all x.
    For $x = 1$, we have

$$e = 1 + 1 + \frac{1}{2!} + \frac{1}{3!} + \cdots = 2.718.$$

Appendix 3

Superposition of a Large Number n of Simple Harmonic Vibrations of Equal Amplitude a and Equal Successive Phase Difference δ

Figure A1 shows the addition of n vectors of equal length a, each representing a simple harmonic vibration with a constant phase difference δ from its neighbour. Two general physical situations are characterized by such a superposition. The first is met in Chapter 6 as a wave group problem where the phase difference δ arises from a small frequency difference, $\delta\omega$, between consecutive components. The second appears in Chapter 12 where the intensity of optical interference and diffraction patterns are considered. There, the superposed harmonic vibrations will have the same frequency but each component will have a constant phase difference from its neighbour because of the extra distance it has travelled.

The figure displays the mathematical expression

$$R \cos (\omega t + \alpha) = a \cos \omega t + a \cos(\omega t + \delta) + a \cos(\omega t + 2\delta)$$
$$+ \cdots + a \cos(\omega t + [n-1]\delta)$$

where R is the magnitude of the resultant and α is its phase difference with respect to the first component $a \cos \omega t$.

Geometrically we see that each length

$$a = 2r \sin \frac{\delta}{2}$$

where r is the radius of the circle enclosing the (incomplete) polygon.

From the isosceles triangle OAC the magnitude of the resultant

$$R = 2r \sin \frac{n\delta}{2} = a \frac{\sin n\delta/2}{\sin \delta/2}$$

Introduction to Vibrations and Waves, First Edition. H. J. Pain and Patricia Rankin.
© 2015 John Wiley & Sons, Ltd. Published 2015 by John Wiley & Sons, Ltd.
Companion website: http://booksupport.wiley.com

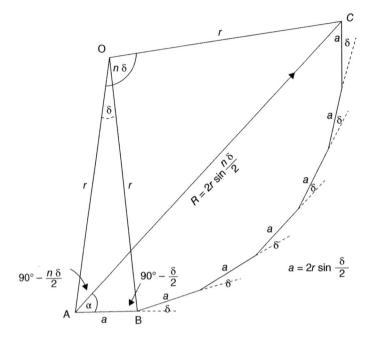

Figure A1   *Vector superposition of a large number n of simple harmonic vibrations of equal amplitude a and equal successive phase difference δ. The amplitude of the resultant $R = 2r\sin\frac{n\delta}{2} = a\frac{\sin n\delta/2}{\sin \delta/2}$ and its phase with respect to the first contribution is given by $\alpha = (n-1)\delta/2$.*

and its phase angle is seen to be

$$\alpha = \hat{OAB} - \hat{OAC}$$

In the isosceles triangle OAC

$$\hat{OAC} = 90° - \frac{n\delta}{2}$$

and in the isosceles triangle OAB

$$\hat{OAB} = 90° - \frac{\delta}{2}$$

so

$$\alpha = \left(90° - \frac{\delta}{2}\right) - \left(90° - \frac{n\delta}{2}\right) = (n-1)\frac{\delta}{2}$$

that is, half the phase difference between the first and the last contributions. Hence the resultant

$$R\cos(\omega t + \alpha) = a\frac{\sin n\delta/2}{\sin \delta/2}\cos\left[\omega t + (n-1)\frac{\delta}{2}\right]$$

We shall obtain the same result later in this chapter as an example on the use of exponential notation.

For the moment let us examine the behaviour of the magnitude of the resultant

$$R = a\frac{\sin n\delta/2}{\sin \delta/2}$$

which is not constant but depends on the value of δ. When n is very large δ is very small and the polygon becomes an arc of the circle centre O, of length $na = A$, with R as the chord. Then

$$\alpha = (n-1)\frac{\delta}{2} \approx \frac{n\delta}{2}$$

and

$$\sin\frac{\delta}{2} \to \frac{\delta}{2} \approx \frac{\alpha}{n}$$

Hence, in this limit,

$$R = a\frac{\sin n\delta/2}{\sin \delta/2} = a\frac{\sin\alpha}{\alpha/n} = na\frac{\sin\alpha}{\alpha} = \frac{A\sin\alpha}{\alpha}$$

The behaviour of $A\sin\alpha/\alpha$ versus α is shown in Figure A2. The pattern is symmetric about the value $\alpha = 0$ and is zero whenever $\sin\alpha = 0$ except at $\alpha \to 0$ that is, when $\sin\alpha/\alpha \to 1$. When $\alpha = 0$, $\delta = 0$ and the resultant of the n vectors is the straight line of length A, Figure A2(b). As δ increases A becomes the arc of a circle until at $\alpha = \pi/2$ the first and last contributions are out of phase $(2\alpha = \pi)$ and the arc A has become a semicircle of which the diameter is the resultant R, Figure A2(c). A further increase in δ increases α and curls the constant length A into the circumference of a circle $(\alpha = \pi)$ with a zero resultant, Figure A2(d). At $\alpha = 3\pi/2$, Figure A2(e), the length A is now 3/2 times the circumference of a circle whose diameter is the amplitude of the first minimum.

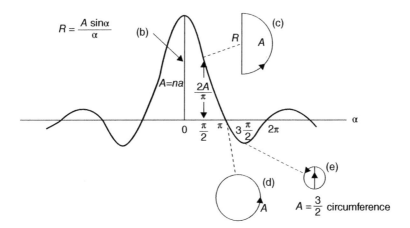

Figure A2 *(a) Graph of A $\sin\alpha/\alpha$ versus α, showing the magnitude of the resultants for (b) $\alpha = 0$, (c) $\alpha = \pi/2$; (d) $\alpha = \pi$ and (e) $\alpha = 3\pi/2$.*

Appendix 4

Superposition of n Equal SHM Vectors of Length a with Random Phase ϕ

When the phase difference between the successive vectors of the last section may take random values ϕ between zero and 2π (measured from the x axis) the vector superposition and resultant R may be represented by Figure A3.

The components of R on the x and y axes are given by

$$R_x = a\cos\phi_1 + a\cos\phi_2 + a\cos\phi_3 \ldots a\cos\phi_n$$

$$= a\sum_{i=1}^{n}\cos\phi_i$$

and

$$R_y = a\sum_{i=1}^{n}\sin\phi_i$$

where

$$R^2 = R_x^2 + R_y^2$$

Now

$$R_x^2 = a^2\left(\sum_{i=1}^{n}\cos\phi_i\right)^2 = a^2\left[\sum_{i=1}^{n}\cos^2\phi_i + \sum_{\substack{i=1\\i\neq j}}^{n}\cos\phi_i\sum_{j=1}^{n}\cos\phi_j\right]$$

Introduction to Vibrations and Waves, First Edition. H. J. Pain and Patricia Rankin.
© 2015 John Wiley & Sons, Ltd. Published 2015 by John Wiley & Sons, Ltd.
Companion website: http://booksupport.wiley.com

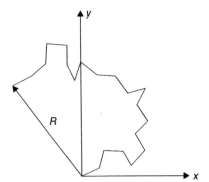

Figure A3 *The resultant $R = \sqrt{n}a$ of n vectors, each of length a, having random phase. This result is important in optical incoherence and in energy loss from waves from random dissipation processes*

The summation $\sum \cos \phi$ for all values

$$-1 \leq \cos \phi \leq +1 = \int_{-\pi}^{0} \cos \phi d\phi = [\sin \phi]_{-\pi}^{0} = 0$$

Similarly, $\sum \sin \phi$ for all values

$$-1 \leq \sin \phi \leq +1 = \int_{-\pi/2}^{\pi/2} \sin \phi d\phi = -[\cos \phi]_{-\pi/2}^{\pi/2} = 0$$

The summation

$$\sum_{i=1}^{n} \cos^2 \phi_i = n \, \overline{\cos^2 \phi}$$

that is, the number of terms n times the average value $\overline{\cos^2 \phi}$ which is the integrated value of $\cos^2 \phi$ over the interval zero to 2π divided by the total interval 2π, or

$$\overline{\cos^2 \phi} = \frac{1}{2\pi} \int_{0}^{2\pi} \cos^2 \phi d\phi = \frac{1}{2} = \overline{\sin^2 \phi}$$

So

$$R_x^2 = a^2 \sum_{i=1}^{n} \cos^2 \phi_i = na^2 \overline{\cos^2 \phi_i} = \frac{na^2}{2}$$

and

$$R_y^2 = a^2 \sum_{i=1}^{n} \sin^2 \phi_i = na^2 \overline{\sin^2 \phi_i} = \frac{na^2}{2}$$

giving

$$R^2 = R_x^2 + R_y^2 = na^2$$

or

$$R = \sqrt{n}a$$

   Thus, the amplitude R of a system subjected to n equal simple harmonic motions of amplitude a with random phases is only $\sqrt{n}a$ whereas, if the motions were all in phase R would equal na.
   Such a result illustrates a very important principle of random behaviour.

Appendix 5
Electromagnetic Wave Equations: Vector Method

For a medium with permeability μ, permittivity ϵ, conductivity $\sigma = 0$ and charge density $\rho = 0$ pages, section 9.2 give Maxwell's equations as

$$\operatorname{curl} \boldsymbol{E} = \nabla \times \boldsymbol{E} = -\frac{\partial}{\partial t}\boldsymbol{B} = -\mu\frac{\partial}{\partial t}\boldsymbol{H} \tag{1}$$

$$\operatorname{curl} \boldsymbol{H} = \nabla \times \boldsymbol{H} = \frac{\partial}{\partial t}\boldsymbol{D} = -\epsilon\frac{\partial}{\partial t}\boldsymbol{E} \tag{2}$$

$$\operatorname{div} \boldsymbol{D} = \nabla \cdot \boldsymbol{D} = 0 \tag{3}$$

$$\operatorname{div} \boldsymbol{B} = \nabla \cdot \boldsymbol{B} = 0 \tag{4}$$

The vector relation, using (3), gives

$$\nabla \times \nabla \times \boldsymbol{E} = \nabla(\nabla \cdot \boldsymbol{E}) - \nabla^2\boldsymbol{E} = -\nabla^2\boldsymbol{E} \tag{5}$$

(1), (2) and (5) give

$$\nabla^2\boldsymbol{E} = \frac{\partial}{\partial t}(\nabla \times \boldsymbol{B}) = \mu\frac{\partial^2}{\partial t^2}\boldsymbol{D} = \mu\epsilon\frac{\partial^2}{\partial t^2}\boldsymbol{E} \tag{6}$$

which, for $E = E_x$ gives the wave equation

$$\frac{\partial^2}{\partial z^2}E_x = \mu\epsilon\frac{\partial^2}{\partial t^2}E_x$$

Introduction to Vibrations and Waves, First Edition. H. J. Pain and Patricia Rankin.
© 2015 John Wiley & Sons, Ltd. Published 2015 by John Wiley & Sons, Ltd.
Companion website: http://booksupport.wiley.com

When $\sigma \neq 0$ an ohmic current of density $J = \sigma E$ flows in addition to the displacement current and equation (2) becomes

$$\nabla \times \boldsymbol{H} = \frac{\partial}{\partial t}\boldsymbol{D} + \boldsymbol{J}$$

which with equation (1), (5) and (6) gives

$$\nabla^2 \boldsymbol{E} = \frac{\partial}{\partial t}(\nabla \times \boldsymbol{B}) = \mu\epsilon\frac{\partial^2}{\partial t^2}\boldsymbol{E} + \mu\sigma\frac{\partial}{\partial t}\boldsymbol{E}$$

which for $E = E_x$ gives

$$\frac{\partial^2}{\partial z^2}\boldsymbol{E}_x = \mu\epsilon\frac{\partial^2}{\partial t^2}\boldsymbol{E}_x + \mu\sigma\frac{\partial}{\partial t}\boldsymbol{E}_x$$

Appendix 6
Planck's Radiation Law

(See also section 10.10.)

Proof that the average energy of an oscillator in Planck's Radiation Law is given by

$$\bar{\epsilon} = \frac{E}{N} = \frac{h\nu}{e^{h\nu/kT} - 1}$$

where E is the total energy, N is the total number of oscillators, k is Boltzmann's constant and h is Planck's constant.

Planck assumed that a large number N of oscillators had energies of 0, $h\nu$, $2h\nu$, $3h\nu$, ..., $nh\nu$, distributed according to Boltzmann's Law.

$$N_n = N_0\, e^{-nh\nu/kT} = N_0 e^{-n\beta\epsilon_0}$$

where we have replaced $h\nu$ by ϵ_0 and $1/kT$ by β for convenience in what follows. Note that the number of oscillators N_n decreases exponentially with increasing n. The total number of oscillators:

$$N = \sum_{n=0}^{\infty} N_n = N_0(1 + e^{-\beta\epsilon_0} + e^{-2\beta\epsilon_0} + e^{-3\beta\epsilon_0} + \cdots) = N_0 \sum_{n=0}^{\infty} e^{-n\beta\epsilon_0}$$

which is a geometric progression with consecutive terms increasing by a factor $e^{-\beta\epsilon_0}$ to give a sum:

$$N_0/(1 - e^{-\beta\epsilon_0})$$

The total energy

$$E = \sum_{n=0}^{\infty} E_n = \sum_{n=0}^{\infty} N_n n\epsilon_0 = N_0 \sum_{n=0}^{\infty} n\epsilon_0 e^{-n\beta\epsilon_0},$$

Introduction to Vibrations and Waves, First Edition. H. J. Pain and Patricia Rankin.
© 2015 John Wiley & Sons, Ltd. Published 2015 by John Wiley & Sons, Ltd.
Companion website: http://booksupport.wiley.com

so the average energy

$$\bar{\epsilon} = E/N = \frac{\sum\limits_{n=0}^{\infty} n\epsilon_0 e^{-n\beta\epsilon_0}}{\sum\limits_{n=0}^{\infty} e^{-n\beta\epsilon_0}}$$

after cancelling N_0,

Examining this expression we see that the numerator is $-\partial/\partial\beta$ of the denominator and writing the denominator as y and using

$$-\frac{\partial}{\partial\beta}\log y = -\frac{1}{y}\frac{\partial y}{\partial\beta}$$

we have

$$\epsilon = \frac{E}{N} = -\frac{\partial}{\partial\beta}\log \sum\limits_{n=0}^{\infty} e^{-n\beta\epsilon_0} = -\frac{\partial}{\partial\beta}\log \frac{1}{1-e^{-\beta\epsilon_0}}$$

because $\sum e^{-n\beta\epsilon_0}$ is a geometric progression. But

$$-\frac{\partial}{\partial\beta}\log \frac{1}{1-e^{-\beta\epsilon_0}} = +\frac{\partial}{\partial\beta}\log(1-e^{-\beta\epsilon_0})$$

because

$$\log N = -\log\frac{1}{N}$$

Using again

$$\frac{\partial}{\partial\beta}\log y = \frac{1}{y}\frac{\partial y}{\partial\beta}$$

with $y = (1-e^{-\beta\epsilon_0})$ we have

$$\bar{\epsilon} = \frac{E}{N} = \frac{\epsilon_0 e^{-\beta\epsilon_0}}{1-e^{-\beta\epsilon_0}}$$

which, multiplying top and bottom by $e^{\beta\epsilon_0}$

$$= \frac{\epsilon_0}{e^{\beta\epsilon_0}-1} = \frac{h\nu}{e^{h\nu/kT}-1}$$

as required.

Appendix 7

Fraunhofer Diffraction from a Rectangular Aperture

The simple approach in section 13.8 took contributions from only the strips along the central axes $x = y = 0$. However, the amplitude at the focal point P in Figure 13.8 is the superposition from all points (x, y) in the aperture with their appropriate phases. The vector $\boldsymbol{k}$, normal to the diffracted wave front, has direction cosines l with respect to the x axis and m with respect to the y axis. The vector $\boldsymbol{r}'$ denotes a typical point (x, y) in the aperture and the phase difference between the contribution from this point and the central point 0 is $r' \, 2\pi/\lambda$ (path difference) $= 2\pi/\lambda(lx + my)$ where $(lx + my)$ is the projection of $\boldsymbol{r}'$ on $\boldsymbol{k}$.

Writing $2\pi l/\lambda = k_x$ and $2\pi m/\lambda = k_y$, we have the phase difference represented by the space part of the incident electromagnetic wave $e^{-i(k_x x + k_y y)}$. Taking the small amplitudes of each point (x, y) as equal to a constant h, the total amplitude in $\boldsymbol{k}$ space is the integration over the area of the aperture as

$$R(k_x{:}k_y) = \frac{h}{(2\pi)^2} \int_{-a/2}^{a/2} \int_{-b/2}^{b/2} e^{-i(k_x x + k_y y)} \mathrm{d}x\mathrm{d}y$$

Integration with respect to y evalues the contribution of a strip of the aperture along the y direction and integrating with respect to x then adds the contribution of all these strips with their appropriate phase relations. Moving from x, y space to k_x, k_y space involves a factor of 2π for each of the coordinates but these, together with h, are incorporated at the end into the value of the maximum intensity I_0 to which all other intensity values in the aperture are *relative*.

The integral

$$\int_{-b/2}^{b/2} e^{-ik_y y} \mathrm{d}y = \frac{1}{-ik_y} \left(e^{-ik_y b/2} - e^{+ik_y b/2} \right)$$

$$= \frac{-2i}{-ik_y} (\sin k_y b/2) = b\frac{\sin \beta}{\beta}$$

Introduction to Vibrations and Waves, First Edition. H. J. Pain and Patricia Rankin.
© 2015 John Wiley & Sons, Ltd. Published 2015 by John Wiley & Sons, Ltd.
Companion website: http://booksupport.wiley.com

where $\beta = k_y b/2 = \pi m b/\lambda$, that is, half the phase difference between the contributions from $y = \pm b/2$. Similarly

$$\int_{-a/2}^{a/2} e^{-ik_x x} dx = a \frac{\sin \alpha}{\alpha}$$

where $\alpha = k_x a/2 = \pi l a/\lambda$, that is, half the phase difference between the contributions from $x = \pm a/2$.

Each term $a \frac{\sin \alpha}{\alpha}$ and $b \frac{\sin \beta}{\beta}$ is a diffraction amplitude and the relative intensities of the diffraction distribution, that is, of the subsidiary maxima, depend on the square of their product

$$a^2 b^2 \frac{\sin^2 \alpha}{\alpha^2} \frac{\sin^2 \beta}{\beta^2}$$

where ab is the area of the aperture.

We write the intensity at P as $I = I_0 \frac{\sin^2 \alpha}{\alpha^2} \frac{\sin^2 \beta}{\beta^2}$

where I_0, the intensity maximum, incorporates the terms a, b, h and $1/2\pi$. The relative values are numerically equal to the products of any two terms from Table 13.1, which are repeated on the left-hand side of Figure 13.9.

Appendix 8

Reflection and Transmission Coefficients for a Wave Meeting a Boundary

Table A1 Waves Incident Normally on a Plane Boundary Between Media of Characteristic Impedances Z_1 and Z_2

Wave type	Impedance Z + for wave in +ve direction − for wave in −ve direction	Boundary conditions	Amplitude Coefficients			
			$\dfrac{Reflected_r}{Incident_i}$ $=\dfrac{Z_1-Z_2}{Z_1+Z_2}$	$\dfrac{Transmitted_t}{Incident_i}$ $=\dfrac{2Z_1}{Z_1+Z_2}$	$\dfrac{Reflected_r}{Incident_i}$ $=\dfrac{Z_2-Z_1}{Z_1+Z_2}$	$\dfrac{Transmitted_t}{Incident_i}$ $=\dfrac{2Z_2}{Z_1+Z_2}$
Transverse on string section 5.5, Reflection and Transmission of Waves on a String at a Boundary	$\dfrac{-T(\partial y/\partial x)}{\dot y}=\rho c=(T\rho)^{1/2}$	$y_i+y_r=y_t$ or $\dot y_i+\dot y_r=\dot y_t$ $T\left[\dfrac{\partial y_i}{\partial x}+\dfrac{\partial y_r}{\partial x}=\dfrac{\partial y_t}{\partial x}\right]$	y and $\dot y$		$-T\dfrac{\partial y}{\partial x}$	
Longitudinal acoustic section 7.6, Reflection and Transmission of Sound Waves at Boundaries	$\dfrac{\rho}{\dot\eta}=\rho_0 c=(B_a\rho)^{1/2}$	$\dot\eta_i+\dot\eta_r=\dot\eta_t$ $p_i+p_r=p_t$	η and $\dot\eta$		p	
Voltage and current on transmission line section 8.4, Reflections from the End of a Transmission Line	$\dfrac{V}{I}=\sqrt{\dfrac{L}{C}}$	$I_i+I_r=I_t$ $V_i+V_r=V_t$	I		V	
Electromagnetic section 9.12, Reflection and Transmission of Electromagnetic Waves at a Boundary	$\dfrac{E}{H}=\sqrt{\dfrac{\mu}{\epsilon}}$	$H_i+H_r=H_t$ $E_i+E_r=E_t$	H		E	

All waves $\dfrac{\text{Reflected intensity}}{\text{Incident intensity}}=\left(\dfrac{Z_1-Z_2}{Z_1+Z_2}\right)^2$ $\dfrac{\text{Transmitted intensity}}{\text{Incident intensity}}=\dfrac{4Z_1Z_2}{(Z_1+Z_2)^2}$

Index

Introduction to Vibrations and Waves, First Edition. H. J. Pain and Patricia Rankin.
© 2015 John Wiley & Sons, Ltd. Published 2015 by John Wiley & Sons, Ltd.
Companion website: http://booksupport.wiley.com

Printed and bound by CPI Group (UK) Ltd, Croydon, CR0 4YY

27/10/2024

14580372-0002